EARTH-SHELTERED BUILDINGS

NAVAL FACILITIES ENGINEERING
COMMAND

Fredonia Books
Amsterdam, The Netherlands

Earth-Sheltered Buildings

by
Naval Facilities Engineering Command

ISBN: 1-58963-968-5

Reprinted from the 1984 edition

Fredonia Books
Amsterdam, The Netherlands
http://www.fredoniabooks.com

ABSTRACT

Design guidance is presented for use by experienced engineers and architects. The types of buildings within the scope of this manual include slab-on-grade, partially-buried (bermed) or fully-buried, and large (single-story or multistory) structures. New criteria unique to earth-sheltered design are included for the following disciplines: Planning, Landscape Design, Life-Cycle Analysis, Architectural, Structural, Mechanical (criteria include below-grade heat flux calculation procedures), and Electrical.

FOREWORD

Earth-sheltered buildings offer an important alternative to conventionally designed facilities. This design manual, which is one of a series, provides criteria for evaluating the habitability and suitability of earth-sheltered space as well as providing new technical design information. Both the aesthetic and technical aspects of this manual represent the most current design practice and engineering methods available for large earth-sheltered buildings.

Many of the technical criteria are the product of on-going research in the scientific community. As research continues in both public and private institutions, improved calculation techniques may become available. The designer is encouraged to make use of new data from professional societies, associations, and institutions. However, deviations from the Design Manual criteria must have prior approval from Naval Facilities Engineering Command Headquarters (Code 04).

This publication is certified as an official publication of the Naval Facilities Engineering Command and has been reviewed and approved in accordance with SECNAVINST 5600.16.

W. M. Zobel
Rear Admiral, CEC, U. S. Navy
Commander
Naval Facilities Engineering Command

TABLE OF CONTENTS

Page

Section 1. REGULATIONS AND GUIDELINES 1.4-1
1. SCOPE 1.4-1
2. RELATED CRITERIA 1.4-1

Section 2. CLIMATE EVALUATION 1.4-3
1. GENERAL 1.4-3
2. SENSITIVITY TO ANALYSIS 1.4-3
a. Infiltration Loads Related to Climate 1.4-3
b. Latent Cooling Loads Related to Climate 1.4-3
c. Energy Costs Related to Climates 1.4-3
d. Internal Heat Gains Related to Climate 1.4-3
e. Ground Temperatures Related to Climate 1.4-4
f. Construction Costs Related to Climate 1.4-4
3. CLIMATE SUITABILITY FOR SMALL BUILDINGS 1.4-4

Section 3. SITE EVALUATION 1.4-7
1. GENERAL 1.4-7
2. SITE INVESTIGATION PROCEDURES 1.4-7
a. Site Records 1.4-7
b. Existing Construction and Utilities. 1.4-8
c. On-Site Investigation 1.4-8
3. SITE INFLUENCES ON THE BUILDING DESIGN 1.4-8
a. Climatic Influences 1.4-8
b. Environmental and Land Use Issues 1.4-9
c. Access and Building Services 1.4-9
d. Sanitary and Storm Sewer 1.4-10
e. Isolation and Protection 1.4-10
f. Topography 1.4-10
g. Ground Conditions/Ground-Water Conditions 1.4-10
h. Excavation Costs 1.4-11

Section 4. BUILDING TYPE EVALUATION 1.4-12
1. GENERAL 1.4-12
2. HIGH INTERNAL HEAT LOADS 1.4-12
3. MODERATE INTERNAL HEAT LOADS 1.4-12
4. BUILDINGS WITH MINIMAL HEATING REQUIREMENTS 1.4-12
5. ROOF-DOMINANT BUILDINGS 1.4-12
6. VENTILATION AND INFILTRATION-DOMINANT BUILDINGS . . . 1.4-13
7. MECHANICAL ZONING OF BUILDING 1.4-13
8. BUILDING SIZE 1.4-14
a. Large Buildings 1.4-14
b. Small Buildings 1.4-14
9. LEVELS OF TECHNOLOGY. 1.4-14
10. SPECIAL PERIMETER REQUIREMENTS 1.4-15
11. LIFE-SAFETY REQUIREMENTS 1.4-15

Section 5. LIFE-CYCLE COST EVALUATION 1.4-16
1. GENERAL 1.4-16
2. INITIAL COSTS 1.4-16
a. Construction Scheduling 1.4-16
b. Land Costs 1.4-16
c. Excavation 1.4-16
d. Earthwork 1.4-16
e. Planting and Ground Cover. 1.4-16
f. Civil 1.4-16
g. Architectural 1.4-16
h. Structural 1.4-17
i. Mechanical 1.4-17
3. FUTURE ADDITIONS 1.4-17
4. MAJOR REPAIRS 1.4-17
5. OPERATIONAL ENERGY COSTS 1.4-18
6. MAINTENANCE AND OTHER OPERATIONAL COSTS 1.4-18
7. GENERAL TRENDS 1.4-18
8. MARGIN OF ERROR 1.4-18
9. INTANGIBLE CONSIDERATIONS 1.4-19

Section 6. SITE CONTEXT 1.4-20
1. GENERAL 1.4-20
2. BASIC RESPONSES TO CONTEXT 1.4-20
a. Nature 1.4-20
b. Reference 1.4-20
c. Coherence 1.4-20
d. Identity 1.4-21
e. Tradition 1.4-21
f. Territoriality 1.4-22
3. EXAMPLES OF RESPONSE TO CONTEXT 1.4-22
a. Earth-Sheltered Building on Wooded Site 1.4-23
b. Earth-Sheltered Building on Unwooded Suburban Site 1.4-24
c. Earth-Sheltered Building on Flat, Open Site . . 1.4-24
d. Approach to Covered Building on Flat Site . . . 1.4-24
e. Earth-Sheltered Building on Shore Front 1.4-28
f. Earth-Sheltered Building Below Commons 1.4-29
g. Earth-Sheltered Building on Corner Site 1.4-29

Section 7. SITE CONDITIONS AND EARTHWORK 1.4-32
1. SOIL CONDITIONS 1.4-32
2. ROCK CONDITIONS 1.4-32
3. GROUND-WATER AND DRAINAGE 1.4-33
a. Identification of Critical Problems 1.4-33
b. Surface Drainage Techniques 1.4-33
c. Subsurface Drainage 1.4-34
4. EXCAVATION, STABILITY, AND DEWATERING 1.4-37
5. BACKFILLING AND COMPACTION 1.4-37
6. EARTH-COVERED ROOFS 1.4-37

Section 8. PLANTING AND IRRIGATION DESIGN 1.4-39
1. PLANT MATERIALS SELECTION AND LAYOUT 1.4-39
a. Suitability to Climate 1.4-39
b. Adaptabilty to Berm Slopes 1.4-39
c. Adaptability to Limited Soil Depths. 1.4-39
d. Ease of Establishment and Maintenance 1.4-43
e. Functional Considerations 1.4-43
2. IRRIGATION SYSTEMS 1.4-43
a. Manual 1.4-43
b. Spray Heads 1.4-44
c. Trickle Systems 1.4-44
3. INSTALLATION 1.4-44
a. Application on Roofs 1.4-44
b. Application on Berms 1.4-46
4. MAINTENANCE 1.4-47
a. Weeding 1.4-47
b. Mowing 1.4-47
c. Irrigation 1.4-47

Section 9. SPACE PLANNING AND PROGRAMMING 1.4-48
1. ACCESS 1.4-48
2. FIRE-FIGHTING ACCESS 1.4-49
3. TRAVEL DISTANCES AND EXITS 1.4-49
4. MECHANICAL ZONING 1.4-49
5. STRUCTURAL BAY WITH EARTH-COVER 1.4-49
6. OPTIMIZATION OF EARTH-CONTACT 1.4-49
a. Earth-Sheltered Buildings with Positive Year-Round Heat Flux 1.4-49
b. Earth-Sheltered Buildings with Negative Year-Round Heat Flux 1.4-50
c. Earth-Sheltered Buildings with Both Heating and Cooling Benefits 1.4-50
7. NATURAL LIGHT AND VIEWS 1.4-50
8. MAINTENANCE AND SECURITY 1.4-51

Section 10. ENVIRONMENTAL PSYCHOLOGY 1.4-52
1. EFFECTS OF PHYSICAL CHARACTERISTICS 1.4-52
a. Natural Light and Views 1.4-52
b. Horizontal Access 1.4-52
c. Internal Spaciousness 1.4-52
d. Internal Environmental Conditions 1.4-52
e. Spatial Orientation 1.4-53
2. EFFECT OF WORK CONDITIONS 1.4-53
a. Internal Activity 1.4-53
b. Length of Stay 1.4-53
c. Need for Underground Location 1.4-53
d. Job Satisfaction 1.4-53
e. Noise 1.4-53

Section 11. NATURAL LIGHTING, SKYLIGHTS, AND PASSIVE SOLAR 1.4-54
1. METHODS OF DAYLIGHTING EARTH-SHELTERED SPACE 1.4-54
a. Beamed Daylighting 1.4-54
b. Diffuse Daylighting 1.4-54
2. PASSIVE SOLAR APPLICATIONS FOR EARTH-SHELTERED BUILDINGS 1.4-54
a. Direct Gain Systems 1.4-55
b. Indirect Gain Systems 1.4-55
c. Isolated Gain Systems 1.4-55
3. CALCULATION PROCEDURES (GENERAL) 1.4-55

Section 12. FIRE PROTECTION AND LIFE SAFETY 1.4-56
1. AIR-HANDLING SYSTEMS 1.4-56
a. Negative Pressures 1.4-56
b. Positive Pressures 1.4-56
c. Fire/Smoke Areas 1.4-56
2. SMOKE AND HEAT VENTING 1.4-56
3. FIRE DEPARTMENT ACCESS 1.4-57
4. REQUIREMENT FOR AUTOMATIC FIRE-EXTINGUISHING SYSTEM 1.4-57
5. FIRE SEPARATION OF STRUCTURES 1.4-57
6. DRAINAGE 1.4-57
7. LIFE SAFETY 1.4-58
a. Refuge Areas 1.4-58
b. Goal-Oriented Exit Design 1.4-58
c. Exit Enclosures 1.4-59
d. Elevators 1.4-59

Section 13. WATERPROOFING 1.4-60
1. SOURCES OF MOISTURE 1.4-60
2. EVALUATION OF WATERPROOFING SYSTEMS 1.4-60
a. Dampproofing Products 1.4-60
b. Built-Up Asphalt and Pitch Membranes 1.4-62
c. Cementitious Materials 1.4-63
d. Liquid-Applied Waterproofing Systems 1.4-65
e. Modified Bitumens 1.4-68
f. Vulcanized and Plastic Sheets 1.4-70
g. Bentonite Clay Products 1.4-73
3. WATERPROOFING SELECTION PROCEDURE 1.4-77

Section 14. INSULATION 1.4-85
1. EFFECT OF INSULATION ON THERMAL PERFORMANCE 1.4-85
a. Reduction of Heat Flow 1.4-85
b. Thermal Stability 1.4-85
2. AMOUNT OF INSULATION 1.4-86
a. Climate 1.4-86
b. Internal Heat 1.4-86
c. Ground Moisture 1.4-86
d. Ease of Adding Insulation 1.4-88

3. DISTRIBUTION OF INSULATION AROUND THE BUILDING ENVELOPE 1.4-88
   a. Optimum Distribution for Minimizing Heat Loss . 1.4-88
   b. Optimum Distribution for Maximizing Cooling. . . 1.4-90
   c. Compromises 1.4-90
   d. Insulation as a Protection Board for the Waterproofing 1.4-91
   e. Passive Solar Gain Areas 1.4-91
   f. Insulation as a Protection Against Cold Floors and Condensation 1.4-91
   g. Insulation for Thermal Breaks 1.4-92
4. LOCATION OF INSULATION 1.4-92
5. INSULATION TYPE 1.4-93
6. INSTALLATION 1.4-94
   a. Method of Adhering Insulation 1.4-94
   b. Backfilling 1.4-94
   c. Prior to Backfilling 1.4-94
   d. Thermal Short Circuits 1.4-94

Section 15. ROOF AND WALL DETAILS 1.4-95
1. GENERAL 1.4-95
2. RECOMMENDED DETAILS 1.4-95
   a. Area Drain at Paved Roof--Pedestrian Traffic--Minimal Freeze-Thaw Cycles 1.4-95
   b. Drain at Paved Roof--Frequent Freeze-Thaw Cycles--Pedestrian Traffic 1.4-97
   c. Area Drain at Concrete Paved Roof--Light Vehicular Traffic 1.4-98
   d. Drain at Paved Roof--Frequent Freeze-Thaw Cycles--Pedestrian Traffic 1.4-99
   e. Area Drain at Earth-Covered Roof 1.4-100
   f. Pipe Penetration at Paved Roof 1.4-102
   g. Expansion Joint at Earth-Covered Roof 1.4-103
   h. Earth-Covered Roof at Masonry Cavity Wall . . . 1.4-104
   i. Expansion Joint at Earth-Covered Roof and Existing Structure 1.4-105
   j. Earth-Covered Roof at Skylight with Concrete Curb Exposed to Interior 1.4-106
   k. Parapet at Earth-Covered Roof with Thermal Break 1.4-107
   l. Back Wall Drainage at Earth-Covered Roof 1.4-108
   m. Window Sill at Berm 1.4-108
   n. Edge of Conventional Roof at Berm 1.4-109
   o. Juncture of Retaining Wall and Exposed Concrete Wall (Shown in Plan) 1.4-110
   p. Juncture of Retaining Wall and Masonry Cavity Wall (Shown in Plan) 1.4-111
   q. Thermal Break and Waterproofing at Retaining Wall (Shown in Plan) 1.4-112
   r. Foundation Near Water Table 1.4-112
   s. Foundation Below Water Table--Hydrostatic Pressure Relieved by Drainage 1.4-113
   t. Foundation Below Water Table--Structure Designed to Withstand Hydrostatic Pressure . . . 1.4-114

Section 16. STRUCTURE 1.4-116
1. GENERAL 1.4-116
2. DESIGN LOADS 1.4-116
a. Roof Loads 1.4-116
b. Wall Loads 1.4-116
c. Seismic Loads. 1.4-117
d. Blast Loads 1.4-118
3. CHECKLIST OF LOADING CONDITIONS 1.4-118

Section 17. MECHANICAL AND ELECTRICAL SYSTEMS 1.4-119
1. AIR-HANDLING SYSTEMS 1.4-119
2. VENTILATION REQUIREMENTS 1.4-119
3. WATER SYSTEMS 1.4-120
4. SANITARY SYSTEMS 1.4-120
5. LIGHTING SOURCE 1.4-120
6. LIGHTING CONTROLS 1.4-120
7. PSYCHOLOGICAL RESPONSE TO LIGHTING 1.4-121
8. EMERGENCY LIGHTING 1.4-121
9. ELECTRICAL COOLING ENERGY AND FAN ENERGY 1.4-121
10. TRANSFORMER LOCATION 1.4-121
11. SERVICE LOCATION 1.4-122
12. EMERGENCY POWER 1.4-122
a. Auxiliary Power Source 1.4-122
b. Emergency Generators 1.4-122
c. Uninterruptible Power Supply 1.4-123
13. FIRE ALARM 1.4-123
14. COMMUNICATIONS 1.4-123

Section 18. ENERGY CALCULATIONS 1.4-124
1. GENERAL
2. OVERVIEW OF HEAT TRANSFER PRINCIPLES IN BELOW-GRADE STRUCTURES 1.4-124
a. Thermal Conductivity of Soils. 1.4-124
b. Temperature Regime Around an Underground Structure 1.4-124
c. Simplified Heat Conduction for Undisturbed Earth 1.4-125
3. CALCULATION PROCEDURES 1.4-126
a. Existing Earth-Contact Methods 1.4-126
b. Computer Model 1.4-129
c. Hourly Weather Data - SOLMET 1.4-132
4. SUMMARIES AND HEAT FLUX TABLES 1.4-132
a. Description of Earth-Sheltered Building Configurations Modeled 1.4-132
b. Input Parameters 1.4-134
c. Heat Flux Tables 1.4-134
d. Results and Discussion 1.4-138
5. SAMPLE LOAD CALCULATION PROCEDURES 1.4-140
a. Sizing of Heating and Cooling Systems 1.4-140
b. Integrated Annual Energy Consumption 1.4-149
6. EARTH-COVERED ROOFS 1.4-151
7. SENSITIVITY OF BUILDING THERMAL PERFORMANCE TO SOIL TYPE 1.4-151

Section 19. DESIGN EXAMPLE 1.4-155
1. GENERAL 1.4-155
2. PROGRAM REQUIREMENTS 1.4-155
3. TOPOGRAPHIC FEATURES 1.4-155
4. CONTEXTUAL CONSIDERATIONS 1.4-158
5. SITE ACCESS 1.4-158
6. BUILDING ORIENTATION 1.4-158
a. Alignment with the Slope 1.4-159
b. Alignment with the Road 1.4-160
c. Alignment with Retained Buildings and Site Features 1.4-161
d. Preliminary Geotechnical Report 1.4-162
7. ORIENTATION FOR PASSIVE SOLAR BENEFIT 1.4-162
8. PROTECTION FROM WINTER WIND 1.4-162
9. EQUALIZATION OF FLOOR AREAS IN THE FULL-BERM CONFIGURATION 1.4-162
10. ALLOCATION OF FUNCTIONS BY MECHANICAL REQUIREMENTS . 1.4-162
11. OPTIMIZATION OF SURFACE-TO-VOLUME RATIO AS A FUNCTION OF MECHANICAL ZONES 1.4-169
12. SURFACE-TO-VOLUME RATIO AS A FUNCTION OF CONSTRUCTION COST 1.4-169
13. SURFACE-TO-VOLUME RATIO LIMITATIONS 1.4-169
14. DETERMINATION OF DEPTH OF EARTH-SHELTERING AS A FUNCTION OF THERMAL LAB 1.4-170
a. Zones with Heating Only 1.4-170
b. Zones with Cooling and Heating Without Thermostat Setback 1.4-170
c. Zones with Both Cooling and Heating and Temperature Setback 1.4-170
15. DETERMINATION OF DEPTH OF EARTH-SHELTERING AS A FUNCTION OF CONSTRUCTION COSTS 1.4-170
16. UTILITIES 1.4-171
17. STRUCTURAL SYSTEMS 1.4-171
18. DIAGRAMMATIC PLAN FOR 2-STORY FULL-BERM BUILDING 1.4-171
a. Full-Berm Configuration 1.4-171
b. Half-Berm Configuration 1.4-173
19. AREA TAKEOFF AND UNIT PRICING 1.4-176
20. ENERGY PREDICTIONS AND MECHANICAL SYSTEMS COSTS . . 1.4-176
21. LIFE-CYCLE COST FACTORS 1.4-177
22. RESULTS OF LIFE-CYCLE ANALYSIS 1.4-178
a. Elimination of Sprinkler System 1.4-178
b. Elimination of Skylight and Arcade 1.4-178
c. Elimination of Earth-Cover on the Half-Berm Building 1.4-178
d. Final Adjusted Life-Cycle Costs for Full-Berm Alternatives 1.4-182
23. REFINEMENTS TO INSULATION THICKNESS AND DEPTH . . . 1.4-182
24. EVALUATION OF INTANGIBLE DIFFERENCES 1.4-182
a. Security 1.4-182
b. Natural Light 1.4-182
c. Exterior Noise Reduction 1.4-182
d. Radiation Protection 1.4-182
25. FINAL SCHEMATIC DESIGN 1.4-184

References Reference-1

Bibliography Bibliography-1

Appendix A - Heat Flux Tables A-1

Appendix B - Boundary Conditions at Ground Surfaces B-1

Appendix C - Calculation Worksheets C-1

Appendix D - Earth Thermal Conductivity Factors D-1

Criteria Sources . . . . Criteria Sources-1

Index Index-1

FIGURES

Figure No.	Title	Page
1.	Climatic Regions for Small, Naturally-Ventilated Earth-Sheltered Buildings	1.4-4
2.	Contrast to Natural Setting	1.4-23
3.	Integration with Natural Setting	1.4-23
4.	Submerged Building with Enclosed Atrium	1.4-24
5.	Incidental Relationship to Parking	1.4-25
6.	Hierarchical Relationship to Parking	1.4-25
7.	Entrance Approach with Poor Image	1.4-26
8.	Entrance Approach with Implied Forecourt	1.4-27
9.	Entrance Approach with Literal Forecourt	1.4-27
10.	Obstruction of View on Harbor Site	1.4-28
11.	Accommodation of View on Harbor Site	1.4-28
12.	Obstruction of Existing Commons	1.4-30
13.	Preservation of Existing Commons	1.4-30
14.	Disruption of Street Order	1.4-31
15.	Reinforcement of Street Order	1.4-31
16.	Backfilling Around Structure in Clay Zone	1.4-34
17.	Drainage at Footing	1.4-35
18.	Drainage at Roof Edge	1.4-35
19.	Climatic Map for Plant Hardiness	1.4-39
20.	Waterproofing Flowchart--Ground-Water Conditions	1.4-78
21.	Waterproofing Flowchart--Building Use Parameters	1.4-79
22.	Waterproofing Flowchart--Environmental Conditions	1.4-80
23.	Waterproofing Flowchart--Structural Conditions	1.4-81
24.	Waterproofing Flowchart--Construction Process	1.4-82
25.	Waterproofing Flowchart--Product Reliability and Contractual Relationships	1.4-83
26.	Waterproofing Flowchart--Product Performance and Cost	1.4-84
27.	Heat Flow from Buried Structure in Winter	1.4-85
28.	Heat Flow From Buried Structure in Summer	1.4-85
29.	Optimum Insulation Pattern for a Buried Structure (Heating Conditions Only)	1.4-89

30. Simulated Below-Grade Heat Flux (February, Minneapolis) . . 1.4-89
31. Optimum Insulation Placement (Heating Conditions Only) . . 1.4-90
32. Areas Requiring Extra Insulation 1.4-90
33. Thermal Break Locations 1.4-92
34. Insulation Within Soil Mass and Detached From Building . . 1.4-92
35. Drain at Brick-Paved Roof--Bituminous Setting Bed 1.4-95
36. Drain at Brick-Paved Roof--Sand Setting Bed 1.4-97
37. Drain at Concrete-Paved Roof 1.4-99
38. Drain at Paved Roof--on Pedestals 1.4-99
39. Drain at Earth-Covered Roof 1.4-100
40. Pipe Penetration at Paved Roof 1.4-102
41. Expansion Joint at Earth-Covered Roof 1.4-103
42. Earth-Covered Roof at Masonry Cavity Wall 1.4-104
43. Expansion Joint at Earth-Covered Roof and Existing Structure 1.4-105
44. Earth-Covered Roof at Skylight with Concrete Curb Exposed to Interior 1.4-106
45. Parapet at Earth-Covered Roof with Thermal Break 1.4-107
46. Backwall Drainage at Earth-Covered Roof 1.4-108
47. Window Sill at Berm 1.4-109
48. Edge of Conventional Roof at Berm 1.4-109
49. Juncture of Retaining Wall and Exposed Concrete Wall (Shown in Plan) 1.4-110
50. Juncture of Retaining Wall and Masonry Cavity Wall (Shown in Plan) 1.4-111
51. Thermal Break and Waterproofing at Retaining Wall (Shown in Plan) 1.4-112
52. Foundation Near Water Table 1.4-113
53. Foundation Below Water Table (Hydrostatic Pressure Relieved in Drainage) 1.4-113
54. Foundation Below Water Table (Structure Designed to Withstand Hydrostatic Pressure) 1.4-115
55. Isotherm Distribution Around an Uninsulated Underground Building for Typical February Conditions in Minneapolis . . 1.4-125
56. Isotherm Distribution Around an Uninsulated Underground Building for Typical August Conditions in Minneapolis . . . 1.4-125
57. Monthly Average Ground Temperatures (Ottawa, Ontario) . . . 1.4-126
58. Radial Isotherms and Circular Heat Flow Lines Assumed by ASHRAE/Boileau and Latta 1.4-127
59. Comparison of ASHRAE and Computer-Predicted Heat Flux for an Uninsulated Basement Wall 1.4-128
60. Comparison of ASHRAE and Computer-Predicted Heat Flux for Basement Wall with R-21 Insulation 1.4-128
61. Computer-Predicted Temperature Profiles Around an Uninsulated Basement for Typical February in Minneapolis . 1.4-128
62. Computer-Predicted Temperature Profiles Around Basement with R-21 Insulation for Typical February in Minneapolis . 1.4-128
63. Distribution of Two-Dimensional Temperature Nodes and Boundary Conditions for Full-Berm, 3-Story Building 1.4-130
64. SOLMET Tape Format 1.4-131
65. Schematic Cross-Section for Slab-on-Grade Building 1.4-133
66. Schematic Cross-Section for Half-Berm Building 1.4-133
67. Schematic Cross-Section for Full-Berm Building 1.4-133
68. Schematic Cross-Section of Full-Berm, 3-Story Building . . 1.4-133

69. Floor Heat Loss for January, Minneapolis 1.4-136
70. Annual Floor Heat Loss Minneapolis, Slab-on-Grade 1.4-137
71. Annual Floor Heat Loss/Gain Miami, Slab-on-Grade Building . 1.4-137
72. Annual Wall and Floor Heat Loss Minneapolis, Full Berm, 1-Story Building 1.4-138
73. Annual Wall and Floor Heat Loss/Gain Miami, Full Berm, 1-Story Building 1.4-139
74. Annual Wall and Floor Heat Loss Minneapolis, Full Berm, 3-Story Building 1.4-141
75. Annual Wall and Floor Heat Loss/Gain Miami, Full Berm, 3-Story Building 1.4-142
76. Isometric of Building Used for Example Problem 1 . . . . . 1.4-144
77. Diagrammatic Plan of Building Used for Example Problem 1 --Exit Level 1.4-144
78. Diagrammatic Plan of Building Used for Example Problem 1 --Below-Grade Level 1.4-144
79. Existing Site for Design Example 1.4-160
80. Half-Berm Building Aligned with Slope 1.4-161
81. Full-Berm Building Aligned with Slope 1.4-161
82. Half-Berm Building Aligned with Road 1.4-161
83. Full-Berm Building Aligned with Road 1.4-161
84. Schematic Section of Training and Administrative Floors . . 1.4-163
85. Space Allocation by Program Requirements 1.4-164
86. Space Allocation to Equalize Floor Areas 1.4-165
87. Instantaneous Below-Grade Heat Flux with Office on Earth-Contact Wall 1.4-166
88. Instantaneous Below-Grade Heat Flux with Storage on Earth-Contact Wall 1.4-167
89. Comparison of Energy Costs for Different Functions Located on Earth-Contact Wall 1.4-168
90. Preliminary Site Plan for Full-Berm Building 1.4-172
91. Preliminary Schematic Plan of Full-Berm Building --Upper Level 1.4-173
92. Preliminary Schematic Plan of Full-Berm Building --Lower Level 1.4-174
93. Preliminary Site Plan for Half-Berm Building 1.4-175
94. Preliminary Schematic Plan of Half-Berm Building --Upper Level 1.4-176
95. Preliminary Schematic Plan of Half-Berm Building --Lower Level 1.4-177
96. Final Schematic Site Plan 1.4-183
97. Final Schematic Plan of Upper Level 1.4-184
98. Final Schematic Plan of Lower Level 1.4-185
99. Axonometric of Circulation Space 1.4-186
100. Perspective Looking West 1.4-187
101. Perspective Looking East 1.4-187
102. 7-Zone Climate Classification of the United States and Representative Cities A-1

TABLES

Table No.	Title	Page
1.	Plant Species by Climate	1.4-40
2.	Weights of Planting Materials	1.4-45
3.	Suggested Resistances for Insulation	1.4-87
4.	Influence of Moisture Boundary Conditions on the 100-Day Moisture Gain of Various Insulations	1.4-93
5.	Building and Soil Properties Used in Computer Simulation	1.4-134
6.	Combined Surface Convection & Linearized Radiation Heat Transfer Coefficients Used in Computer Model	1.4-134
7.	Specified Building Interior Dry-Bulb Air Temperatures Used in Computer Model	1.4-135
8.	Constant Deep Ground Temperatures Used in Computer Model	1.4-135
9.	Heat Loss Calculation for Example 1	1.4-145
10.	Cooling Load Calculation for Example 2	1.4-148
11.	Sensitivity of Slab Thermal Performance to Soil Type, Slab-on-Grade, Minneapolis	1.4-152
12.	Sensitivity of Slab Thermal Performance to Soil Type, Slab-on-Grade, Phoenix	1.4-152
13.	Sensitivity of Slab and Wall Thermal Performance to Soil Type, Full Berm, 1-Story, Minneapolis	1.4-153
14.	Sensitivity of Slab and Wall Thermal Performance to Soil Type, Full Berm, 1-Story, Phoenix	1.4-154
15.	Space Allocation by Mechanical, Natural Light, and Access (Training Functions)	1.4-156
16.	Space Allocation by Mechanical, Natural Light, and Access (Administrative Functions)	1.4-157
17.	Program Area Tabulation for Vehicular Maintenance Facility	1.4-158
18.	Summary of Program by Zones	1.4-159
19.	Summary of Life-Cycle Cost Analysis	1.4-179
20	Adjustments to Life-Cycle Cost Analysis	1.4-181
21.	Monthly Average and Peak Heat Fluxes for Minneapolis with No Insulation	A-2
22.	Monthly Average and Peak Heat Fluxes for Boston with No Insulation	A-3
23.	Monthly Average and Peak Heat Fluxes for Seattle with No Insulation	A-4
24.	Monthly Average and Peak Heat Fluxes for Kansas City with No Insulation	A-5
25.	Monthly Average and Peak Heat Fluxes for Albuquerque with No Insulation	A-6
26.	Monthly Average and Peak Heat Fluxes for Phoenix with No Insulation	A-7
27.	Monthly Average and Peak Heat Fluxes for Miami with No Insulation	A-8
28.	Monthly Average and Peak Heat Fluxes for Minneapolis with R-10.4 Insulation	A-9
29.	Monthly Average and Peak Heat Fluxes for Boston with R-10.4 Insulation	A-10
30.	Monthly Average and Peak Heat Fluxes for Seattle with R-10.4 Insulation	A-11

31. Monthly Average and Peak Heat Fluxes for Kansas City with R-10.4 Insulation A-12
32. Monthly Average and Peak Heat Fluxes for Albuquerque with R-10.4 Insulation A-13
33. Monthly Average and Peak Heat Fluxes for Phoenix with R-10.4 Insulation A-14
34. Monthly Average and Peak Heat Fluxes for Miami with R-10.4 Insulation A-15
35. Monthly Average and Peak Heat Fluxes for Minneapolis with R-20.8 Insulation A-16
36. Monthly Average and Peak Heat Fluxes for Boston with R-20.8 Insulation A-17
37. Monthly Average and Peak Heat Fluxes for Seattle with R-20.8 Insulation A-18
38. Monthly Average and Peak Heat Fluxes for Kansas City with R-20.8 Insulation A-19
39. Monthly Average and Peak Heat Fluxes for Minneapolis with R-10.4 (Deep) Insulation A-20
40. Monthly Average and Peak Heat Fluxes for Boston with R-10.4 (Deep) Insulation A-21
41. Monthly Average and Peak Heat Fluxes for Seattle with R-10.4 (Deep) Insulation A-22
42. Monthly Average and Peak Heat Fluxes for Kansas City with R-10.4 (Deep) Insulation A-23

Section 1. REGULATIONS AND GUIDELINES

1. SCOPE. This manual contains design criteria for earth-sheltered buildings. The design criteria include site planning, architectural, civil, structural, mechanical, and electrical disciplines. The mechanical design data for the evaluation of below-grade heat flux, heating and cooling loads, and peak loads is restricted to buildings over 15,000 square feet. (See Section 18, Energy Calculations, for additional restrictions.)

This manual does not apply to housing (single-family dwellings and dormitories).

The term "earth-sheltered" does not necessarily mean "underground." The term applies to any building that has some portion of its exterior walls below ground. By this definition, "earth-sheltering" includes a building that has an earth berm up to the window sills. It also includes any conventional building that has a usable climate-controlled basement.

To demonstrate the use of the criteria contained in this manual, a step-by-step example is included in Section 19.

2. RELATED CRITERIA. Certain criteria related to earth-sheltered construction appear elsewhere in the DM series. Refer to the following sources:

Subject	Source
Site Planning	NAVFAC DM-1
Waterproofing, Dampproofing and Condensation Control	NAVFAC DM-1
Space Planning	NAVFAC DM-1
Construction Requirements by Climate	NAVFAC DM-1
Structural Engineering General Requirements . . .	NAVFAC DM-2.1
Structural Loads	NAVFAC DM-2.2
Steel Structures	NAVFAC DM-2.3
Concrete Structures	NAVFAC DM-2.4
Timber Structures	NAVFAC DM-2.5
Aluminum, Masonry, Composite Structures and Other Structural Materials	NAVFAC DM-2.6
Mechanical Engineering	NAVFAC DM-3
Plumbing Systems	NAVFAC DM-3.1
Heating, Ventilating, Air-conditioning and Dehumidifying Systems	NAVFAC DM-3.3
Electrical Design Considerations	NAVFAC DM-4.1
Electrical Power Distribution Systems	NAVFAC DM-4.2
Electrical Utilization Systems	NAVFAC DM-4.4
Wire Communication and Signal Systems	NAVFAC DM-4.7

Subject	Source
Hydrology and Hydraulics	NAVFAC DM-5.2
Drainage Systems	NAVFAC DM-5.3
Pavements	NAVFAC DM-5.4
General Provisions and Geometric Design for Roads, Streets, Walks, and Open Storage Area	NAVFAC DM-5.5
Soil Conservation	NAVFAC DM-5.11
Fences, Gates, and Guard Towers	NAVFAC DM-5.12
Soil Mechanics, Foundations, and Earth Structures .	NAVFAC DM-7 Series
Fire Protection Engineering	NAVFAC DM-8
Cold Regions Engineering	NAVFAC DM-9
Site Evaluation of Waterfront Structures	NAVFAC DM-25

Section 2. CLIMATE EVALUATION

1. GENERAL. This section provides general guidelines for evaluating the suitability of earth-sheltered buildings for various climates.

2. SENSITIVITY TO ANALYSIS. Climate is the most sensitive index to life-cycle performance of earth-sheltered buildings. As a general rule, earth-sheltered buildings will tend to have a better life-cycle performance in cold climates than in hot climates. Earth-sheltered buildings in mild climates, such as San Diego, usually will not have a life-cycle benefit over conventional construction. These general rules are subject to qualification by the variables discussed below.

a. Infiltration Loads Related To Climate. Placing exterior building surfaces below-grade reduces infiltration. Because of the larger indoor/outdoor temperature differential in cold climates, energy savings due to reduced infiltration will be greater in cold climates than in hot climates.

b. Latent Cooling Loads Related to Climate. Even though the indoor/outdoor temperature differentials in hot climates may not be as great as those in cold climates, the latent component of cooling loads attributable to infiltrating air will tend to improve the comparative feasibility of earth-sheltered buildings in hot and hot-humid climates.

c. Energy Costs Related to Climates. Because of the different kinds of energy sources available, the variation in local energy costs, and the range of operating efficiencies for different mechanical systems, it is not possible to correlate energy costs with climate in general. In some instances, the relative advantage of earth-sheltering in harsh, cold climates will be reinforced by fuel costs, particularly if fuel oil is used for heating. For example, the cost of fuel oil per million Btu's of heating load would be $9.50 ($32.42/million Wh) based on $1 per gallon ($.26/L) of fuel oil and a boiler efficiency of 75 percent. By comparison, electricity for cooling would cost $5.86 per million Btu's ($20/million Wh) of cooling load based on $.05 per kilowatt-hour and a coefficient of performance of 2.5. On this basis, there is a positive correlation between energy costs and colder climates. On the other hand, if natural gas is used as an energy source for heating at the rate of $2.50 per 1,000 ft^3 ($0.088/m^3) of gas, the correlation between cold climates and fuel costs can be negative. At $2.50 per 1,000 ft^3 ($0.088/m^3) of gas and with a boiler efficiency of 75 percent, heating would cost $3.33 per million Btu's ($11.36/million Wh) of load as compared with $5.86 ($20) for cooling as calculated above.

d. Internal Heat Gains Related to Climate. Internal heat gain from equipment, lights, and people will reduce heating requirements. In cold and temperate climates, earth-sheltering will reduce the number of hours in a year where heating loads attributable to heat loss, ventilation, and infiltration exceed the internal heat gains. Thus earth-sheltering, while reducing the overall heating load, will also reduce the amount of usable internal gains. In climates where there is a dominant cooling load year-round, the internal gains will not be significantly affected by earth-sheltering.

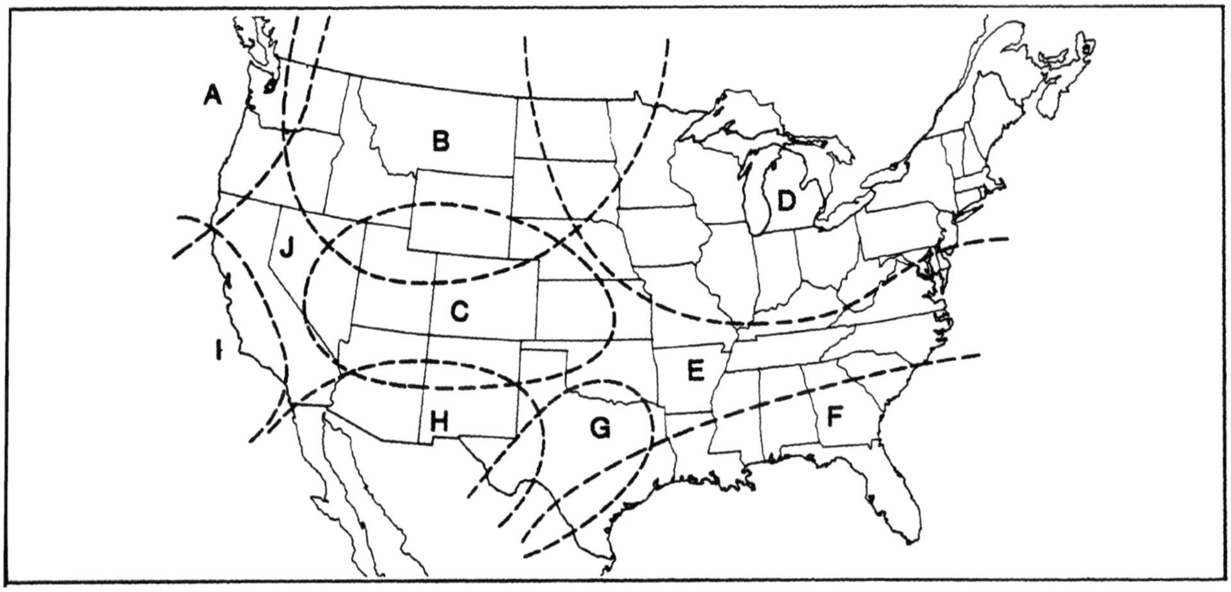

FIGURE 1
Climatic Regions for Small, Naturally-Ventilated Earth-Sheltered Buildings

e. Ground Temperatures Related to Climate. Ground temperatures have a positive correlation to air temperature but the correlation is not proportional. With an earth-contact configuration, the improvement in the indoor/outdoor temperature differential is much greater in cold climates than in hot climates. Further, earth-sheltered buildings in cold, temperate climates can take significant advantage of thermal lag. In effect, thermal lag can provide some heat during the early heating season and provide cooling benefits during the cooling season. In hot climates, solar radiation tends to increase ground temperatures. Warm ground temperatures do not contribute to cooling. In hot, arid climates, however, the relatively steady ground temperatures will benefit the energy requirements by removing the building from exposure to diurnal temperature extremes.

f. Construction Costs Related to Climate. Local price differences notwithstanding, construction costs for conventional buildings tend to be higher in cold climates than in hot climates. This is mainly due to frost footings and greater insulation thicknesses for conventional buildings in cold climates. Earth-sheltering tends to eliminate the frost footing. In addition, insulation costs are reduced since below-grade insulation can be less expensive than above-grade insulation. Earth-sheltered construction is thus more competitive in cold climates than in hot climates.

3. CLIMATE SUITABILITY FOR SMALL BUILDINGS. Because small buildings are more sensitive to passive energy techniques than large buildings, it is possible to map out the suitability of small earth-sheltered buildings by climatic region. This has been done by Kenneth Labs (See Reference 1, Regional Analysis of Ground and Above-Ground Climate by Kenneth Labs). The conclusions of this report pertain only to small, skin-dominated buildings or portions of buildings that are not hermetically sealed and are not mechanically ventilated. The comfort criteria which determine the suitability of earth-sheltering are derived from climatic data which was

cross-checked with comfort zones indicated in a bio-climatic chart. Figure 1 and the following text, which refers to Figure 1, are based on that report.

Region A. Good Summer and Winter Suitability. Cool to cold, cloudy winters maximize the value of earth-tempering as a heat conservation measure. Cool soil and dry summers favor subgrade placement and earth-covered roofs, with little likelihood of condensation.

Region B. Excellent Summer and Winter Suitability. Severely cold winters demand major heat-conservation measures, even though more sunshine is available here than on the coast. Dry summers and cool soil favor earth-covered roofs and ground coupling.

Region C. Strong Summer and Winter Benefits. Good winter insolation somewhat lessens the need for extraordinary winter heat conservation, but the value of the summer benefit is more important here than in the zone above. Earth-covering is advantageous, the ground offers some cooling, condensation is not a problem, and ventilation is not a major necessity.

Region D. Major Winter Benefit, Summer Mixed Blessings. Cold and often cloudy winters place a premium on heat conservation. Low summer ground temperatures offer a cooling source, but with the likelihood of condensation. High summer humidity makes ventilation the leading summer climate control strategy. An above-ground, superinsulated structure designed to maximize ventilation is an important alternative.

Region E. Marginal Winter and Summer Benefit. Generally, good winter sun and minor heating demand reduce the need for extreme heat conservation measures. The ground offers protection from overheated air, but not major cooling potential as a heat sink. The primacy of ventilation and the possibility of condensation compromise summer benefits. Quality of design will determine the actual benefit realized here.

Region F. Insignificant Winter Benefit, Small or Negative Summer Benefit. Ground temperatures can actually be higher than the indoor ambient temperature at times during the summer months with a 65°F (18.1°C) indoor temperature. Although a slight cooling benefit can exist for a 78°F (26°C) indoor temperature, especially for large buildings, small buildings can obtain better passive cooling by ventilation. Earth-sheltering would reduce the envelope heat gain for both large, mechanically-ventilated buildings and small buildings. However, even in this respect, small buildings are better off using passive ventilation and shading. Persistent high humidity levels negate the value of roof mass and establish ventilation as the only important summer cooling strategy. Any design that compromises ventilation effectiveness without contributing to cooling is considered counterproductive.

Region G. Insignificant Winter Benefit, Marginal Summer Benefit. This is a transition area between zones F and H. Comments for zones F and H apply here in degree. The value of earth-tempering increases moving westward through this zone, and diminishes moving southward.

Region H. Insignificant Winter Benefit, Useful Summer Advantage. Summer ground temperatures are high, but relatively much cooler than air. Aridity favors roof mass, reduces the need for ventilation, and eliminates concern for condensation. The potential for integrating earth-tempering with other passive design alternatives is high.

Region I. Insignificant to Marginal Winter and Summer Benefit. Extraordinary means of climate control are not required due to the relative moderateness of this zone. Earth-tempering is compatible with other winter and summer strategies, with no strong argument for or against it.

Region J. Insignificant to Useful Winter Benefit, Marginal to Useful Summer Advantage. Latitude and topography cause wide variation in local climate and soil temperatures. Generally, comments for zones C and H apply here in degree.

Section 3. SITE EVALUATION

1. GENERAL. This section deals with site-evaluation issues affecting the design of earth-sheltered structures. The emphasis is on those site issues which are more important or more pertinent for earth-sheltered structures than for conventional structures.

2. SITE INVESTIGATION PROCEDURES. In an earth-sheltered building, the underground work may represent the major portion of the project. Thus, poor site investigation practices leave a large margin of uncertainty in the total project costs.

Checklists of site investigation procedures are given below:

a. Site Records. Existing records (legal and historical) should be the starting point for a serious investigation of a site. Aside from providing the legal description of a site together with any existing agreements restricting its use, such a survey can reveal what is likely to be encountered on the site in terms of existing foundations, utilities, a past history of water problems, subsidence, and so forth. Types of information that should be included:

- o Original plat/survey information
- o Easements (horizontal and vertical extent)
- o Mineral rights (together with any agreements as to access from the property)
- o Existing topographic information
- o Existing utility and building information (discussed below)
- o Records of seismic activity or zone clarifications for seismic design
- o Existing surficial and bedrock geologic information
- o Existing foundation or soil reports
- o Existing percolation test reports
- o Existing agricultural or planning studies
- o Existing information on ground water conditions and surface drainage
- o Soil temperature

The best sources of this information can vary, but the usual sources include:

- o Local, city, or county engineer's office
- o City or county records department
- o Aerial survey companies
- o State geological survey information
- o United States Geological Survey maps and reports
- o United States Department of Agriculture soils information (usually shallow only)
- o Previous records from nearby construction
- o Well-drilling records
- o Drilling records of public agencies
- o Technical papers on local geology, microclimate, soil conditions, and so forth (in nearest university library)
- o Experienced local soil-testing firms and consultants (as paid service)

- o Government agency responsible for adjacent roads or nearby facilities
- o Local weather reporting agency

b. Existing Structures and Utilities. Existing below-grade structures can include the following:

- o Storm and sanitary sewer lines, manholes
- o Water mains
- o Gas lines
- o Electric cables and vaults
- o Telephone cables and vaults
- o Tanks, septic tanks, cisterns
- o Steam or hot-water lines
- o Industrial waste lines
- o Fire lines or firetanks
- o Pipelines
- o Tunnels
- o Existing/abandoned pilings
- o Old building or equipment foundations
- o Age, condition, and depth of adjacent structures
- o Buried road beds
- o Filled-in basements of old buildings
- o Fortifications
- o Dump-site locations
- o Toxic agents
- o Buried explosives

c. On-Site Investigation. The on-site investigation should include the following:

- o Evidence of existing construction
- o Location and mapping joints on any rock outcrops
- o The existing canopy and understory vegetation, location of mature trees, and other natural features to be preserved
- o Changes in vegetation indicating changes in soil type, depth or ground water conditions
- o Erodibility of soil on any steep slopes
- o Potential views
- o Obstructions to access
- o Evidence of surface water drainage patterns
- o Adjacent structures
- o Verification of surface utility features such as manholes or power poles

Based on the site inspection, information from existing maps and records, and the anticipated siting of the building, a soil-boring program should be designed as outlined in NAVFAC DM-7 Series.

3. SITE INFLUENCES ON THE BUILDING DESIGN. The following site conditions are evaluated in terms of their relative impact on predesign decisions.

a. Climatic Influences. Since energy efficiency is a major criteria for evaluating earth-sheltered buildings, it is important to be aware of the general climate of the region in which the site is located (see Section 2,

Climate Evaluation). It is also important to establish the likely microclimate of the site, as determined by the existing topography, vegetation, and surrounding structures. These issues have been analyzed in detail by others. (See Reference 2, Design with Climate, by Olgyay; Reference 3, Man, Climate and Architecture, by Givoni; and Reference 4, Plants, People and Environmental Quality, by Robinette.)

For earth-sheltered buildings, the range in air temperature will indicate the moderating potential of the earth. High air temperatures indicate some earth-cooling potential. Although total energy savings from an underground location are usually less in warmer climates, earth-contact cooling may be one of the few passive techniques available. Wind speeds and humidity in combination with air temperature indicate the availability of comfort conditions in warm climates through ventilation strategies.

These are not always practical for large buildings or where dust and humidity must be controlled. The ground-temperature information gives a baseline comparison with outside conditions but does not represent the conditions in the ground adjacent to an operating building. The solar information should indicate the importance of incorporating passive or active solar strategies into the building design.

b. Environmental and Land Use Issues. Environmental or land use considerations very often are complementary to the development of an earth-sheltered design. Occasionally, these factors will suggest an earth-sheltered building as the only appropriate solution, especially when the preservation of natural surroundings is of primary importance. In non-urban areas, environmental issues generally relate to the potential of earth-sheltered buildings to preserve existing ecological systems.

c. Access and Building Services. A major problem in the site evaluation of an earth-sheltered or underground facility is the need to provide adequate and efficient access to the building. This usually conflicts with the desire to blend the building with the context of the site and to maintain the maximum amount of earth-sheltering. (See Section 9, Space Planning and Programming, for access requirements.)

A further design problem, from an aesthetic point of view, is that the building elements which provide the service and access functions are usually the least attractive elements of a building and will, in an earth-sheltered structure, be among the most visible. Solutions to this problem imply careful site and building design to allow as unobtrusive a service access to the building as possible.

A checklist of building access and service items to be considered is given below:

- Pedestrian access to the building
- Pedestrian access through or across the building
- Access for the disabled
- Service access
- Any significant materials-handling requirements
- Equipment access/removal
- Exterior storage areas
- Fire department access
- Overhead utilities

- o Surface utilities, for example, meters, transformers, junction boxes, fire hydrants
- o Relative heights of the lowest building level and the storm and sanitary sewers
- o Site drainage

d. Sanitary and Storm Sewer. The item on which a lot of attention has been placed in the discussion of major underground facilities is the potential location of a portion of the building below the level of the community's sanitary and storm sewer system. This is not particularly uncommon in building design and the situation is handled by providing a sump and submersible pumps to lift any drainage water or sewage from the lowest levels of the building. Sewage and water can be pumped to considerable heights. It is customary to provide standby power equipment for critical installations and also to provide duplicate equipment where servicing would exceed allowable shut-down times. See Section 17, paragraph 4 for further requirements. A standby pump could be designed to come in service in conditions such as drainage for firefighting. Outside access for pump suction pipes should be considered.

e. Isolation and Protection. In general, an earth-sheltered or underground building will provide improved protection in the following cases

- o Exterior noise protection
- o Low vibration requirements
- o Building security
- o Vandalism protection
- o Blast protection against exterior blasts
- o Protection to surrounding buildings from interior explosions
- o Tornado and hurricane protection
- o Earthquake resistance
- o Fallout protection

f. Topography. The most desirable sites for minimum cost of construction will generally be gently sloping sites of a large size in relation to the building. Deep buildings on flat sites will inevitably involve substantial net excavation, requiring either hauling or redistribution of the excess on the same site. Steep slopes should not always be construed as disadvantageous for earth-sheltered buildings. In fact, a series of narrow earth-sheltered buildings can make effective use of steep slopes which would otherwise be almost unbuildable.

g. Ground Conditions/Ground-Water Conditions. Ground conditions will determine the ease of excavation, the bearing capacity of foundations, lateral pressures for design, and the suitability of the excavated material for backfilling. Special consideration must be given to the effects of earthen load relative to potential consolidation. Larger loaded areas will transmit pressures to deeper depths.

Sand or gravel is usually recommended for backfilling around an underground structure for three reasons: (1) it is easy to compact, (2) it provides good drainage, and (3) it provides more predictable and lower lateral pressure when drained. If sand and gravel are not available on the site, they can be brought to the site in limited quantities for the immediate backfilling of the walls. If this is not economical owing to very great hauling distances, the walls may need to be designed for higher lateral pressures and drainage provided adjacent to the wall using a drainage mat. See Section 7, paragraph 3 for further discussion.

Rock conditions within the excavation will necessitate a change in excavation method. Rock ripping or drilling and blasting are both usually substantially more expensive than soil excavation. To some extent blasting costs can be offset by the elimination of temporary retaining walls, which are not usually required for excavation in competent rock. This aspect can be significant for deep buildings on small sites but blasting will usually be subject to restrictions to protect adjacent structures. The larger types of excavating equipment available today are able to rip most weathered rock and even jointed layers of quite competent rock. Ripping and blasting are both more feasibile on large than on small sites. A bedrock layer that slopes across the site will require careful foundation design to ensure even settlement of building foundations.

The presence of ground-water levels within the planned building depth will always complicate the building design and construction and, hence, increase its cost. There are two main options: (1) draining the soil and drawing down the water level, or (2) completely waterproofing the foundation and floor.

If the building is to be fully waterproofed, it must be designed to counter buoyancy and to withstand the lateral water and soil pressures. Buoyancy may be countered by one of the following: (1) providing a thick, mass concrete floor, (2) tying the building down to a rock layer below using drilled anchors, or (3) by using an extension of the floor slab beyond the building walls to mobilize the weight of the soil mass above this portion of the slab.

h. Excavation Costs. Several factors affect the cost of excavation and the temporary and permanent retaining walls required. Excavation is cheapest when the excavated material can be loaded directly by large capacity equipment onto trucks or, better still, stockpiled (on a large site) for use as backfill. Costs increase when conditions are not ideal, as indicated by the following examples:

- Costs are higher for deep excavations where trucks cannot drive to the bottom of the excavation and a large portion must, therefore, be excavated by drag line or bucket.

- An excavation where the progress is too slow to allow continuous operation of trucks may cause stockpiling and double handling of material.

- Excavation sites with insufficient adjacent-site area may result in inefficient use of equipment due to access problems and a lack of storage area.

- Costs are higher for excavations requiring extensive temporary retaining walls.

- Costs are higher for excavations which extend below the foundation levels of nearby structures.

Section 4. BUILDING TYPE EVALUATION

1. GENERAL. Some building types are more suitable to earth-sheltering than others. This section identifies the relative advantages and disadvantages of earth-sheltering for various types and sizes of buildings.

2. HIGH INTERNAL HEAT LOADS. Building functions that tend to have high internal heat loads include data-processing rooms, communication facilities, food-preparation facilities, and similar functions. Often, the heat-producing portions of these buildings will require cooling year round. Where these functions are located in separate mechanical zones away from the perimeter of the building, earth-sheltering will not affect the energy requirements for those portions of the building significantly. If the heat-producing zones are located on the perimeter, energy savings can be achieved for those zones. However, even in the most optimal design considerations--a zone with maximum earth-contact and cold ground temperatures--heat loss from the earth-contact surfaces may be one or two orders of magnitude smaller than the internal heat generated by people, lights, and equipment. If the cooling-load profile is variable in nature (for example, a conference room which is not always occupied), the heat loss to the earth-contact surfaces will serve to dampen the peak cooling load, thus reducing cooling equipment capacities, leading to lower first costs in mechanical equipment and higher operating efficiencies. It is essential that not only the integrated annual energy performance, but also the instantaneous energy performance of the building be considered when quantifying the thermal benefits available from earth-sheltering.

3. MODERATE INTERNAL HEAT LOADS. Buildings characterized as having moderate internal heat loads include offices, housing, medical facilities, and educational facilities. Buildings of this type are all conducive to earth-sheltering depending on other variables such as climate.

4. BUILDINGS WITH MINIMAL HEATING REQUIREMENTS. Buildings with minimal heating requirements are best represented by dry-storage warehouses. A typical ambient air temperature for such facilities may be 55°F (12.8°C). Such buildings do not easily achieve long-term economy by earth-sheltering. This is because the lower the indoor ambient temperature, the lower will be the energy savings in the heating seasons. Also, dry-goods warehouses are usually not air conditioned; thus, the cooling benefits associated with earth-sheltering will not serve to offset the operating costs of that kind of building.

Refrigerated warehouses have the potential to benefit from earth-sheltering in several ways. First, the earth acts as a temperature moderator, reducing heat gains to the refrigerated spaces. As a result, cooling loads will be reduced and refrigeration equipment will operate under more stable cooling loads, resulting in higher operating efficiencies. Second, in the event of a power or equipment failure, the warehouse will maintain its design temperature for a longer period of time than would a lightweight structure, thus minimizing the need for backup cooling equipment

5. ROOF-DOMINANT BUILDINGS. Roof-dominant buildings have roof to wall area ratios greater than unity. From the point of view of life-cycle cost,

roof-dominant buildings have less to gain by earth-sheltering than buildings with small roof areas. When land costs are not a factor in the life-cycle cost analysis, life-cycle savings will not normally be achieved by earth-sheltering the roof. In this case, depths of earth cover that would have a thermal mass close to the thermal mass obtainable at the wall, are not economically achievable due to the high structural costs for supporting the additional weight. Consequently, the mechanical savings per unit of construction cost of the building envelope will be much lower for roof areas than for wall areas.

In northern climates, an earth-sheltered alternative to a conventional single-story building will have better life-cycle results as a three-story building than as a one-story building. Even better life-cycle results will be obtained for earth-sheltering a building that would otherwise be multistory in its conventional form. In southern climates, a multistory earth-sheltered building will normally not be a life-cycle improvement over a one-story conventional building.

6. VENTILATION AND INFILTRATION-DOMINANT BUILDINGS. In conventional buildings, uncontrolled infiltration is often the single largest factor contributing to the heating loads. Infiltration occurs at building joints, roof edges, windows, and doors. Infiltration rates can be improved by appropriate detailing and by the selection of products with better infiltration performance. Infiltration at entrances can be decreased by using revolving doors and vestibules instead of single-swing doors. To a large extent, however, infiltration through the building envelope is a matter of quality control. Earth-sheltering effectively eliminates infiltration without regard to quality control. Where the amount of fenestration can be reduced below normal requirements, infiltration reduction by earth-sheltering will often be the most significant contribution to the life-cycle savings for buildings in northern climates. Buildings with low heating loads, and therefore buildings that are not infiltration-dominant, will not achieve as significant a reduction in energy costs.

The heat losses or gains due to the introduction of outside air into an occupied space for ventilation purposes usually comprise a major portion of the total load. Any energy saving potential of underground space can easily be offset by ventilation requirements of the building. It is important that the ventilation load (amount of outside air introduced into the building) be minimized. Refer to the recommended levels of ventilation air as set forth by American Society of Heating, Refrigerating and Air-Conditioning Engineers Standard 62-1981 (See Reference 5, ASHRAE Standard 62-1981). Minimizing ventilation requirements will certainly enhance the energy saving potential of underground space. Energy recovery systems have the potential to recover 30 to 60 percent of the energy lost through ventilation.

7. MECHANICAL ZONING OF THE BUILDING. The mechanical zoning of a building has an important effect on the relative energy performance of an earth-sheltered building. Buildings having discrete mechanical zones with little or no thermal interaction between zones cannot be earth-sheltered to reduce energy loads on interior zones. Consequently, such buildings can make the most of earth-sheltering by locating the zones with the highest energy requirements on the perimeter of the building. It may also be

possible to recycle air from zone to zone to achieve greater thermal interaction between earth-sheltered zones and non-earth-sheltered zones, though, the parasitic power costs associated with moving air from zone to zone may offset the energy savings. The use of heat pumps between zones having opposite cooling or heating requirements may also be a consideration.

8. BUILDING SIZE. The size in total floor area of a building affects at least four other variables. These are: roof-dominance, life-safety costs, perimeter requirements, and mechanical zoning arrangement. In general, the smaller and more compact the building, the more these variables will work to improve the life-cycle economy of earth-sheltering. Generally, a smaller earth-sheltering building with a smaller roof area will be more effectively earth-sheltered than a building with a larger roof area.

a. Large Buildings. As the size of a building increases, the life-safety requirements for the number of exits increases. Further, if the depth from a windowless wall is increased beyond certain limits, additional fire access panels or an automatic fire-extinguishing system may be required. Large buildings will tend to have more zones without earth contact than smaller buildings. The ratio of earth-contact zones to non-earth-contact zones can be increased, however, by increasing the length-to-width ratio of the building. The economic trade-off will be the increased initial costs for the additional wall area. In large buildings, earth contact will have little impact on the heat flux through the floor for zones located away from the perimeter.

b. Small Buildings. If a building is small enough, fenestration on one side (which may be used for passive solar) may adequately serve most of the building area. Smaller buildings will have a smaller volume of independent interior mechanical zones. Thus, more of the zones will be able to take advantage of earth-sheltering. In very small structures without interior zones, the earth-sheltered building can be arranged to provide all exiting and natural light requirements on a south-facing wall which also takes advantage of passive solar. Passive solar and earth contact in residential-size buildings will affect all zones of the building. At this scale, the structural bay becomes small enough to make earth-cover on the roof less expensive. However, in warm humid climates where small conventional buildings would normally take advantage of passive cooling, earth-sheltering will not have an advantage. In such climates, the most effective means of passive cooling will be by cross-ventilation rather than by earth contact.

9. LEVELS OF TECHNOLOGY. Building size normally correlates with the type of construction. At one end of the scale, large buildings will normally be constructed out of reinforced concrete or structural steel; at the other end, the building may be based on hammer-and-nail technology. A rough line divides these buildings into what can be termed "institutional" and "domestic" types of construction. Domestic construction is inherently less expensive both in terms of labor and materials. Environmental controls are less exact. Small earth-sheltered projects which would normally be associated with a less sophisticated construction are often drawn into an institutional type of technology. Thus, on very small projects, the possible costs of shifting to a different technology must be considered in the life-cycle cost analysis.

10. SPECIAL PERIMETER REQUIREMENTS. Buildings that have special requirements for access, such as warehouses, may not be able to take advantage of earth-sheltering on all sides. The requirement for natural light and/or views at exterior walls also conflicts with earth-sheltering. In some instances, the sill height can be raised above berm height. Skylights and atriums, which are provided to make up for natural light lost at the exterior walls, must be carefully analyzed for energy loss and additional construction cost. Interior planning should include the possible use of borrowed lights to more economically distribute available natural light.

11. LIFE-SAFETY REQUIREMENTS. Windowless and underground space is at a disadvantage from a life-safety point of view. Most of the exterior walls will not be accessible for firefighting. Exiting in an upward direction is more difficult. Without proper mechanical measures, smoke will tend to accumulate in the upper portions of the building and clog exit ways. In order to offset these disadvantages and maintain the same degree of safety, automatic fire-extinguishing systems, refuge areas, compartmentalization, pressurization, and smoke removal by mechanical means often must be provided. These requirements may increase the initial construction cost. Buildings which would ordinarily have such features as the automatic fire-extinguishing system and pressurized stairwells, will consequently be more cost effective as earth-sheltered buildings than facilities which would not otherwise have these features (see Section 12, Fire Protection and Life Safety).

Section 5. LIFE-CYCLE COST EVALUATION

1. GENERAL. An evaluation of the economy of earth-sheltering must be based on a life-cycle cost analysis which conforms to NAVFAC P-442, Economic Analysis Handbook (see Criteria Sources). This section identifies the analysis-sensitive variables and overall life-cycle trends for various types of earth-sheltered facilities.

2. INITIAL COSTS. Initial costs include all one-time expenditures incurred for financing construction and any related costs such as land acquisition.

a. Construction Scheduling. Consideration must be given to the duration of the construction phase, seasonal timing, and mechanical systems start-up.

b. Land Costs. Occasionally, where a value can be assigned to land, and where there is a programmatic requirement or a future requirement for open space, the relative cost for land must be included in the life-cycle analysis.

This can be either a negative or a positive factor. For instance, if there are no present or future requirements for open space (parking, parade fields, parks, and so forth), a bermed building plus the related site work may require more land area than the conventional alternative. If, on the other hand, the program requires a parade field which can be located on the roof, the total land area for the project may be reduced.

c. Excavation. Excavation costs for earth-sheltered buildings can increase significantly over conventional buildings. Particular attention should be given to possible shoring and underpinning costs. Excavation below a water table will result in additional costs for dewatering. Determine if material will have to be hauled from the site.

d. Earthwork. Buildings with extensive berming and/or earth cover will involve moving greater volumes of soil. Due to the method of placement and compaction requirements, unit costs for placing soil against the building will be higher than normal grading unit costs. Unit costs for placing soil on the roof will also be higher. In addition to the placement of soil, materials such as soil separators, gravel, and, in some instances, irrigation systems, must be accounted for. Determine what materials and quantities will have to be hauled to or from the site.

e. Planting and Ground Cover. Ground cover will be more expensive to install than grass, but, the maintenance costs may be lower.

f. Civil. Consideration should be given to costs for additional paving required to meet existing road elevations, area drains, sumps, additional storm sewer, and retaining walls.

g. Architectural. The possible need for parapet walls, guardrails, and special security measures (at skylights, for example) should be considered. Roof edge details may be more costly for structures bermed only to the roof. Reduction of above-grade exterior finishes and windows will result in significant initial savings depending on the relative cost of exterior wall systems.

Consider the possibility of increased insulation thickness over the amounts required for the conventional structure. Because insulation is usually more economically installed below-grade than above-grade, additional amounts of below-grade insulation may have a life-cycle cost benefit that is not as attainable above-grade.

Waterproofing will normally be a more significant factor in earth-sheltered buildings than in above-grade buildings. A distinction should be made between "dampproofing" and waterproofing. Determine if waterproofing is required below the floor slab.

Consider additional costs for an automatic fire extinguishing system and smoke removal systems. Requirements for compartmentalization and refuge areas will increase architectural costs.

h. Structural. Bermed single-story buildings may achieve reductions in the depth of frost footings. Earth-covered roofs will increase the structural costs for roof framing, columns, and footings. Lateral pressure will affect the thickness and reinforcing of perimeter walls. Structures designed for buoyancy will have additional foundation and floor-slab costs. Column spacing may be different for buildings with earth-covered roofs. Perimeter columns may be fewer or even eliminated with cast-in-place concrete walls. Unconventional structural systems such as long-span concrete barrel vaults may be especially cost-effective in a below-grade application with earth cover on the roof. See Section 16, Structure.

i. Mechanical. Equipment and distribution-system costs will be lower based on lower peak design loads. Reduced mechanical costs typically represent a major savings for earth-sheltered buildings. In cases where the sanitary sewer is higher than the lowest floor, additional costs may include sewage ejection.

3. FUTURE ADDITIONS. If the facility requires future expansion, the costs of expansion will likely be higher for an earth-sheltered building. The extent of the additional work involved in below-grade connections must be estimated. (This amount must be increased for inflation to the probable year of the addition and then discounted to present worth.)

4. MAJOR REPAIRS. Major repairs for both the conventional building and the earth-sheltered building must be inflated and discounted in the same way. Roof repairs for an earth-covered building will be higher than repairs for conventional roofs. However, if the roof membrane is guaranteed by the contractor for a number of years and if the repair costs borne by the contractor include removal and replacement of the earth-cover, then repair costs will likely be higher for the conventional roof. Once leaks have been discovered and repaired within the guarantee period, the membrane roofing on an earth-sheltered building should perform better than the conventional built-up roof over a 25-year period. Further, the built-up roof will normally have to be replaced within the life-cycle cost period. The replacement year will vary according to the type of the roof construction and exposure, but in any case, replacement of the built-up roof should constitute a major life-cycle advantage for the earth-covered roof. Replacement of metal coping, flashing, scuppers, and downspouts should also be taken into account. Replacement of mechanical and electrical equipment

will tend to favor earth-sheltering. Tuckpointing, painting, caulking, and patching should be included for above-grade walls if applicable.

5. OPERATIONAL ENERGY COSTS. Energy costs must be calculated for both the conventional and earth-sheltered alternative. Costs for heating and cooling will normally be substantially lower for the earth-sheltered building. Electrical costs for lighting can be considered to be the same as for the conventional building unless it can be demonstrated that lighting levels in the conventional structure would be reduced due to the availability of more natural light. Electricity costs for fan energy may or may not be significantly reduced, depending on ventilation requirements and the type of air handling system. All energy costs must be reduced to cumulative discounted worth.

6. MAINTENANCE AND OTHER OPERATIONAL COSTS. Maintenance costs are annually recurring expenses and must be reduced to cumulative discounted worth. Often the greatest differences in maintenance costs will be for irrigation water and mowing, pruning, and fertilization for grass and ground cover. Other affected costs include window cleaning. Operational costs for security may be reduced depending on surveillance requirements. Insurance costs will not normally be applicable to Navy facilities.

7. GENERAL TRENDS. Buildings in northern climates are economically bermed half-way up the exterior wall. This eliminates foundation work between the floor slab and the frost footing. These construction savings may be offset, however, if the portion of the wall below grade must be designed as a cantilever retaining wall.

Earth-covered buildings that prove to be economical will usually achieve life-cycle economy by energy savings attributable to the earth-contact of the walls. An earth-covered building that uses conventional structural systems may not be more economical than a fully bermed building. There are obvious exceptions, including those buildings that achieve a reduction in land costs by utilization of the roof area.

Compact buildings with low roof-to-wall area ratios will tend to be more economically earth-sheltered than buildings with large roof-to-wall area ratios (See Section 4, Building Type Evaluation).

Earth-sheltered buildings in very mild climates are usually not justifiable on an economic basis alone. The less energy at stake, the less likely that earth-sheltering will have a life-cycle advantage.

The inclusion of an automatic fire-extinguishing system in an earth-sheltered building will be a significant set back to life-cycle savings if the conventional building (used as a basis in the analysis) does not have such a system. Construction below a water table will have a similar impact on the life-cycle results.

If, in a northern climate, the life-cycle basis is a building that has high rates of uncontrolled infiltration, major savings will usually be realized by earth-sheltering.

8. MARGIN OF ERROR. Regardless of the accuracy of the cost estimate for

the alternatives under consideration, these values will be combined with gross estimates for fuel and electricity escalation rates and discount factors (see NAVFAC P-442). Over a 25-year period, a difference of 3 percent in the chosen discount factor will make a difference of as much as 30 percent in energy savings and this is also true of fuel escalation rates.

9. INTANGIBLE CONSIDERATIONS. In comparing earth-sheltered buildings with above-grade buildings, a number of factors to be considered that are not readily translated into cost. These include exterior noise reduction, blast and vibration resistance, and aesthetic and psychological considerations. Thus benefits may be realized that are not specifically required by the building program.

All life-cycle cost comparisons involving earth-sheltered buildings must be accompanied by a qualitative description of the aesthetic, programmatic, psychological, and lighting implications of the proposed design. The criteria in Sections 6, 9, 10, and 11 should be used as a checklist for this report.

Section 6. SITE CONTEXT

1. GENERAL. This section discusses the visual relationship of earth-sheltered buildings and their surroundings. Selected examples of situations that are common to earth-sheltered buildings are included.

2. BASIC RESPONSES TO CONTEXT. The following conditions must be considered in response to context.

a. Nature. Earth-sheltered buildings are readily integrated with the natural surroundings. An extreme version of this is the entirely submerged building. Where it is desirable to eliminate all exposed walls, an atrium may be necessary for natural light, ventilation, and exits. An atrium allows nature to be reintroduced at the interior of the building to replace the views lost at the perimeter.

b. Reference. A large earth-sheltered building can contribute to the reference system by creating a well-defined open space. For instance, the roof of an earth-sheltered building can be designed as a central parade ground.

With regard to visual hierarchy, earth-sheltered buildings are initially at a disadvantage. This is obvious in terms of the facade--often, an earth-sheltered building will have fewer exterior elements for manipulation.

Without proper attention, the building services and mechanical features of earth-sheltered buildings will tend to be more prominent. Exterior access for the replacement of mechanical equipment, access for trash removal, mechanical louvers, meters, vents, and boiler flues must all be located above grade. Such features should not be allowed to become more dominant than the building entrance.

A similar problem can occur with the parking lot. If an unobtrusive earth-sheltered building is directly juxtaposed to the parking, the building can visually become an accessory to the parking rather than the other way around. Hierarchical articulation of the building form and visual reinforcement of the approach to the building will mitigate this problem.

c. Coherence. Coherence is best explained as a condition that complies with an expected arrangement. If a new facility is added to a base, a degree of coherence can be maintained by incorporating similar building materials.

If the building is earth-sheltered, it is quite possible that the form of the building will not have coherence in terms of the way it meets the ground. For the above-grade portions of the building, however, the designer has recourse to using materials that are similar to the materials used on existing adjacent buildings.

In an urban environment, coherence is maintained with respect to the streets and lot lines. A conventional building which would normally exceed the established building height or bulk if built entirely above-grade can be lowered by providing a number of sub-grade levels. This strategy not only takes advantage of earth-sheltering but allows the building cornice line to match the existing cornice lines. It is imperative that an earth-sheltered building follow the established setback pattern with its perceived above-grade mass. By doing so, the new building will contribute to the definition of the street and will appear to be conventional in terms of its street image. The above-grade portion will complete the street while the below-grade portion will take full advantage of earth-sheltering.

In general, coherence with adjacent buildings is less critical in non-urbanized environments, but, there are circumstances where it should be addressed. For instance, along a shoreline or other natural boundary, it will normally be desirable to preserve the continuity of this natural edge. An earth-sheltered building has a strong potential in this regard. Further, a submerged building may allow a view that would otherwise be blocked. In some instances the roof of the building may serve as an overlook or an esplanade.

d. Identity. Positive identities should be reinforced rather than disrupted. Where a group of buildings have characteristics that promote a positive identity, the above-grade portions of an earth-sheltered building can be designed to reinforce that identity. In the process, any negative associations of underground space will be improved by this external identity

e. Tradition. Certain traditional building features, such as fenestration, porticos, and cornice lines indirectly communicate the status of the building, how it is to be approached, and how it is to be used. In terms of the approach to a building, two features--stairs and the foundation--have particular significance for earth-sheltered buildings.

An ascending flight of stairs has a strong subconscious impact. Ascent before entering has a preparatory effect. Raising the entrance above the ground lends importance to the building. Most important of all, it correlates with the direction that has a positive psychological identity.

In much modern architecture, the exterior flight of stairs is omitted--in part because of such practical considerations as the economy of slab-on-grade construction and handicap access. Without proper design, earth-sheltered buildings can compound this problem. Not only is the preparatory ascent missing, but the ultimate direction is now downward.

Often, on sloping sites, a horizontal approach can be easily accommodated. For a multistory, earth-sheltered building, a horizontal approach is usually simple to achieve, especially when all or part of the top story is above-grade. On flat sites, this condition can be obtained by designing the building so that the downward path is contained within the building.

The expression of the foundation has traditionally served a purpose similar to the exterior stair. A foundation allows the building to be perceived as being special and disassociated from the ground. Premodern buildings, especially, delineated a buffer zone between the main floor and the ground. In much modern architecture, the articulation of the base is omitted, partly due to the economy of construction and partly due to prevailing aesthetic theory. The lack of the base delineation, which can be a perceptual disadvantage in conventional buildings, is usually critical to the perception of earth-sheltered buildings. Normally, earth-sheltered buildings cannot meet the ground through the transition of a base; instead they comprise what was traditionally the base itself. The earth-sheltered building, consequently, becomes aethestically and psychologically associated with the base and with basements.

There are two basic strategies for correcting this condition.

First, in a context where coherence with adjacent buildings is not crucial, the building can be arranged so as to break up the references to the rectilinear base entirely. Alternate building features which do not connote a base can be employed. These associations include pavilions (as entrances), greenhouses (as skylights), and pitched or shed roofs (as clerestories). Exterior materials such as glass, metal, and tile can be used instead of exposed concrete to further develop these images. Color may also be used to advantage. Preferably, the approach would be horizontal and would terminate at an above-grade entrance feature.

The second strategy involves articulating a traditional facade. This will be particularly effective in an urban context where the expression of a conventional facade is an appropriate response to existing buildings. Alternatively, the earth-sheltered building can be located completely below-grade with access and mechanical service from an existing building. Additional required exits and natural light can be provided through open recessed atriums or courtyards, depending on the site.

f. Territoriality. The term "territoriality" refers to the need for psychologically defined and defensible personal space. The extent to which a person will physically invade a space is related to the degree to which territoriality is defined, often subconsciously, by commonly understood conventions. The territoriality of open spaces between buildings is improved by visually defining public, private, and semi-private domains. This involves orienting the buildings and utilizing landscape features to avoid ambiguous residual spaces. In urban environments, or large military installations, especially, where multiple housing is involved, good territorial definition has the practical benefit of reducing crime and vandalism and, in the case of housing, fostering pride and voluntary maintenance.

Earth-sheltered buildings which are substantially covered or bermed cannot define exterior space as well as conventional buildings. Further, lack of fenestration overlooking exterior spaces inhibits voluntary surveillance. Greater attention must, therefore, be given to landscape design to visually define such spaces. Slight changes in grade elevation, terraces, changes in planting and ground cover, shrub and tree placement, and special paving materials at private areas and entrances can provide effective psychological barriers.

3. EXAMPLES OF RESPONSE TO CONTEXT. The following examples are of buildings in temperate climates. In general, though, the basic principles illustrated are valid for other climates as well.

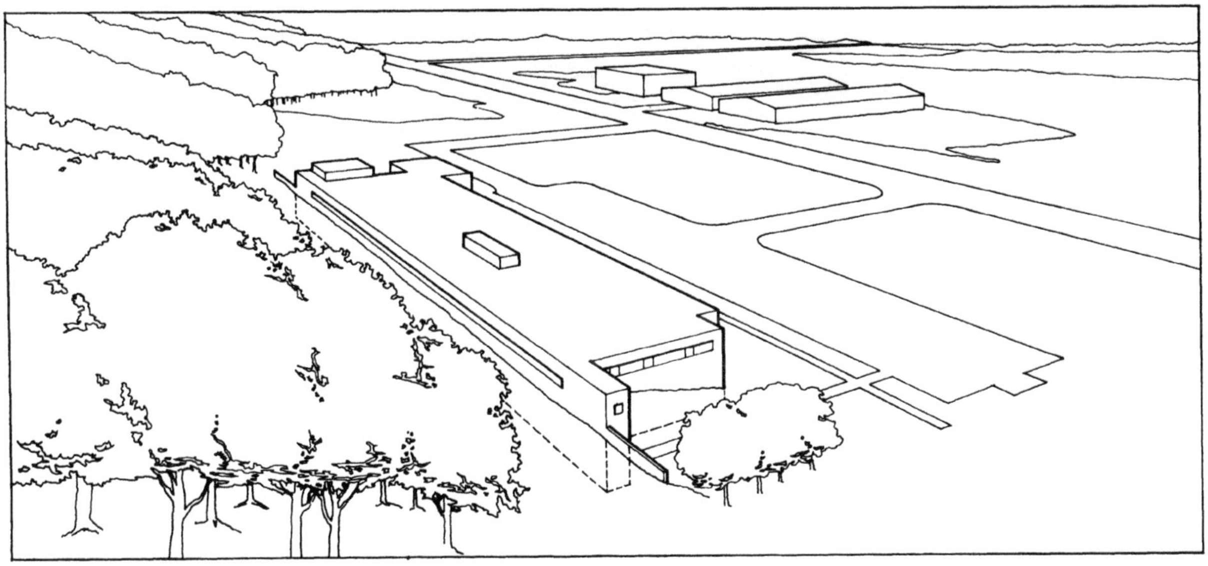

FIGURE 2
Contrast to Natural Setting

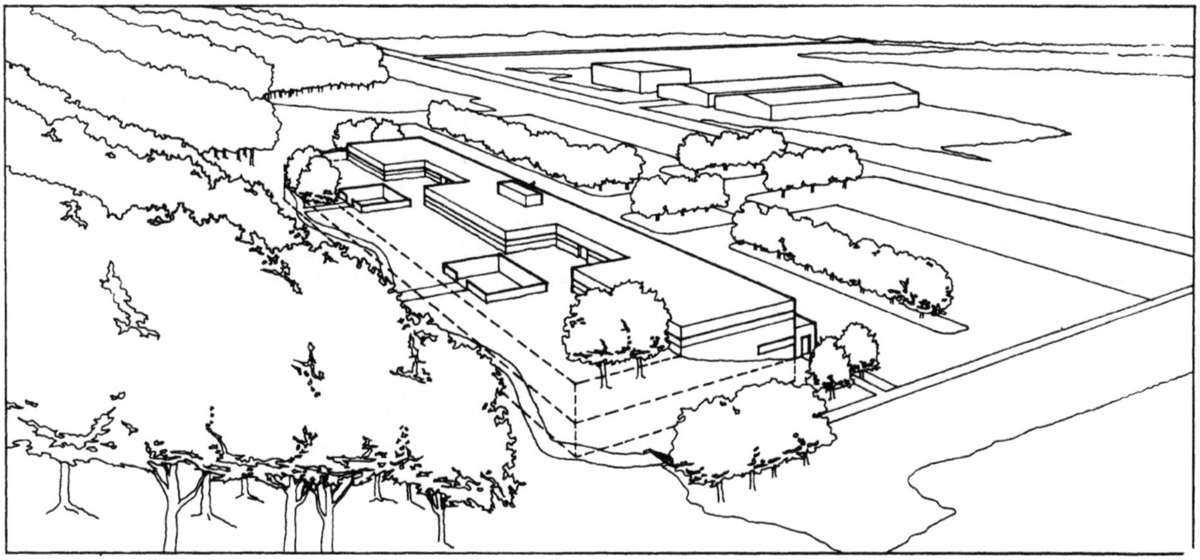

FIGURE 3
Integration with Natural Setting

a. Earth-Sheltered Building on Wooded Site. The site illustrated in Figure 2 is partially wooded with a buildable open area adjacent to access roads. Site planning suggests that the parking be located in the open area with the building located partially in the woods. The building illustrated in Figure 2 is conceived as a juxtaposition of the man-made and natural environment. In Figure 3, the building occupies the same location. In this case, however, the concept is to avoid the confrontation between building

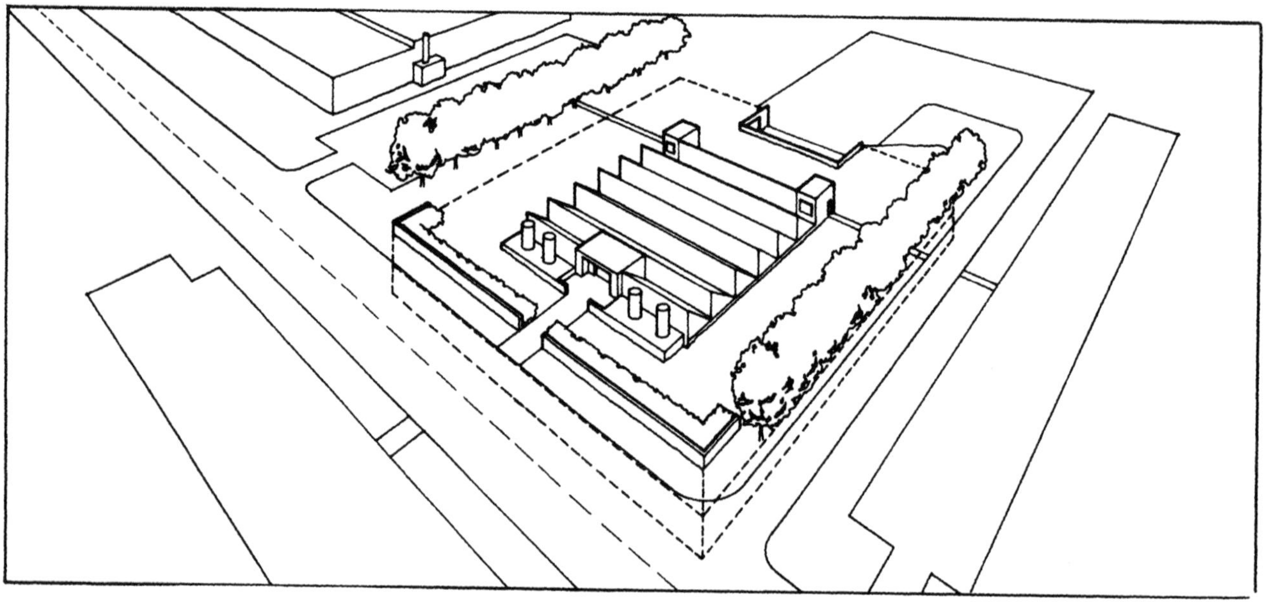

FIGURE 4
Submerged Building with Enclosed Atrium

and nature. The form of the building and the planting are modified to accomplish this. In addition, the parking area is more extensively planted.

b. Earth-Sheltered Building on Unwooded Suburban Site. The site illustrated in Figure 4 is adjacent to a highway. The land is moderately flat without any exceptional views. The surrounding area is zoned for light industrial and businesses. The earth-sheltered building shown is bermed and covered. To take the place of perimeter windows, a skylighted atrium is provided. Service access is slightly below grade at the first-floor elevation at the rear of the building. Mechanical shafts are located to reinforce the importance of the main entrance.

c. Earth-Sheltered Building on Flat, Open Site. The site illustrated in Figure 5 is surrounded by roads on three sides, with the main road on the east side of the site. The facility illustrated is a single-story administration/warehouse building. The administration half of the building is earth-covered and is located on the south for passive solar advantage. Notice that the relationship between the main street, the parking, and the main entrance is awkward--visually the main entrance seems to be incidental to the parking. To correct this situation, the building can be turned so that the windows face east. This is justified by the relatively insignificant effect of orientation on solar gain for a large earth-sheltered building in a temperate climate. Figure 6 illustrates this solution. The importance of the entrance is heightened by a hierarchical articulation of the facade and the planting and parking layout visually extend the entrance to the street.

d. Approach to Covered Building on Flat Site. The approaches and entrances to buildings which have few exposed exterior areas must be carefully designed with regard to hierarchical and traditional references.

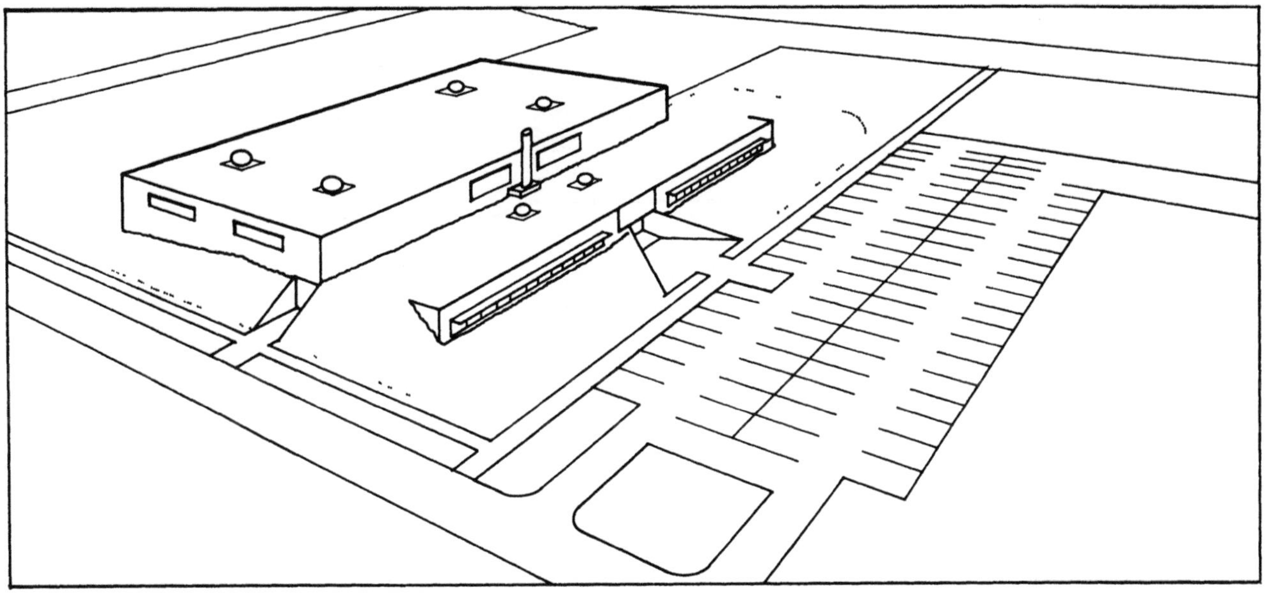

FIGURE 5
Incidental Relationship to Parking

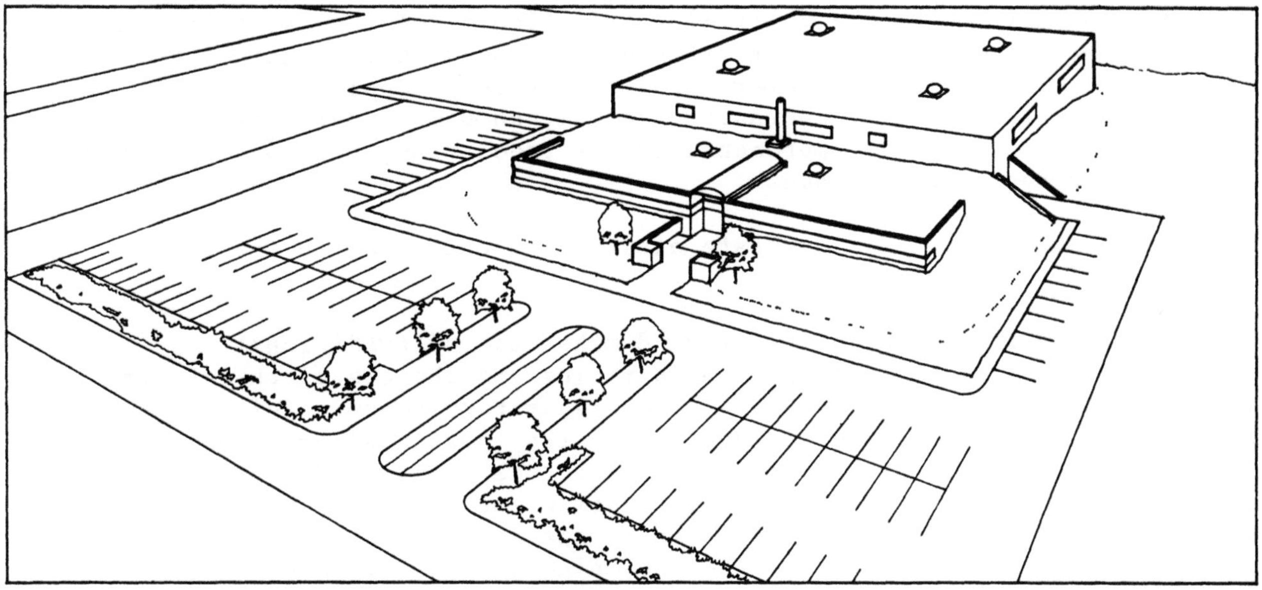

FIGURE 6
Hierarchical Relationship to Parking

Figure 7 illustrates a poorly designed approach to a covered single-story building on a flat site. The traditional references in this example have negative images. In particular, the sloped concrete retaining walls connote tunnel or culvert construction. Further, the broad downward sloping ramp with the trench drain at the bottom is an image that is customarily associated with parking garages. The only references to the fact that this is a pedestrian entrance are the handrails and the entrance doors themselves. From the parking lot, however, these clues are visually lost.

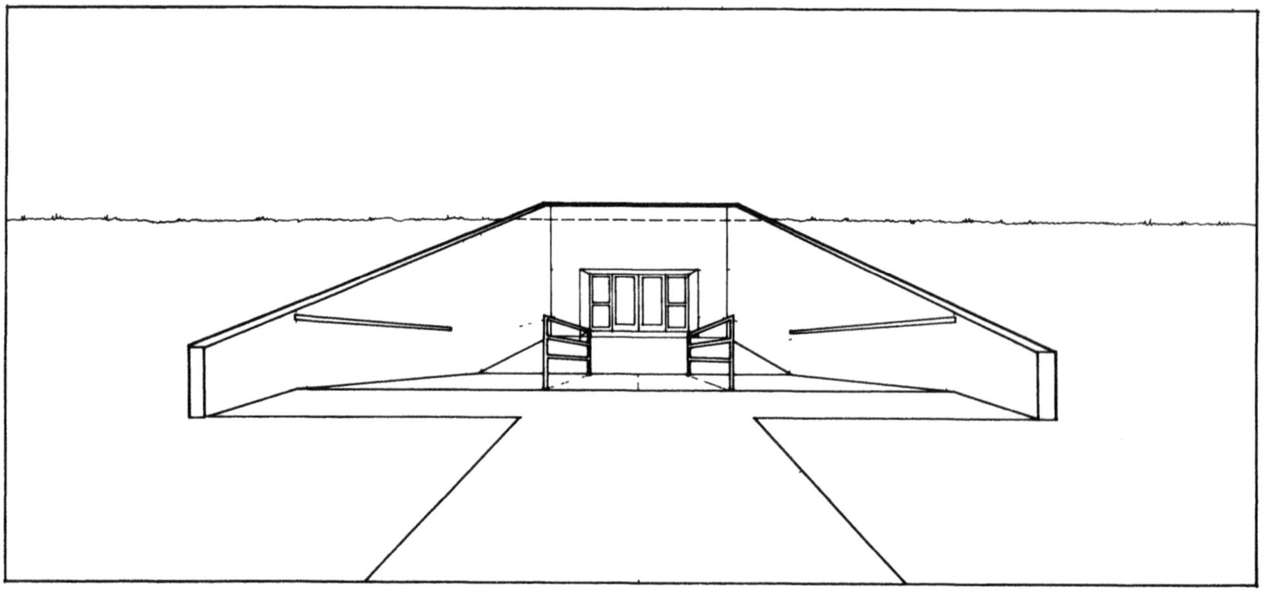

FIGURE 7
Entrance Approach with Poor Image

In addition to the misleading traditional references, this example lacks a hierarchical sequence of references. The visual impact of the parking lot will easily overpower a building with an approach like the one illustrated. The building entrance, instead, should be perceived as a goal. As the entrance is approached by the pedestrian, he should encounter a reassuring sequence of images.

Figures 8 and 9 illustrate alternative entrances to the same building.

The building shown in Figure 8 makes use of traditional architectural elements. Between the building and the toe of the sloping approach walk, an area is reserved as a preparatory space. This has the same psychological function as approach stairs, porticos, or canopies in traditional architecture. In order to make this landing effective, it is finished with a paving material that is different from the surrounding paving. The paving pattern also forms strips across the approach path that form psychological thresholds. The retaining wall is allowed to step down in a manner similar to a side-parapet on traditional stairs. The bottom half is detailed as a traditional rusticated base by nailing horizontal strips to the inside of the concrete formwork. The top half of the retaining wall is sheathed in masonry, thus establishing a material that is commonly used above-grade. The incorporation of window openings and a mullion grid at the entrance reinforces the human reference. Finally, a symmetrical skylight is added over the lobby and main corridor area. In addition to providing direct and indirect natural light to the interior, the gable end references a traditional entrance pediment.

The entrance in Figure 9 accomplishes a similar result without conventional use of traditional features. The entrance is recessed back from the building line in this case. The retaining walls are inflected

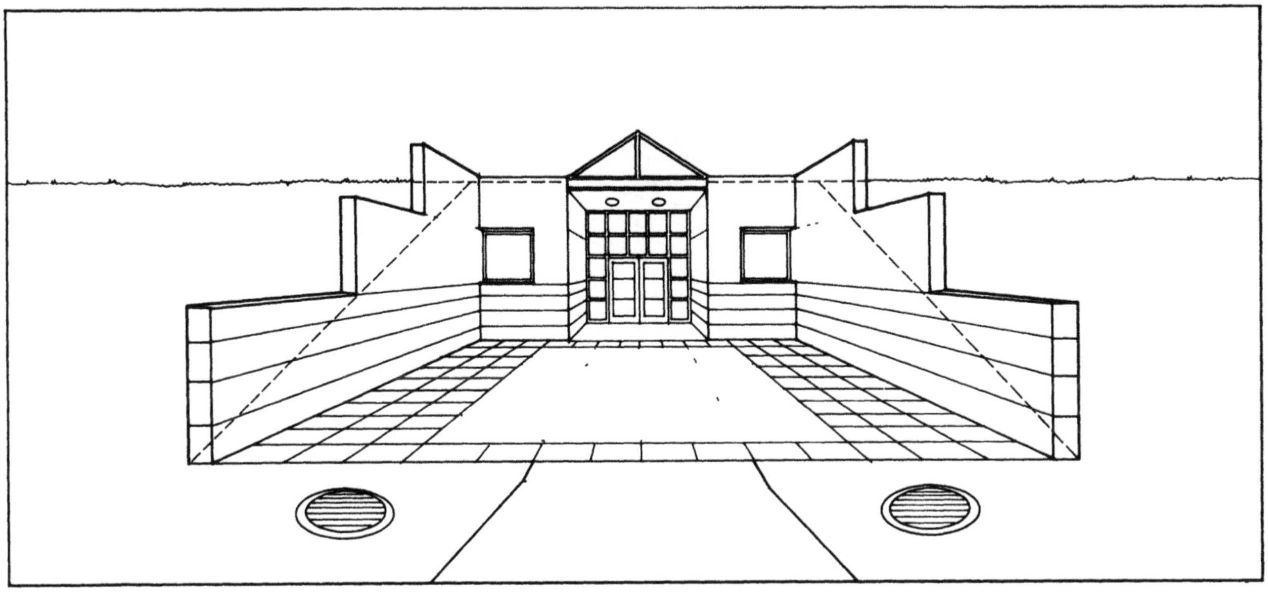

FIGURE 8
Entrance Approach with Implied Forecourt

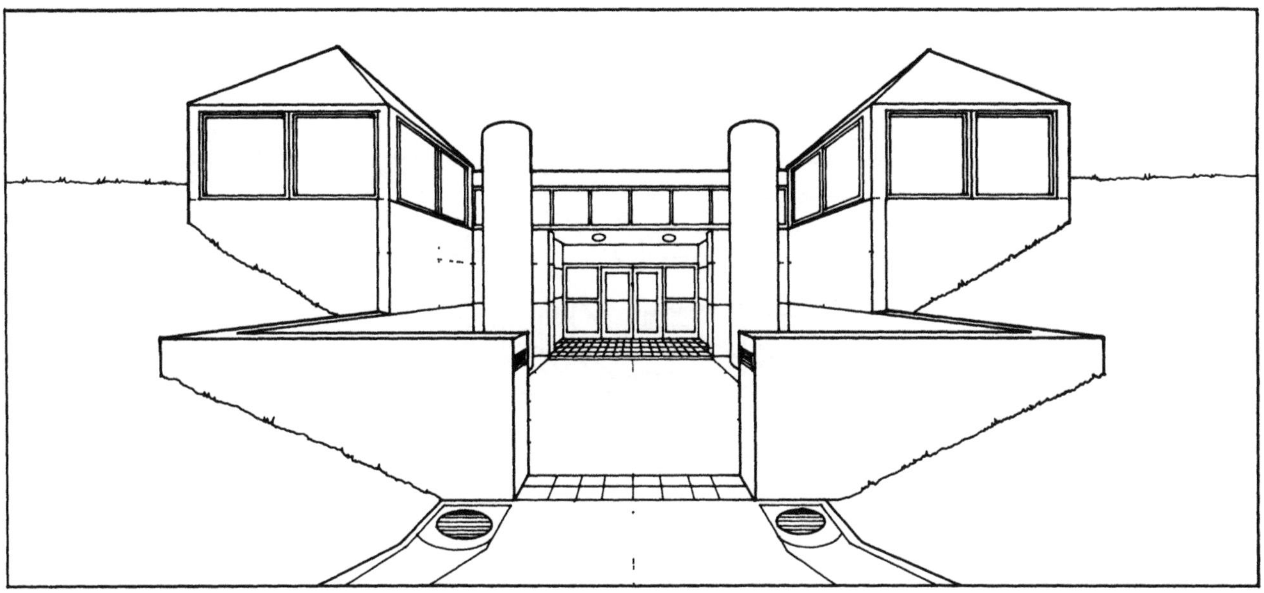

FIGURE 9
Entrance Approach with Literal Forecourt

towards the approach path. This results in a preparatory space that is like an exterior room or forecourt. In the process, the references to the traditional base and its association with the ground are suppressed. Thresholds are formed at both ends of the forecourt by using special paving blocks. The clerestory band continues over the entrance, giving the impression of light and airy construction. The symmetrical placement of the clerestories with the hip roof identifies the entrance from the vehicular approach and the parking area before the small-scale features come into view. The fresh-air intake ducts serve to visually reinforce the entrance.

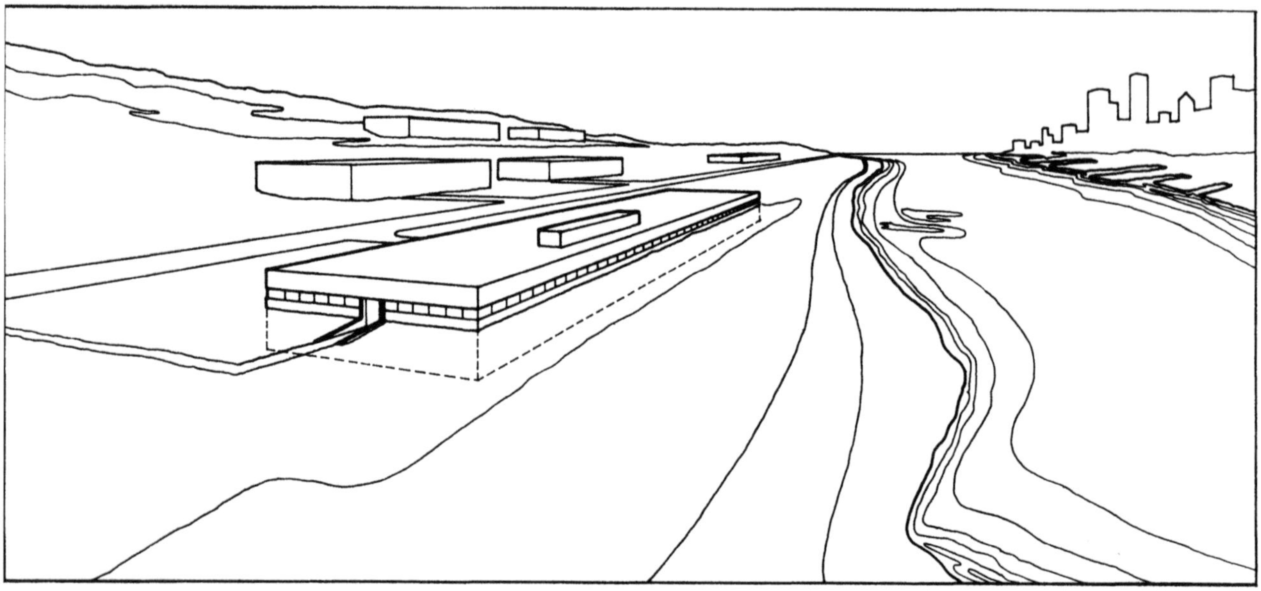

FIGURE 10
Obstruction of View on Harbor Site

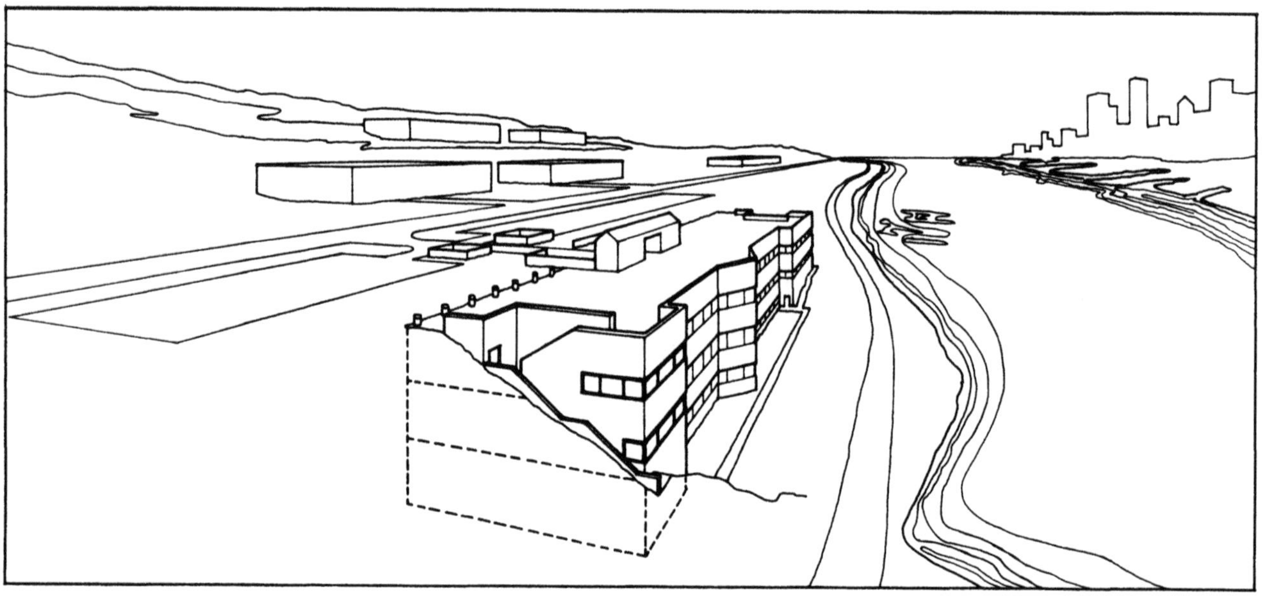

FIGURE 11
Accommodation of View on Harbor Site

e. Earth-Sheltered Building on Shore Front. Figure 10 illustrates a two-story, half-bermed building on a site with a view of a harbor and city skyline. With the berm up to the second floor level, only half of the office spaces on the harbor side have a view of the harbor and skyline. Also, the view is blocked for other buildings across from the site. Both of these problems can be solved by setting the building below grade near the top of the slope. Figure 11 illustrates this solution. The roof of the building serves as an overlook. Though the pavilion entrance occupies a small portion of the roof, the landscaping and building form are designed to visually reinforce the entrance. Otherwise, such a diminutive entrance would be out of scale with the size of the facility. The facade facing the harbor accommodates additional required fire exits as well as affording views of the harbour.

f. Earth-Sheltered Building Below Commons. On sites where other buildings are arranged around a commons or quadrangle, it is important to maintain the existing site relationships. If there is available land around the commons for a new building, then an above-grade portion of the building can take its place with the others as a conventional building. A bermed building adjacent to the commons would not be suitable. Not only would such a building weaken the definition of the commons area, but the building would preclude appropriate construction in the future. A totally submerged building could be designed with structural provisions for an above grade addition. Or, a new earth-sheltered building may be used to complete the third side of a partially enclosed yard by matching the adjacent buildings. Any remaining excess bulk of the building could be placed below grade.

Figures 12 and 13 illustrate a similar site. In this case, however, there is no available land immediately adjacent to the yard. Yet the program calls for the function housed in the building on the left to be doubled. The solution shown in Figure 12 is unacceptable because the normal elevational relationships between the buildings and the yard are upset. The shallow berm and the mechanical appurtenances all block the view of the base of the surrounding buildings. The grade elevation is, in effect, raised so that the approaches and entrances to the existing buildings are awkward. On any other site, a 5-foot-high wall (1500 mm) or berm might be called unobtrusive, but in this case it is completely incompatible with the context. Figure 13 shows a more appropriate solution. The building is submerged below the existing level of the yard. The principal pedestrian and mechanical access is below grade from the building on the left.

g. Earth-Sheltered Building on Corner Site. The earth-sheltered building illustrated in Figure 14 is on a site once occupied by an open parking lot. There are other existing gaps in the urban structure formed by such parking lots. The existence of such temporary gaps, however, does not justify a below-grade building which constitutes a permanent gap in the coherence of the urban fabric. Totally submerged earth-sheltered buildings are appropriate only where a permanent open space is allowed by proper site planning criteria such as the criteria outlined in the first half of this section and NAVFAC P-960, Installation Design (see Criteria Sources).

Figure 15 illustrates a proper response to the context. This building is a conventional building with three below-grade levels. The design benefits the context in the following ways: the setback, second-floor line, and cornice are coherent with the adjacent buildings; the capacity of the street as a primary reference is reinforced; special identity is imparted to the block by the presence of the new building; and, finally, the improved territoriality provides a safer, more liveable street.

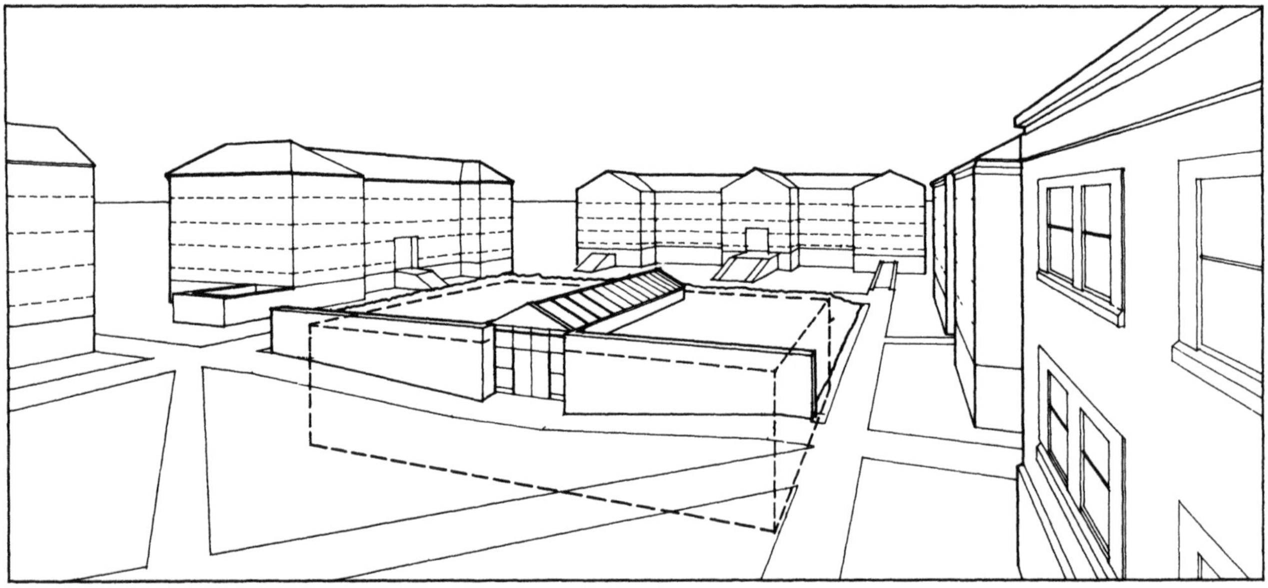

FIGURE 12
Obstruction of Existing Commons

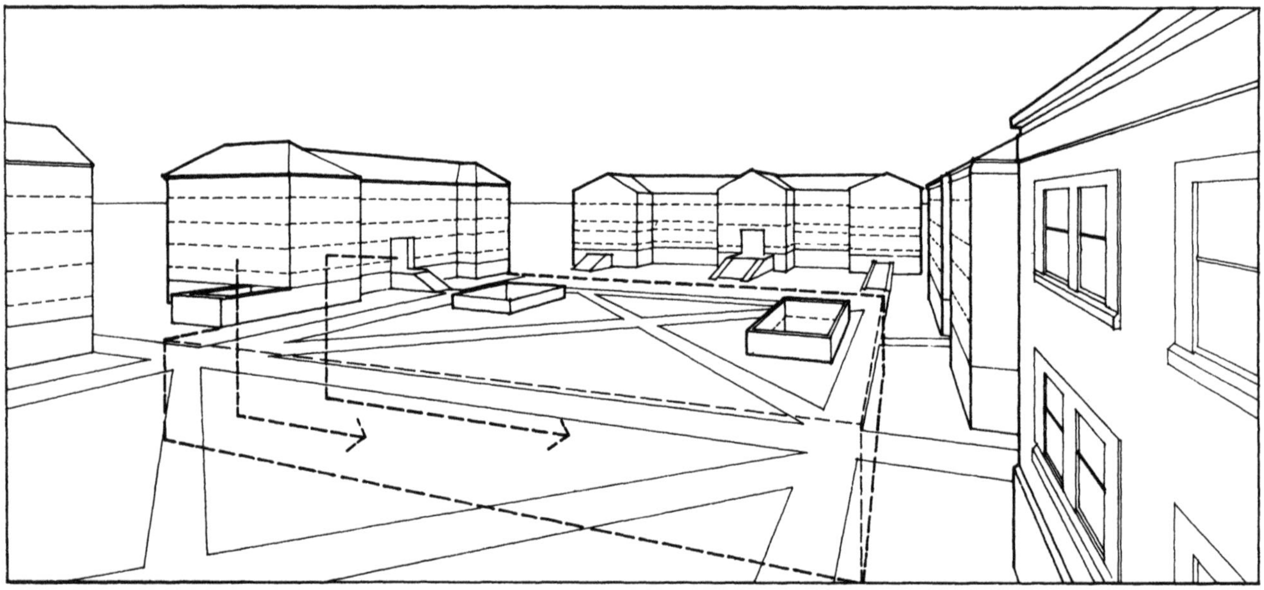

FIGURE 13
Preservation of Existing Commons

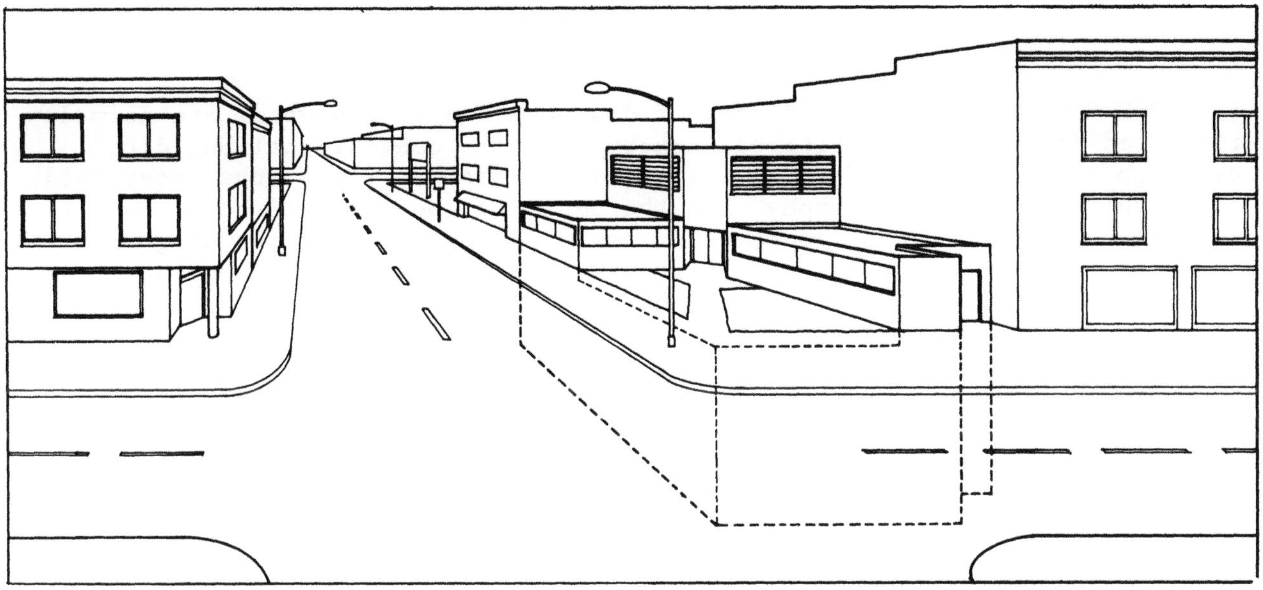

FIGURE 14
Disruption of Street Order

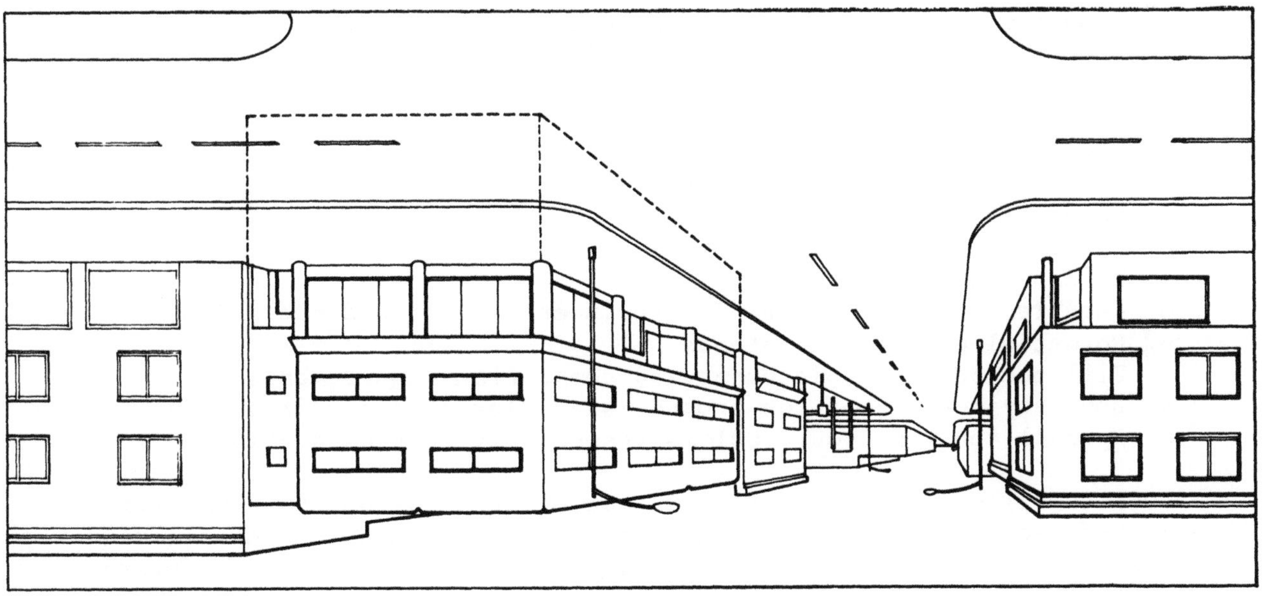

FIGURE 15
Reinforcement of Street Order

Section 7. SITE CONDITIONS AND EARTHWORK

1. SOIL CONDITIONS. The problems of drainage and ground water are normally increased for an earth-sheltered structure over a conventional building although the magnitude of this difference depends heavily on the water table location, the local topography, and the depth of the proposed building. Bearing conditions may be improved or worsened depending on the weight of any soil cover on the roof and the increase in foundation-bearing capacity with increasing depth into the ground.

Major problems with deep excavations usually relate to surface and subsurface water or soft ground conditions. Inflows to excavation can cause erosion or quick conditions in soils or loose sands. Soft clays can flow beneath retaining walls and cause heaving in the bottom of the excavation. Loss of ground into the excavation or movement of retaining walls must be minimized if the foundations of adjacent buildings are to be protected.

A geotechnical exploration testing program shall be used to determine the following:

- Drainage characteristics or permeability
- Water table
- Probable variation in water table over annual cycle
- Potential volume change upon wetting or drying
- Frost heave potential
- Potential re-use of native soil for fill
- Foundation bearing capacity
- Potential settlement
- Permissible construction slopes
- Maximum permanent slope angles
- Erodability
- Required site conditions for equipment access and operation
- Potential excavation methods
- Fill and backfill placement and compaction requirements
- Boundary conditions for ground-water flow analysis if needed
- Special construction problems (especially deep excavations)

2. ROCK CONDITIONS. Several factors can mitigate or balance the extra expense of excavating in rock. For instance, if excavation in soil requires temporary retaining walls, a rock excavation that requires no temporary supports becomes more cost effective. If the rock is of high enough quality and fissures do not have significant water flows, the rock can be exposed as an internal surface of the building, saving the cost of a permanent wall and waterproofing. (An internal drain is used with this system.)

Geotechnical information required on rock conditions for earth-sheltered buildings should include:

- Rock quality evaluation
- Bearing capacity
- Spacing, attitude, and condition of joints
- Changes in rock condition upon exposure (for example, shales)
- Rippability or potential alternate excavation methods

- Presence of water in joints or seams of rock
- Potential flows through rock joints or seams
- Controls or prohibitions on blasting

3. GROUND WATER AND DRAINAGE.

a. Identification of Critical Problems. The following conditions require careful analysis of the drainage provisions for an earth-sheltered building.

(1) Sunken Courtyards. Buildings using sunken courtyards below the surrounding ground level must be designed so that water entering the courtyard is collected and pumped to a storm sewer or to the surface. Sunken courtyards should not be used where there is a danger of surface flooding.

(2) Drainage Gullies. Buildings should not be constructed across surface or subsurface drainage paths without adequate provision for drainage to prevent a build-up of water on the upstream face of the building.

(3) Lowering Water Tables. Buildings extending below the water table should only be designed to be drained if the permeability of the surrounding soil is low enough to restrict the quantities of water to be pumped to a reasonable level. Back-up pumps or a blow-out panel should be incorporated to protect the structure.

(4) Provision of Free-Draining Backfill. If the natural soil on the site will not freely drain, then the necessity of providing a drainage material adjacent to the wall must be considered even in conditions without a water table. Sand or gravel are usually recommended for backfilling when available because they are easy to compact, provide good drainage, and will exert lower lateral pressures (when drained) than other soils.

However, if such free-draining material is surrounded by poorly draining soils to the extent that a natural sump would be formed, then the capacity of the perimeter drainage system will have to be increased. Figure 16 illustrates a recommended alternative to merely increasing the perimeter drainage capacity at the footings.

b. Surface Drainage Techniques. Surface drainage techniques involve contouring the land to divert water away from the building. This will aid in both surface and subsurface water control. Surface drainage requirements include the following:

- The ground surface should have a slope 2 percent or greater away from the building.
- Soil erosion must be avoided.
- Water flow should be distributed evenly along the remainder of the site where possible. This increases infiltration and reduces run off and erosion.
- Earth berms steeper than 3 (horizontal) to 1 (vertical) should be avoided.
- Swales can be used to slow surface water runoff on long steep slopes.

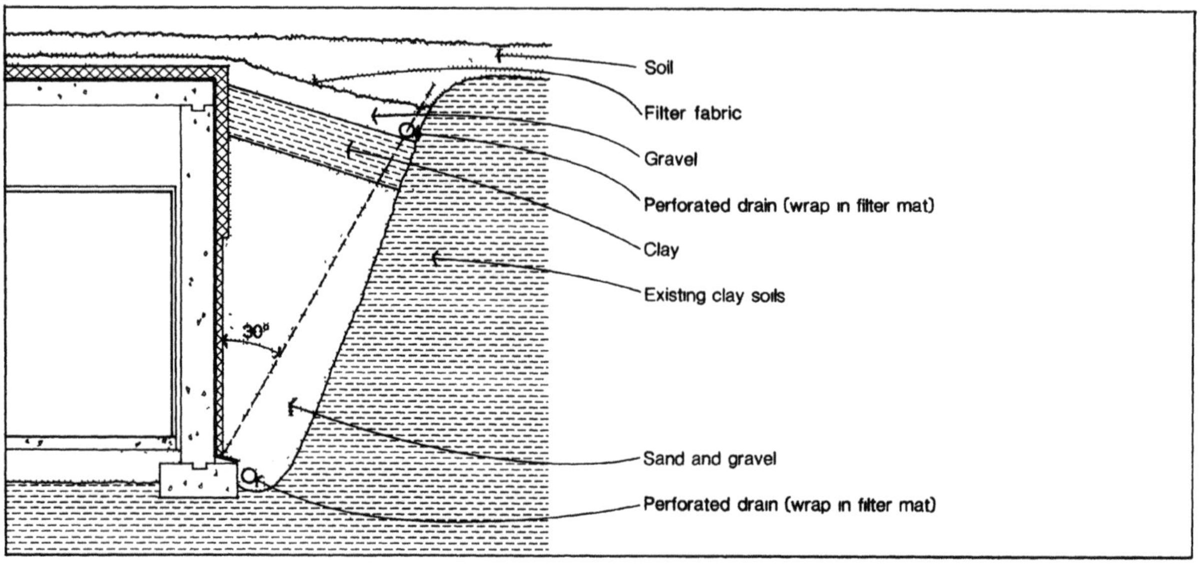

FIGURE 16
Backfilling Around Structure in Clay Zone

- Runoff should not be concentrated onto erodable slopes.
- Junctions between smooth-surfaced walls and earth berms or banks which are particularly susceptible to erosion must be designed for durability.
- Gravel cut-off trenches can trap minor surface water flows.
- Drainage channels should be designed to handle the anticipated peak flows without erosion.

For additional requirements refer to NAVFAC DM-5.11, Soil Conservation.

c. Subsurface Drainage. The function of subsurface drainage around an earth-sheltered building is to drain water which penetrates the soil cover over and around the building and to keep any permanent ground-water conditions from affecting the building.

(1) Roof Drainage. An earth-covered roof must include a drainage layer to allow the water entering the soil layer to move laterally to the edge of the building or to another drain point. Soil above the drain layer should be capable of holding sufficient moisture for plant survival. A filter mat is necessary to prevent clogging of the drainage layer.

(2) Wall Drainage. If a granular material is used for wall backfill it should extend beyond a 30 degree angle from the base of the wall to allow the soil pressures appropriate for the backfill to be used (see Figure 16). If a granular backfill is not readily available, a smaller quantity can be used with a perimeter drain tile and a proprietary wall drainage mat as shown in Figure 17.

If significant quantities of water are expected to drain from the roof drainage layer to the wall drainage layer, a drainage pipe should

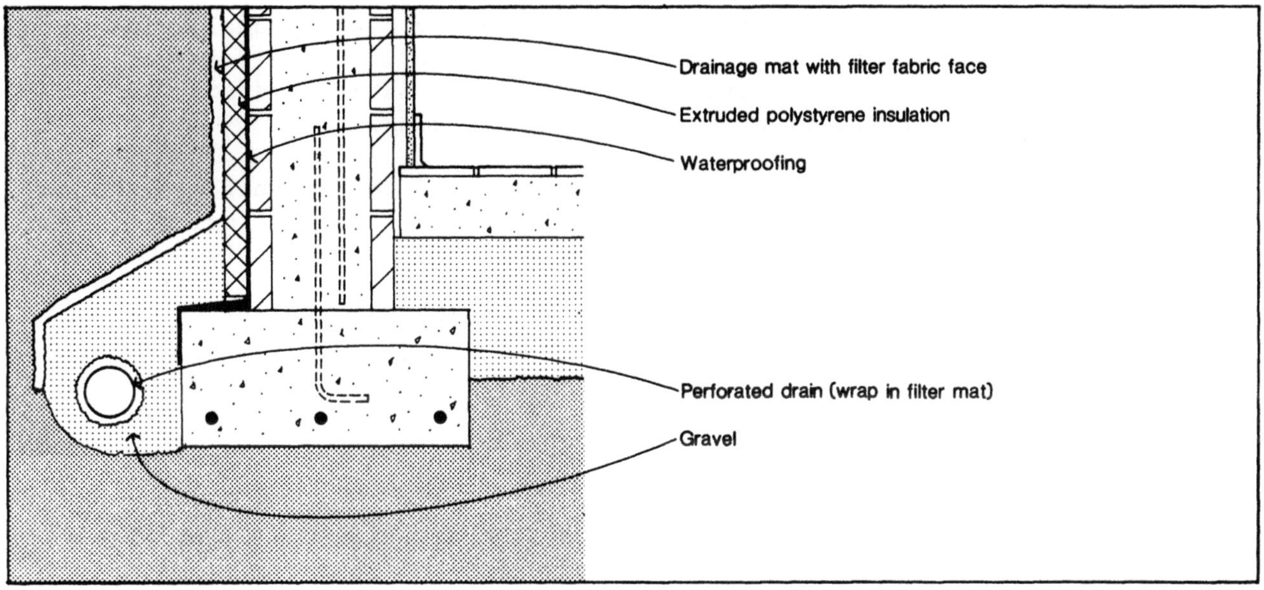

FIGURE 17
Drainage at Footing

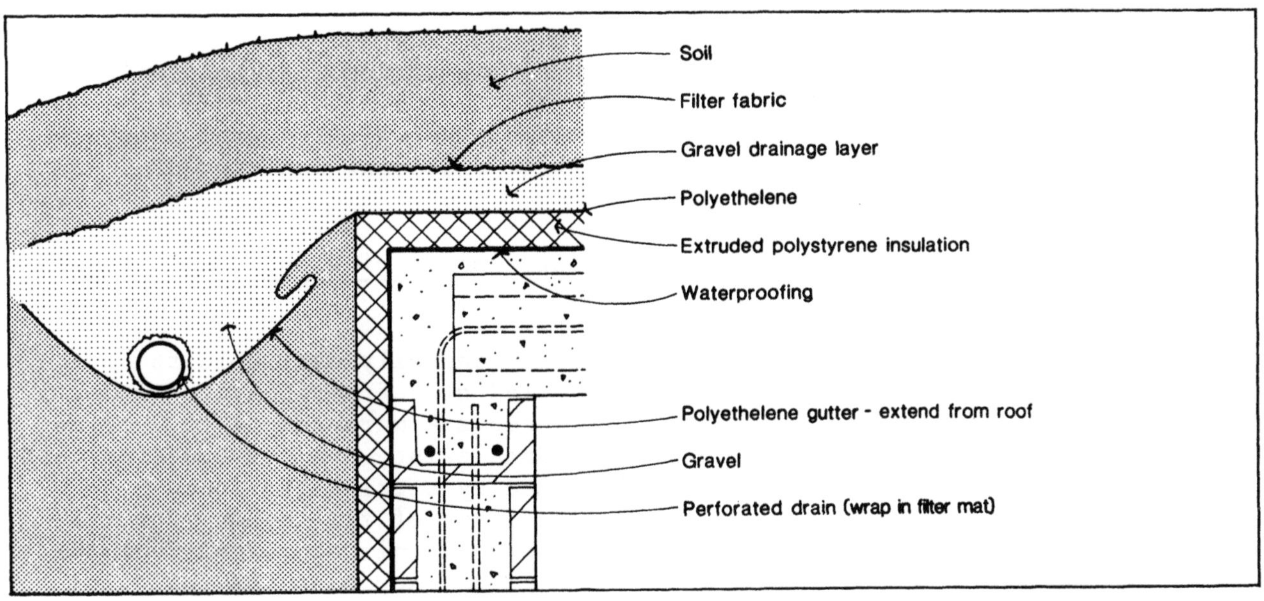

FIGURE 18
Drainage at Roof Edge

be installed at the top of the wall as shown in Figure 18. This reduces water pressure on the walls. Further, the thermal performance of the building will be improved by drier backfill.

(3) Floor Drainage. In designs above the water table, the floor is not usually waterproofed but a drainage layer is incorporated to halt capillary rise in the soil and to drain any temporary pressures caused by heavy rains. Connection should be provided between this drainage layer and the perimeter drain tile. This can be accomplished by a separate drain-tile

system beneath the floor (for large buildings) or by casting small pipes through the foundation at intervals of 3 to 4 feet (914 to 1219 mm) to provide a connection to the exterior drain system. Exterior drain tiles should always be below the floor level of the building.

(4) Drainage Pipes. Drainage pipes should be designed for the life of the building. They must resist the soil pressure and degradation in the soil environment. The minimum diameter should be 4 inches (100 mm). All drain tile for permanent buildings should be protected by filter fabrics or by an appropriately designed soil filter system. Filter fabrics are preferred for reasons of simplicity and quality control. Provision for an occasional clean-out of the drainage system should also be made. Minimum slopes for a drain tile should be 1 inch in 20 feet (25 mm in 6097 mm) except in small buildings where they may be laid horizontally. Vermin protection must be provided.

(5) Filters. All drainage layers or drain tile must be protected from clogging by fine particles from the adjacent soil material. Geotextile fabrics are now in wide use for filtration and drainage. Requirements for geotextile fabrics to be used in filtration/drainage applications are:

(a) The fabric must be (and remain) more permeable than the adjacent soil.

(b) The fabric must prevent piping of the adjacent soil.

(c) The fabric must have enough physical strength, puncture resistance, and abrasion resistance to survive placement and provide adequate in-service performance.

See Reference 6, Use of Engineering Fabrics in Transportation-Related Applications by Haliburton.

(6) Cut-and-Fill Quantities. It is generally desirable to minimize the amount of soil that must be hauled away from the site since this is an expensive item if the haul distances are more than 1 to 2 miles (1.61 to 3.22 km). When appropriate conditions exist, the construction cost for earth-sheltered buildings can be minimized by setting the foundation elevation of the building such that the quantities of fill needed for the berms and earth cover are equal to the quantity of soil excavated from the lower portions of the building. This balancing presupposes the following conditions:

- The site is large enough to store the excavated material before its final replacement.
- The excavated material is suitable for backfill (and earth cover, if used).

Cut and fills are most easily balanced on sloping sites. Buildings of large area extent on flat sites will not have balanced cut-and-fill quantities unless they are kept very shallow and bermed extensively. Where berming is not possible on flat sites due to lack of site area it will not usually be possible to develop an earth-sheltered design without net excavation and hauling.

4. EXCAVATION, STABILITY, AND DEWATERING. Details of excavation stability, dewatering procedures, and seepage and drainage analysis are contained in NAVFAC DM-7 Series, Soil Mechanics, Foundations, and Earth Structures. Temporary or permanent dewatering must include an analysis of the effect of dewatering on the settlement or stability of any adjacent moisture-sensitive soil materials. Earth berms or final grading surrounding earth-sheltered buildings should be landscaped as soon as practicable to prevent erosion.

5. BACKFILLING AND COMPACTION. All fill placed under or adjacent to earth-sheltered buildings should be adequately compacted. Insufficient compaction can cause settlement of the backfill. Settlement of wall backfill causes drag forces on the insulation and waterproofing and can alter surface drainage patterns, trapping water adjacent to the building. Backfilling must be carried out carefully to avoid damage to the building insulation and waterproofing. Large equipment should not be allowed on a building roof or adjacent to the walls unless the structure is adequately designed for this. Bulldozing earth fill across the roof of an earth-sheltered building should not be permitted since this can shift previously installed insulation and waterproofing. Backfill should not be concentrated in one spot on the roof before grading since this represents a more severe design loading condition than the uniform load.

Free-draining granular soil backfills cannot only significantly lower lateral pressures, but can also divert water from the structure. It is recommended that the soil used for backfilling side walls be compacted to 90 percent of the maximum density as determined by ASTM D1557. The soil should be compacted only enough to ensure that no major settlements occur. Overcompaction of the soil around the side walls can induce additional lateral pressure on the walls. The backfill should be placed and compacted in 8-inch (200-mm) lifts. Compaction requirements, methods, and equipment are given in NAVFAC DM-7 Series.

Fill required beneath floor slabs should have a low plasticity index (below 13), a low liquid limit (below 36) and should be compacted to a dry density of at least 95 percent of the maximum dry density as determined by ASTM D1557. Compaction of this fill should be accomplished by placing the fill in 6- to 8-inch (150- to 200-mm) thick lifts. Soil placed on the roof should be only moderately compacted. Vegetation on the roof will play an important role in how the soil will behave over time.

6. EARTH-COVERED ROOFS. Earth-covered roofs have unique properties which offer energy benefits. The large thermal mass associated with earth-covered roofs not only serves to dampen the continuous fluctuations in the external air temperature, but also has the capability to reflect the summer sun. Radiant heat gains are also countered by latent energy losses due to evapotranspiration.

Based on computer simulation, earth-covered roofs are more than capable of countering radiant heat gains (solar) if plant cover is provided and well irrigated. In fact, in northern (Minneapolis) and southern climates (Jacksonville and Phoenix), earth-covered roofs can provide net cooling benefits during the cooling season; more heat is lost from the roof than is gained. These benfits are significant, but typically will represent less than 1 percent of the total building cooling load (commercial buildings).

The important distinction to be made here is that an earth-covered roof may provide cooling benefits instead of adding to the building's total cooling load, as a conventional roof would.

During the heating season, an earth-covered roof's annual energy performance will be similar to that of a well-insulated conventional roof. The difference in energy performance is directly dependent on the depth of earth cover, and the amount of insulation applied to the conventional roof. Winter peak design loads of the two roof types will be similar unless large amounts of earth cover are applied to the roof (over 12 inches (300 mm)). It is difficult to justify analyzing an earth-covered roof with small amounts of earth cover (6 inches (150 mm)) on a transient basis, since a roof of this type will nearly reach steady-state heat loss conditions over periods when cold fronts exist for days or weeks at a time. Earth-covered roofs with large amounts of earth cover should be analyzed by considering the thermal mass when determining the peak heating loads of the roof component.

If extremely accurate thermal performance data is required, hour-by-hour computer simulations must be carried out to predict the peak and annual energy performance of the roof component (heating and cooling).

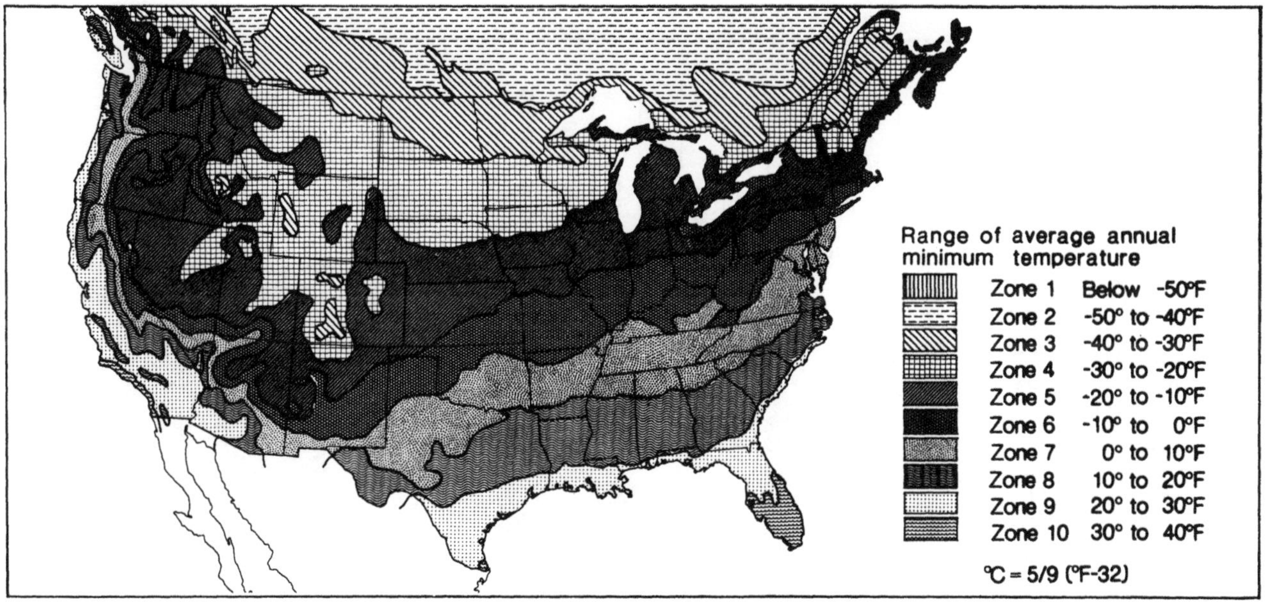

FIGURE 19
Climatic Map for Plant Hardiness

1. PLANT MATERIALS SELECTION AND LAYOUT. The following criteria should be identified for each species considered in a planting design.

a. Suitability to Climate. Table 1 contains a partial list of species based on the climatic zones shown in Figure 19. These species are suitable for berm roof-top planting. There are, of course, many more species suitable for use than can be documented here. The landscape designer should research the species native to the project's specific climatic zone.

b. Adaptability to Berm Slopes. For the purposes of moisture protection and the reduction of soil pressure, most berms are constructed of materials that drain rapidly. Species of plants selected for use on berms should be those that can tolerate long periods of drought and rapid fluctuations from wet to dry conditions. Species that have extensive root systems to hold and maintain slopes should also be of primary consideration.

c. Adaptability to Limited Soil Depths. Due to the structural costs required to carry even moderate soil depths on roofs, plant selection for roofs should be limited to species that are capable of surviving on a limited soil depth. To assist moisture protection of the building, the soil and drainage layer construction is usually designed to provide rapid drainage. These conditions result in extreme fluctuations of soil moisture content. During dry spells, plants that develop extensive root systems will seek moisture near the building's waterproofing membrane. To avoid damage to the waterproofing, select species with less aggressive root systems.

TABLE 1
Plant Species by Climate (Continued on Page 41)

PLANT NAME	Deciduous	Evergreen	Vines	Ground Cover	Shrubs	Trees	Flowering	Bermed Area	Rooftop Planting	Climatic Zone 1	2	3	4	5	6	7	8	9	10
Ampelopsis Blueberry Climber	X		X				X	X						X	X	X	X	X	X
Ampelopsis Cordata Heart-Leaf Ampelopsis	X		X				X		X				X	X	X	X	X	X	X
Campsis Radicans Common Trumpet Creeper	X		X				X		X					X	X	X	X	X	
Euonymus Fortunei Big-Leaf Winter Creeper		X	X				X		X					X	X	X	X		
Hedera Ivy		X	X					X	X					X	X	X	X	X	X
Hydrangea Anomata Climbing Hydrangea	X		X				X		X					X	X	X	X	X	
Loniceria Japonica Japanese Honeysuckle		X	X				X		X					X	X	X	X	X	X
Parthenocissus Tricuspidata Boston Ivy	X		X					X	X					X	X	X	X	X	X
Parthenocissus Quinquefolia Virginia Creeper	X		X					X	X			X	X	X	X	X	X	X	X
Polygonum Anbertii Silver Lace Vine	X		X				X	X	X					X	X	X	X	X	X
Vinca Minor Dwarf Periwinkle		X	X				X		X				X	X	X	X	X	X	X
Ajuga Reptans Ajuga		X		X			X		X				X	X	X	X	X	X	X
Anthemis Nobilis Chamomile		X		X					X					X	X	X	X	X	X
Arctostaphylos Uva-ursi Bearberry, Kinnikinnick		X		X					X			X	X	X	X	X	X	X	X
Cerastium Tomentosum Snow-in-Summer		X		X			X		X				X	X	X	X	X	X	X
Cotoneaster Dammeri Bearberry Cotoneaster		X		X			X		X					X	X	X	X	X	X
Cotoneaster Horizontalis Rock Cotoneaster	X			X				X	X					X	X	X	X	X	X
Euonymus Fortunei Radicans Common Winter Creeper		X		X				X	X					X	X	X	X	X	
Gaultheria Procumbens Wintergreen, Checkerberry		X		X				X	X			X	X	X	X	X	X	X	
Hedera Helix English Ivy		X		X				X	X					X	X	X	X	X	X
Hedera Helix 'Baltica' Baltic Ivy		X		X				X	X					X	X	X	X	X	X
Hypericum Calycinum Aaron's Beard		X		X					X					X	X	X	X	X	X
Iberis Sempervirens Evergreen Candytuft		X		X				X	X					X	X	X	X	X	X
Juniperus 'Bar Harbor' Bar Harbor Juniper		X		X				X	X				X	X	X	X	X		
Juniperus Chinesis Sargentii Sargent Juniper		X		X				X	X				X	X	X	X	X		
Juniperus Conferta Shore Juniper		X		X				X	X					X	X	X	X		
Juniperus Horizontalis Creeping Juniper and varieties		X		X				X	X			X	X	X	X	X	X		
Juniperus 'Plumosa' Andorra Juniper		X		X				X	X				X	X	X	X	X		
Juniperus Sabina Tamariscifolia Tam Juniper		X		X				X	X					X	X	X	X		
Lonicera Japonica Halliana Hall's Honeysuckle		X		X			X	X	X				X	X	X	X	X	X	X
Pachysandra Terminalis Pachysandra		X		X				X	X					X	X	X	X		
Phlox Subulata Moss Pink		X		X			X	X	X			X	X	X	X	X	X		

TABLE 1
Plant Species by Climate (Continued from page 40)

PLANT NAME	Deciduous	Evergreen	Vines	Ground Cover	Shrubs	Trees	Flowering	Bermed Area	Rooftop Planting	Climatic Zone 1	2	3	4	5	6	7	8	9	10
Vinca Minor Dwarf Periwinkle		X		X			X	X	X				X	X	X	X	X	X	X
Amelanchier Serviceberry/Juneberry					X		X	X	X		X	X	X	X	X	X	X	X	X
Amorpha Canescens Lead Plant					X				X				X	X	X	X	X	X	X
Artemesia Sage					X		X		X					X	X	X	X	X	X
Atriplex Canescens Four-winged Saltbrush					X		X		X					X	X	X	X	X	X
Chrysothamnus Rabbit Bush	X				X		X		X			X	X	X	X	X	X	X	X
Comptornia Peregrina Sweet Fern	X				X				X				X	X	X	X	X	X	X
Cornus Baileyi Bailey Dogwood	X				X		X	X	X		X	X	X	X	X	X	X		
Cornus Mas Corneliancberry Dogwood	X				X		X	X	X		X	X	X	X	X	X	X		
Cornus Racemosa Gray Dogwood	X				X		X	X	X		X	X	X	X	X	X	X		
Cornus Stolenifera Redosier Dogwood	X				X		X	X	X		X	X	X	X	X	X	X		
Cornus Stolenifera Flaviramea Redosier Flaviramea Dogwood	X				X		X	X	X		X	X	X	X	X	X	X		
Epigaea Repens Mayflower		X			X		X	X						X	X	X	X	X	X
Gaultheria Shallum Salal		X			X		X		X				X	X	X	X	X	X	X
Gaylussacia Huckleberry		X			X		X	X	X		X	X	X	X	X	X	X	X	X
Hamamelis Witchhazel	X				X		X		X				X	X	X	X	X	X	X
Ilex Holly, Dahoon		X			X		X		X				X	X	X	X	X	X	X
Junipers Common Junipers		X			X			X	X	X	X	X	X	X	X	X			
Leiophyllum Buxifolium Box Sandmyrtle		X			X		X		X			X	X	X	X	X	X	X	X
Lyonia Mariana Staggerbush Lyonia	X				X		X					X	X	X	X	X	X	X	X
Myrica Pennsylvanica Northern Barberry	X				X		X		X		X	X	X	X	X	X	X	X	X
Picea Spruce		X			X			X		X	X	X	X	X	X	X			
Pirus Pine		X			X			X			X	X	X	X	X	X	X		
Prunus Western Sandcherry	X				X		X		X			X	X	X	X	X	X	X	X
Rhus Sumac	X				X				X		X	X	X	X	X	X	X	X	X
Rubus Odoratus Flowering Raspberry	X				X		X		X		X	X	X	X	X	X	X	X	X
Sheperdia Buffaloberry	X				X				X		X	X	X	X	X	X			
Symproriaarpos Albus Common Snowberry	X				X		X		X		X	X	X	X	X	X	X	X	X
Vaccinium Blueberry	X				X				X		X	X	X	X	X	X	X	X	X
Acer Ginnala Amur Maple	X					X		X	X		X	X	X						
Acer Negundo Box Elder	X					X		X	X		X	X	X						

TABLE 1
Plant Species by Climate (Continued from page 41)

PLANT NAME	Deciduous	Evergreen	Vines	Ground Cover	Shrubs	Trees	Flowering	Bermed Area	Rooftop Planting	Climatic Zone 1	2	3	4	5	6	7	8	9	10
Betula Davurica Dahurian Birch	■					■		■	■				■	■					
Betula Populifolia Gray Birch	■					■		■	■			■	■						
Broussorietia Papyrifera Paper Mulberry	■					■		■							■				
Celtis Occidentalis Hackberry	■					■		■					■	■	■	■			
Cornus Alternifolia Pagoda Dogwood	■					■		■	■			■	■						
Cupressus Macrocarpa Monterey Cypress		■				■		■								■			
Elaeagnus Angustifolia Russian Olive	■					■		■			■	■	■						
Eucalyptus Eucalyptus		■				■		■										■	■
Ficus Carica Fig		■				■		■							■				
Fraxinus Pennsylvanica Lanceolata Green Ash	■					■		■			■	■	■						
Ginko Biloba Ginko	■					■	■	■			■	■	■						
Gleditsia Triacanthos Honey Locust	■					■		■				■	■	■					
Gymnocladus Dioicus Kentucky Coffee Tree	■					■		■					■	■	■				
Leptospermum Laevigatum Australian Tea Tree		■				■	■	■										■	
Maclura Pomifera Osage-Orange	■					■	■	■					■	■					
Melaleuca Leucadenara Cajeput Tree		■				■	■	■											■
Olea Europaea Common Olive		■				■		■	■									■	
Ostrya Virginiana Ironwood	■					■	■	■					■						
Phellodendron Amurense Amur Corktree	■					■	■	■				■	■	■					
Pinus Banksiana Jack Pine		■				■		■			■								
Pinus Carariensis Canary Pine		■				■		■									■		
Pinus Torreyana Torrey Pine		■				■		■									■		
Pinus Virginiana Virginia Pine		■				■		■					■	■					
Populus Alba White Poplar	■					■		■				■	■						
Populus Fremontii Fremont Poplar	■					■		■								■			
Populus Tremulodies Quaking Aspen	■					■		■	■	■	■	■							
Quercus Kelloggii California Black Oak	■					■		■								■			
Quercus Marilandica Black-Jack Oak	■					■		■							■				
Quercus Prinus Chestnut Oak	■					■		■					■	■					
Robinia Pseudoacacia Black Locust	■					■		■					■						
Ulmus Siberica Siberian Elm	■					■		■			■	■	■						

d. Ease of Establishment and Maintenance. Reduced operating costs and protection of the waterproofing are best obtained by the establishment of healthy plants that require little or no maintenance. Cutting, digging, or the operation of maintenance machinery can damage the waterproofing at building edges or roof openings, especially when roof-top soil layers are thin. The need for permanent establishment of the initial planting without reworking is essential.

e. Functional Considerations. Certain functional considerations can be enhanced or accomplished by the proper selection and layout of planting materials for earth-sheltered buildings.

Species that form dense, tightly packed masses or appear to bear thorns can be an effective security barrier. Such species may be used in place of guardrails and fences at retaining walls, parapets, roof openings, mechanical equipment, and skylights to provide safety barriers as well as protection from accidental or intentional damage.

Proper selection of plant materials can have a significant impact on reduction of mechanical cooling loads within the structure by shading roof top soils and by holding moisture over longer periods of time. Shading reduces solar gain and moisture retention promotes evaporative cooling. Shading can also be useful in reducing heat gain through windows or skylights. Deciduous plants will accomplish this without blocking beneficial solar gain in winter. Proper plant selection can also control wind movement in and around buildings. Plants can serve as buffers to reduce the direct wind velocity at the building, thereby reducing convective and infiltrative losses. Wind breaks will also limit plant desiccation that can drain soil moisture and damage the health and growth of the plants.

The proper selection of plant materials can impact the installation image. A carefully designed native association can assist the building in blending with the surrounding landscape when integration with the natural environment is required. Screening with plant materials can be effective in reducing the visual presence of the building, parking areas, or on-grade mechanical or electrical equipment. On the other hand, planting may be used to visually reinforce the building entrance. In urban contexts, appropriate planting design may call for geometric or formal patterns.

2. IRRIGATION SYSTEMS. Because roof-top and berm drainage systems are generally designed to carry moisture away from the building, extremely droughty conditions can arise which may test the hardiness of even the most suitable plant selections. There are three basic options for irrigation: manual, spray heads, or trickle systems.

a. Manual. The use of manual irrigation in these situations should be restricted to installations of modest or small size which can be reliably observed and maintained. Consideration should be given first to the selection of plant materials that require the least amount of irrigation. For instance, durable shrub species that shade the soil and help hold moisture will survive better than turf. These species should be capable of tolerating extremes of flooding or drought. The layout of available hose bibbs should be closely coordinated with the probable equipment available for providing the manual irrigation so that no area of planting is inaccessible. Manual irrigation should also be limited to installations where inadvertent over-irrigation will not endanger the construction due to hydrostatic pressure.

b. Spray Heads. The standard method of laying irrigation lines with appropriately spaced spray heads is the most desirable from the point of view of maintaining plant materials. Spray-head application simulates actual rain conditions, thereby washing away unwanted dust or pollutants and protecting the plants from desiccation. An automatic spray-head installation will reduce the danger of over-irrigation, although, miscalculation of the rate of flow can result in extensive mineral leaching or erosion on slopes.

c. Trickle Systems. Trickle systems are a unique compromise for providing appropriate soil moisture. The systems themselves are lower in cost and require a lower water pressure for operation. Normally, a dispensing head is located at each significant plant to provide a slow trickle of water for keeping the soil moist. The system is time clock controlled and is normally installed just beneath a 3- to 4-inch (75-mm to 100-mm) mulch layer, or just under the top of the soil. Providing moisture with a trickle system reduces soil leaching, eliminates streaking or water spotting on windows that may be caused by spray heads and reduces the chance of extensive over-irrigation due to the smaller volumes of water distributed over time. These advantages must be balanced with the negative effect of not being able to wash the plant materials of dust or pollutant materials. In areas with average or significant rainfall, however, plant washing by the irrigation system should not be necessary.

3. INSTALLATION. Some of the most important considerations in the landscaping of earth-sheltered structures are the design and installation of drainage layers and the placement of plants.

a. Application on Roofs. The design of soil layers and installation of plants on roof-top situations is a complicated and sensitive process which should take into account all of the following considerations.

(1) Structural Integrity. Table 2 indicates general weights of commonly used materials in roof-top planting situations. In the conceptualization of landscape design, the landscape architect should work carefully with the building structural engineer to provide the necessary structural capacity and to ensure proper location and distribution of weights. It is critical to remember that the installation process must be accomplished by hand, since most road vehicles and heavy machinery will exceed the efficient design loads for an underground structure. It is advantageous that the landscape design in some way prohibit the future inadvertent wandering of large vehicles or maintenance equipment onto a structural roof deck that is not designed to carry their load.

(2) Protection of Waterproofing. For any underground structure, the protection of the integrity of the waterproofing is critical. The layered drainage and soil installation should include layers that prevent the long-term accumulation of standing water and potential damage from plant roots. Care must be taken to protect the waterproofing, especially at parapets or around roof openings. Staking methods for large shrubs and trees should be such that it is impossible for the stakes to damage the waterproofing.

TABLE 2
Weights of Planting Materials

TREES

Ball Size	Caliper	Weight
20 in. (508 mm)	3/4 - 1 in. (19-25 mm)	180 lbs (80 kg)
22 in. (559 mm)	1-3/4 in. (44 mm)	260 lbs (120 kg)
24 in. (610 mm)	2 in. (51 mm)	340 lbs (150 kg)
28 in. (711 mm)	2-1/2 in. (64 mm)	560 lbs (250 kg)
32 in. (813 mm)	3 in. (76 mm)	880 lbs (400 kg)
38 in. (965 mm)	3-1/2 in. (89 mm)	1100 lbs (500 kg)
42 in. (1067 mm)	4 in. (102 mm)	1500 lbs (680 kg)
48 in. (1219 mm)	4-1/2 in. (114 mm)	2000 lbs (910 kg)
54 in. (1372 mm)	5 in. (127 mm)	2600 lbs(1180 kg)
60 in. (1524 mm)	5-1/2 in. (140 mm)	3100 lbs(1400 kg)
68 in. (1727 mm)	6-7 in. (152-178 mm)	3800 lbs(1720 kg)

SHRUBS

Ball Size	Ht/Spread	Weight
18-24 in. (457-610 mm) Evergreen		24 in. spread (610 mm)
60-70 lbs (27-32 kg)		
18 in. (457 mm) Evergreen	18 in. spread (457 mm)	50-60 lbs (20-27 kg)
Deciduous 18 ft.	2 ft. height (610 mm)	60-70 lbs (27-32 kg)
18-24 in. (457-610 mm)	3 ft. height (914 mm)	70-90 lbs (32-40 kg)

SOILS MIXTURES AND MATERIALS

Material by Volume	lbs/ft^3	(kg/m^3)
Clay Soil	75	(1200)
Sand	92	(1472)
Peat, semidry	4 - 13	(64 - 192)
Perlite	5 - 8	(80 - 128)
Vermiculite	4	(64)
Baked clay particles	37	((592)
1/3 soil, 1/3 sand, 1/3 peat	55-67	(880 - 1072)
1/2 peat, 1/2 sand	52	(832)
1/2 peat, 1/2 perlite	10	(160)
1/3 soil, 1/3 perlite, 1/3 peat	37	(592)
Loam (Dry)	7590	(12,1440)
Compact Peat Moss	15-18	(240-288)
Loose Peat Moss	3-5	(48-80)
Dry Sand	95-110	(1520-1760)
Wet Sand	120-130	(1920-2080)
Wet Clay	105-120	(1680-1920)
Dry Clay	90-110	(1440-1760)
Granite	170	(2720)
Gravel	120-135	(1920-2160)
Dry-Loose Loam	90-100	(1440-1600)
Wet Silt	130-145	(2080-2320)

Note: These are initial weights. Use full grown weight for structural design.

(3) Soil Layers. The rigid insulation assists in protecting the membrane. A polyethylene layer over the insulation is recommended to improve drainage and prevent moisture saturation which may reduce thermal resistance. If concrete is used over the waterproofing (such as at planters), the waterproofing must be covered completely with protection board first. The primary drainage layer should be composed of pea gravel. A filter fabric over the drainage layer controls some leaching of materials from the planting soil layer and also prevents invasion of plant roots.

As a rule of thumb, soil depth for shrubs and trees should be 1 foot (300 mm) plus rootball depth. Soil depth for lawn and ground cover should be 1.5 feet (450 mm), or 1 foot (300 mm) if irrigated.

(4) Containers. The container and its location must be coordinated with the structural engineer. Special soil mixtures which reduce the soil weight are often incompatible with the plant's natural soil requirements. Experience has shown that such mixes reduce the growth rate of the plant and result in more rapid leaching of minerals in the highly irrigated environment of a container. If lightweight soil mix must be used, it should be formulated by a soil and plant lab. In cold climates, it is absolutely necessary to provide rigid insulation between the soil in the container and the container wall to prevent soil temperatures from dropping below the point of tolerance developed by native plants growing in a normal ground environment.

b. Application on Berms. Most planting installations for berms are similar to conventional installations on slopes with the exception that berms along building walls are generally formulated for more rapid drainage and have a smaller mass for holding moisture due to the close proximity of the building wall.

(1) Plant Placement. In establishing plant layouts on sloping conditions, the designer must calculate the appropriate spacing based on the actual dimension of the slope rather than the horizontal projection of the slope onto a plan. Detailing should also indicate how shrubs and trees are to be installed and staked so that their growth habit is in appropriate alignment to grade, rather than angled in an unnatural way. Plant layouts must also take into account the danger of placing hand-excavated holes too close to the top of the berm at the building wall. This increases the chance of inadvertently damaging the waterproofing.

(2) Erosion Control and Establishment. In establishing plantings on slopes of greater than 4:1 by seeding or transplantation, one must use proper erosion control during construction and early-growth periods. A variety of erosion control mats are commercially available, the straw-fabric mat being the most common and inexpensive. Straw-fabric mats will hold the soil in place and help preserve moisture while the plants are becoming established, but will gradually deteriorate and disappear over time. Wood-chip or rock mulch will not remain on the slope during erosive rains. Plant type and spacing should be selected so as to fill in rapidly without the need for rock or wood-chip mulch. Moisture holding can be enhanced by the application of additional loose straw mulch or hydro-sprayed systems.

(3) Sod Placement. Sod intended for mowing should not be placed on slopes greater than 3:1 due to the inability of large mowing machines to operate at that angle. Sod on slopes of 2-1/2:1 or greater runs the risk of movement or sliding in extremely wet conditions. On slopes between 4:1 and 3:1, sod should be laid perpendicularly to the slope from the lowest point and moving uphill. Joints should be staggered to avoid erosion during irrigation or rain. Ends of sod strips should be pegged. Pegs should be located at each upper corner and in the center of a sod strip. Always drive pegs vertically, not perpendicularly, to the slope.

4. MAINTENANCE. Early and careful maintenance of a planting installation on berms or slopes is necessary for proper establishment and reduced future maintenance.

a. Weeding. During the first few seasons of establishment for berm or roof-top planting installations, regular weeding is necessary. Weeding should be done by trained personnel who can distinguish the plants to remain from the plants that should be removed. The designer should verify that the waterproofing selected is not susceptible to damage from chemical herbicides or fertilizers.

b. Mowing. Due to the tendency of berm and roof conditions to become droughty situations, mowing should be minimized to assist in the preservation of soil moisture and turf durability. The option of allowing certain areas to grow without mowing should be considered, as it will do the most to enhance the growing environment of the grasses. Heavy mowing equipment cannot be used on roofs unless the structure is designed to accommodate such loads. Consideration should be given to the availability of hand-mowers on the roof-top to minimize undesirable passages through the building interior.

c. Irrigation. Although the most effective use of irrigation to promote plant establishment is to provide water on a time-clock controlled basis, it is advisable to provide a system of manual override to prevent irrigation when natural rainfall has already saturated the soil and to provide for additional irrigation during extremely droughty conditions.

1. ACCESS. One of the major problems in the siting and site planning of earth-sheltered facilities is the need to provide adequate and efficient access to the building. This usually conflicts with the goal of maintaining a maximum amount of earth-contact. A large number of access points may reduce the amount of earth-sheltering and often requires retaining walls at bermed areas.

In climates which have snow and ice, ramps used for pedestrian access and exiting that have a gradient exceeding 5 percent should be provided with a system for protection from snow and ice.

Consideration should be given in design to the maintenance of depressed entrance areas--especially snow removal (where required). Space for stockpiling snow should be provided within the depressed area when feasible. To avoid expensive hand labor for cleaning paved areas, access by snow-removal equipment should be considered.

Access for vehicles, including warehouse docking facilities, can be difficult to design in an earth-sheltered building. Service and vehicular access functions, which are frequently the least attractive, will be among the most prominent. Some or all of the following vehicular access functions may have to be accommodated, both in terms of appearance and site engineering.

A means of removing and servicing mechanical equipment must be provided. If possible, this access point should be grouped with storage and material-handling functions away from pedestrian entrances. However, to maximize the utility of exposed wall areas, the designer should consider the possibility of locating a required fire exit here, rather than opening up another wall area. In this case, special provisions may be necessary to keep the exit path unobstructed.

Refuse collection and outdoor storage areas should be located away from primary entrances. Depressed storage areas with retaining walls on three sides should be considered as a strategy for concealing the area as well as maintaining a maximum of earth-contact against the building.

If the topography of the site is relatively flat, the elevation of the loading berths places an additional restriction on how low a building can be sited before special provisions are required to handle surface drainage. In addition, greatly depressed loading berths may require inefficient and expensive ramps or runout areas to reach the elevation of an existing road. Because the practical gradients for truck ramping are low, the necessary runout length may not be possible within the available area of the site. The alternative of loading at a mezzanine level may be too inefficient.

Intake air louvers should not be located near service drives or other areas having exhaust fumes.

Allowance must be made for unobtrusive locations for meters, junction boxes, fire hydrants, and so forth. These should be specified and not left to the discretion of installers.

2. FIRE-FIGHTING ACCESS. An automatic fire-extinguishing system may have to be provided (see Section 12, Fire Protection and Life-Safety and NAVFAC DM-8, Fire Protection Engineering).

3. TRAVEL DISTANCES AND EXITS. In order to maximize earth-contact, it is desirable to minimize the number of entrances. Retaining walls may be used at access locations to maximize earth-contact, however, retaining walls can be costly. It is therefore desirable to plan the building so that the number of access points is kept to a minimum. Refer to NAVFAC DM-8 for requirements for travel distances and exits.

4. MECHANICAL ZONING. The location of functions having different areas and different HVAC requirements will affect the energy savings attributable to earth-contact. Large undivided floor areas placed on exterior earth-sheltered walls will allow a greater volume of the building to interact with the exterior walls. The overall amounts of energy saved will usually be improved if the functions having larger energy demands are located on the perimeter. An open-plan arrangement for the perimeter areas may be a workable alternative to relocating functions. On a bermed exterior wall having fenestration, locating open-plan areas on the perimeter will also has the advantage of distributing natural light to a greater portion of the building.

5. STRUCTURAL BAY WITH EARTH COVER. One of the consequences of placing earth on the roof is the additional structural cost. Structural costs can be reduced in some instances by either employing a different structural system, such as barrel vaults, or by reducing the bay sizes (see Chapter 6, Structural). Often the bay size can be reduced without compromising the functioning of the building. For instance, in warehouses where column lines are located in the space between back-to-back palletized storage racks, the number of columns can be increased without interfering with the storage layout.

6. OPTIMIZATION OF EARTH-CONTACT. In most instances, earth-contact will yield greater energy savings than construction which is exposed to the weather. Possible exceptions to this include small buildings or perimeter zones that could otherwise take advantage of either passive solar gain or passive cooling by ventilation. Appendix A provides tables for determination of below-grade heat flux as a function of climate, soil type, insulation, and depth below grade. Construction costs and planning considerations will normally prevail over any attempt to optimize the building shape for thermal performance. However, the following observations will serve as a conceptual guide.

a. Earth-Sheltered Building With Positive Year-Round Heat Flux. In cold climates, large buildings with normal internal gains nearly always experience a positive heat flow (from the building interior to the earth mass). In these buildings energy efficiency is improved by minimizing the perimeter wall area and by maximizing the average distance to the ground surface. Insulation placement should be separately optimized.

Buildings with large internal gains, however, can often be optimized by increasing the earth-contact area. Such buildings in cold and in moderate climates have predominant cooling loads even in winter. High-internal-gain functions placed on the perimeter will experience a year-round cooling benefit. An increased floor-slab area will also contribute to cooling in this case.

b. Earth-Sheltered Buildings with Negative Year-Round Heat Flux. Earth-sheltered buildings in very hot climates, especially those with large internal heat loads, will have little or no heating requirements. In this case, the building may experience heat gain from the surrounding soil, especially at shallow depths in summer months. Such buildings tend to be optimized by minimizing the building wall area and placing the building at a depth near the constant deep-ground temperature. However, in southern climates, earth-sheltering tends to act more as a buffer to the climate than as a positive cooling benefit. Cooling derived from earth-contact surfaces in southern climates tends to be insignificant.

c. Earth-Sheltered Buildings with Both Heating and Cooling Benefits. The majority of earth-sheltered buildings, in particular those in temperate climates, will benefit in both heating and cooling. In general, the goal is to get ground temperatures to offset the direction of seasonal heat flow. Ground temperatures warmed during the cooling season, in which there was a high thermostat setting, will release the stored heat during the heating mode when the thermostat setting is lower. Similarly, the lowered ground temperatures in the heating mode increase the heat sink capacity of the earth in the cooling mode. The extent to which earth contact offsets both negative and positive heat flux and the depths at which this is maximized must be estimated using the criteria provided in Section 18.

The question of wall area at these depths is answered by comparing the net energy usages of perimeter zones and internal zones which are otherwise identical. If the earth-contact perimeter zones use less total energy than the interior zones, then the building configuration can be optimized (in terms of energy savings) by reducing the number of interior zones and increasing the number of exterior zones. This would imply a configuration with maximum wall area at the predetermined depth. If, on the other hand, the internal zones have a better total-energy performance than the earth-contact exterior zones, the optimum configuration would be square in plan.

Caution must be exercised, however, when increasing the earth-contact surface area. The construction costs for additional wall areas can exceed the life-cycle energy savings to be derived from the reduced internal zones and increased earth-contact area.

7. NATURAL LIGHT AND VIEWS. The requirement for natural light in small earth-sheltered buildings is not as problematic as for large earth-sheltered buildings. In small buildings, natural light, views, and passive solar can work together to serve a significant portion of the building. Large earth-sheltered buildings, on the other hand, must often resort to clerestories and atriums to make the building more habitable. The disadvantages tend to include both increased initial cost and higher energy loads.

In large earth-sheltered buildings, windows will normally represent a relatively small portion of the building envelope. Window orientation (unless combined with passive solar systems) should, therefore, be determined primarily by building function and the exterior views. Differences in potential heat gains or losses due to compass orientation should be the second priority.

When exterior views are limited by an earth-sheltered configuration, it is often desirable to provide internal atriums or spaces which provide an alternative visual relief. In this case, life safety must be given special consideration.

8. MAINTENANCE AND SECURITY. Some building elements may be more prone to damage than in conventional buildings because of the ease of access by non-authorized personnel. Examples of such building elements are building vents and skylights which may be quite readily accessible on an earth-covered roof. Design for such building elements should include considerations of building security, privacy and noise or blast protection if necessary.

Earth-cover resists casual vandalism damage to most building surfaces but critical points are building services such as vents, which are accessible on an earth-covered roof, and skylights, which may also be readily accessible. A determined vandal could also cause considerable damage and repair cost by puncturing the roof waterproofing system in a disguised location.

Section 10. ENVIRONMENTAL PSYCHOLOGY

1. EFFECT OF PHYSICAL CHARACTERISTICS. The location of a building slightly below-grade does not preclude providing the amenities of natural light, sunshine, and exterior views. There are, however, psychological barriers to the physical location of a space below grade even if it has the identical physical amenities of a a conventional building. Such fears may be related to a fear of structural collapse, burial, or entrapment by fire or flood. Psychological reactions to an underground location tend to heighten awareness of physical characteristics that may go unnoticed in a conventional building. The following considerations affect the perception of earth-sheltered space.

a. Natural Light and Views. Access to natural light is important to users of a building even if the proportion of daylight to artificial lighting is relatively low. The feeling produced by daylight, its variability and the sense of contact with the outside world are important reasons for its desirability (see Section 11, Natural Lighting, Skylights and Passive Solar).

Windows are desired for their ability to provide a direct view, to monitor weather conditions, provide a sense of contact with the environment and to provide visual relief from the immediate surroundings.

b. Horizontal Access. A horizontal entrance into a below-grade building will reduce the impression of entering an underground space (see Section 6, Site Context).

c. Internal Spaciousness. The fear of tight or enclosed spaces is clearly a potential problem for users of an earth-sheltered building. Building designs that allow natural light, provide a good internal environment, and have a spacious interior will reduce the claustrophobic effect of a below-grade space.

Since the functions of a window, besides admitting light, are to extend the immediate environment of the building occupant and provide visual change and relief, improvement in user satisfaction and contentment can be obtained with more spacious interiors.

Carefully planned interior views which utilize borrowed lights, open-plan arrangements, atriums, and/or skylights will achieve greater visual relief and spaciousness.

d. Internal Environmental Conditions. Windowless- or underground-building users frequently complain of poor temperature and humidity control and a lack of ventilation or stuffiness. These complaints should be no more of a problem for a below-grade or windowless building than for a sealed climate-controlled conventional building. In fact, temperature and humidity control should be easier to obtain than in an above-ground building. In addition to the actual air changes provided, the user perception of ventilation is important. Where a superior internal environmental control of an earth-sheltered building is made apparent, this will be recognized as a compensation by most building users.

Good ventilation should be provided to work stations and it should be felt or observed. Individual temperature control may improve the perception of an adequate mechanical system.

e. Spatial Orientation. Since orientation to surrounding buildings and other external features can be lost underground, disorientation may affect the perception of visitors to an underground facility. Simple building layouts and clear coding or signing of circulation areas is important. Color coding, preferably warm colors, may assist orientation. Courtyards and skylights can also provide occasional chances for orientation

2. EFFECT OF WORK CONDITIONS. The following work conditions can affect the perception of a building.

a. Internal Activity. Internal activity within a building that can offset the lack of external stimuli will normally be beneficial in a work environment, provided that it is not too intrusive in terms of noise or distraction.

b. Length of Stay. An individual's reaction to an underground or windowless environment may be substantially colored by the length of time he or she expects to spend in that environment. Underground facilities primarily used for short-term activities such as indoor sports facilities, restaurants, and shopping centers will thus normally raise fewer objections than an underground office.

c. Need for Underground Location. Employees of underground or windowless facilities appear to be more accepting of their environment if they perceive a rational basis for the location or design of the facility. In other words, since windows are detrimental to the operation of many sports facilities, museums, restaurants and shops, employees working in these facilities may be more resigned to their drawbacks. Similarly, windowless precision laboratory and manufacturing environments provoke less criticism than a windowless office building.

d. Job Satisfaction. Employees who derive considerable satisfaction from their work may be more tolerant of windowless space than employees who work at repetitive tasks.

e. Noise. Reduction of noise levels from sources external to the building may be an important benefit to occupants in a noisy exterior environment--close to an airport or freeway--for instance. Internal noise levels should be controlled to provide a balance between an unnerving silence and a disturbing level of noise.

Section 11. NATURAL LIGHTING, SKYLIGHTS, AND PASSIVE SOLAR

1. METHODS OF DAYLIGHTING EARTH-SHELTERED SPACE. Daylighting is one of the more promising energy conservation strategies for reducing lighting loads, but it needs to be carried out very efficiently in order to achieve the maximum gain and to avoid the danger of increasing the summer cooling and the winter heating loads. In general, daylighting can be divided into two categories: (1) daylighting schemes which rely strictly on the beam component of the sun, and (2) strategies which utilize both the beam and diffuse components.

a. Beamed Daylighting. Beamed daylighting shows the most promise in lighting deep underground space since the light can be projected through the building interior for delivery to target spaces. The optical systems required to do this are very sophisticated. The components for such a system include a set of heliostats (sun trackers), and a system of lenses and mirrors in an arrangement through which light can be projected. A major advantage of this system is the ability to bring large amounts of light through small window openings. Cold mirrors may be incorporated into the system to separate the infrared part of the spectrum from the visible, thus reducing cooling loads. A lighting source such as this will produce one-fourth less heat than a fluorescent fixture providing the same amount of light.

Light shelves are an attractive alternate to the "high tech" solar optics discussed above. This technique is a very workable daylighting solution for single-story earth-sheltered structures. The greatest asset of this system is that no moving parts are required. This system is most suited to ambient lighting.

b. Diffuse Daylighting. Daylighting strategies that rely on the diffuse solar component are much simpler to design than systems which utilize the beam component. These systems do not require moving parts, and are less likely to cause visual comfort problems. Such systems include the use of windows and skylights.

2. PASSIVE SOLAR APPLICATIONS FOR EARTH-SHELTERED BUILDINGS. Passive heating and cooling techniques can be effectively incorporated into earth-sheltered building design to further reduce space conditioning loads. Design optimization requires manipulation of the building configuration, orientation, and construction.

The heat transfer process in passively heated and cooled buildings is variable and of low density. As a result, passive energy performance is difficult to estimate. To date, the only accurate way to estimate energy performance is the use of computer modeling. This technique provides an interactive approach to passive design and enables a parametric analysis to be carried out quickly. At present, however, this technique has not been fully developed and is not universally cost effective to the designer.

Window location may affect the overall building configuration. Buildings oriented to the south can take advantage of direct solar gain. Spaces that are remote from the south side of the building may take advantage of skylights and clerestories. Clerestories will normally be preferred to skylights in earth-sheltered construction since they allow fixed seasonal shading, are typically less expensive and need less maintenance, and accept thermal shuttering more readily. In addition, depending on the configuration, clerestories may allow partial earth-sheltering. Where lack of fenestration may impair the psychological habitability of the building, passive clerestories may be especially appropriate.

a. Direct Gain Systems. Because direct gain techniques may result in a greater range of room temperatures, it is important to carefully assess the needs of the occupants. Some occupants may permit a wide range of temperature fluctuations; others may not. Further, some activities may not be amenable to direct solar gain. In some cases, north light will be preferred.

These conditions will affect the type of glazing to be used. In place of clear glazing, translucent materials such as diffusing glass, fiberglass, and plastic films may be appropriate (refer to NAVFAC DM-8, Fire Protection Engineering, for use of plastic). Reflective glazing may reduce unwanted radiation as well as provide more privacy for the occupants. Glazing that diffuses light will better disperse the radiation to interior surfaces, resulting in more even temperatures.

b. Indirect Gain Systems. As applied to earth-sheltered buildings, thermal storage walls will reduce precious views to the exterior. A second disadvantage is that for solar heating, this system does not take advantage of thermal mass inherent in earth-sheltered construction. The mass associated with earth-sheltering is not only capable of storing heat on a diurnal cycle, but also on an annual cycle.

Attached sunspaces allow wide temperature fluctuations which reduce the amount of thermal mass normally required in a direct gain system. In an earth-sheltered building with abundant mass, this may not be a distinct advantage. For large earth-sheltered buildings, however, sunspaces have the potential to provide visual relief and natural light.

Walls between the sunspaces and the building provide the most effective thermal mass for diurnal cycling. These walls, especially south-facing walls, receive full solar insolation during the day in winter months and reradiate heat into the interior space; the remaining heat warms the sunspace. Such a south-facing wall inside a sunspace can be constructed as a passive or trombe wall. Heat transfer to the interior spaces can be accomplished by direct solar transmission, direct air exchange, conduction through common walls, or by storage in rock beds.

c. Isolated Gain Systems. In an isolated heating system, solar heat is collected and stored in an area set apart from the occupied space. Because the occupied area is not directly exposed to the sun, the amount of heat it receives can be more closely controlled than with other systems. Earth-sheltering does not necessarily afford additional free thermal mass in this application.

3. CALCULATION PROCEDURES (GENERAL). Currently, the only accurate way of predicting thermal performance in earth-sheltered buildings is the use of computerized hour-by-hour simulations. Simplified methods are available for small above-grade buildings (see Reference 7, Passive Solar Design Handbook, by Balcomb).

Section 12. FIRE PROTECTION AND LIFE SAFETY

1. AIR-HANDLING SYSTEMS. Refer to NAVFAC DM-8, Fire Protection Engineering, for criteria for air handling systems. The following general principles apply to the effective use of air-handling systems for pressurization:

a. Negative Pressures. A negative air pressure should be obtained in the fire/smoke area by the shutdown of air supply fans and diversion of return air directly to outside.

b. Positive Pressures. A positive air pressure should be obtained in adjacent areas by the shutdown of return air fans and by using 100 percent outside air.

c. Fire/Smoke Areas. Separate fan and duct systems shall be used for each fire area. Where the building is not divided into fire/smoke areas, all supply air must immediately shut off and all return air exhausted to the exterior.

2. SMOKE AND HEAT VENTING. Refer to NAVFAC DM-8 for design requirements for smoke and heat roof vents. To effectively implement these requirements, the designer should be aware of the considerations discussed below. See also NFPA 101, Life Safety Code, for outside smoke venting requirements.

Smoke venting is used to alleviate smoke and heat build-up in an underground or windowless building by exhausting smoke and fire gases from the building. The current trend in firefighting, however, is to provide venting only after the fire is out. Storage buildings may require extensive venting facilities, whereas buildings with low fire loading, noncombustible construction, and automatic sprinklers may be cleared of smoke through the ventilation system provided for normal changes of air.

Vents are most applicable to large-area one-story buildings lacking adequate subdivision. They are also useful in windowless and earth-sheltered buildings. Vents are not a substitute for automatic sprinkler protection.

Where vents are used on windowless buildings it is necessary to ensure that there is an area of openings near the ground, at least equal to the area of the vents, or the vents will not function effectively.

Vents will not function in earth-sheltered buildings unless provision is made for air to enter the burning compartment near floor level. Fresh air falling through the layer of hot smoky gases will itself become heated and smoky. Sprinklers will cool the gases.

In below-grade areas where gravity venting is not feasible, power ventilators can provide positive heat and smoke venting. If venting is initiated only to clear the smoke after the fire, then power ventilators will be more suitable.

In large, high spaces, smoke can be exhausted with roof-mounted fans only to a height of 50 to 60 feet (15 000 to 18 000 mm). If smoke becomes cool, wet or heavy because of the operation of the sprinkler system, and the ceiling is higher than 55 feet (16500 mm), it is necessary to mechanically introduce air at the floor level in order to remove the smoke through the vents. The supply air should be directed vertically--roughly in line with the exhaust fans. The supply should be 75 percent of the exhaust.

3. FIRE DEPARTMENT ACCESS. Access to a building for manual fire-fighting must be made easy. Normally the usual entrances to a building will provide satisfactory access for fire-fighters, but this may not always be the case with underground buildings. A fire in these spaces often becomes an excessively smoky fire. In addition, heat build-up is usually more intense because heat venting is inadequate. Firefighters are therefore often placed in the position of attacking fires in these spaces in the face of the heat and smoke.

For access requirements for windowless buildings, refer to the provisions of NAVFAC DM-8.

4. REQUIREMENT FOR AUTOMATIC FIRE-EXTINGUISHING SYSTEM. NAVFAC DM-8 requires an automatic sprinkler system for windowless or underground structures, "when the combustibility of the contents, or life-safety features, warrant the protection."

Note, however, that earth-sheltered buildings are not necessarily "underground structures." "Structure" is defined by NFPA 101, Chapter 30, as follows: "A structure in which there is no direct access to outdoors or to another fire area other than by upward travel." It is therefore possible to have an earth-sheltered building where the direction of travel is downward or horizontal. A building built into a hillside with exits on the exposed wall would be an example.

5. FIRE SEPARATION OF STRUCTURES. Refer to NAVFAC DM-8 for separation criteria. Completely below-grade buildings are not considered in the above fire separation criteria and safe separation should be analyzed using available criteria in other publications such as DOD 4270.1-M (see Criteria Sources) and the Uniform Building Code, subject to NAVFAC approval.

NAVFAC DM-8 requires that all grass, weeds, and brush within 25 feet (7500 mm) of all structures be kept trimmed.

6. DRAINAGE. Drainage of water from sprinkler discharge or from hose streams can be a serious problem in underground structures and therefore requires study in the early planning stages. If provisions for drainage are required for reasons other than fire protection, it may be possible to use only one drainage system for all purposes, provided that it is designed to handle the expected maximum flow.

Refer to National Fire Protection Association, Fire Protection Handbook, Section 6, for relevant criteria (see Criteria Sources).

Watertight floors are important in preventing the spread of water damage. Of greater importance is the number and location of floor drains. If interior drains and scuppers are available, salvage teams can remove water effectively with a minimum of damage.

Although gravity flow drainage is the most reliable means of removing water, mechanical means may have to be used in areas below grade. A waste removal pump is best, and it should take suction from a screened sump. Electric power for the sump pump should be taken off ahead of the service disconnecting means so that the building power can be interrupted for firefighting without interfering with the pump operation. Emergency power should be provided for the pump system.

7. LIFE SAFETY. General life-safety criteria are covered in NAVFAC DM-8. In addition to meeting the exit requirements contained or referenced in NAVFAC DM-8, the following practices are suggested:

a. Refuge Areas. Maximum division of areas limits fire extent and severity, and provides areas of refuge for occupants. Temporary refuge areas, stairwells, and elevator shafts are often the prinicipal areas where pressurization and barriers may be utilized effectively. The Federal Building in Seattle, Washington, for example, combines exhausting the fire floor and the use of physical barriers and pressurization of spaces above and below the fire to confine the smoke to the fire area and to protect the occupants in refuge areas within the building.

Stairways may be pressured to help ensure smoke-free conditions. From a theoretical viewpoint, the magnitude of pressure need be only slightly greater than the pressure of the smoky atmosphere. From a practical viewpoint, a positive pressure of 0.05 to 0.10 inches (1.27 mm to 2.54 mm) of water column should be maintained, even with as many as three stairwell doors open.

For the safety of the occupants, every level of a windowless or underground building should be divided into two or more separate areas to provide safe places of refuge within that level. Areas of refuge are not substitutes for the required exits; however, areas of refuge should be considered when life-safety provisions are minimal. Each refuge area or "fire area" should be on an independent air supply and exhaust system. The separation between areas is critical. A possible configuration might locate a vestibule between adjoining areas with the vestibule vented as would be required for a mechanically-operated smokeproof enclosure. If pipe or conduit goes through the separation wall which establishes the refuge area, penetrations must be tightly caulked to prevent passage of smoke. The use of listed and tested transition devices is suggested. Ideally, there should be no ducts through that wall. Other penetrations should be limited to the connecting horizontal exit.

b. Goal-Oriented Exit Design. Goal-oriented (dynamic evaluation) exit design is discussed in the NFPA *Fire Protection Handbook* although it is not identified in either the *Uniform Building Code* or NFPA 101, *Life Safety Code*, (see Criteria Sources) as an alternate to the more specific exit requirements.

Fatigue is judged to become important after five minutes of travel in a downward direction or one minute of travel in an upward direction. To help reduce fatigue, stairs can be made more gradual. Instead of the 7-inch (178-mm) rise and 10-inch (254-mm) run stairs, a 7-inch (178-mm) rise with an 11-inch (279-mm) run would significantly reduce the amount of energy a person would have to expend in climbing. It can take about 30 percent less energy to climb up the more shallow stairs.

It is suggested that the following performance criteria be applied in addition to meeting the criteria cited in NAVFAC DM-8:

(1) All occupants exposed to the fire environment should be able to evacuate to a safe area within 90 seconds of alarm.

(2) A portion of this time, not to exceed approximately 15 seconds, can be involved in traveling in a direction towards the fire.

(3) All occupants should reach an area of refuge within five minutes of downward vertical travel or within one minute of upward vertical movement (fatigue is judged to be important after these travel times).

c. Exit Enclosures. Exits from underground buildings which involve upward travel, such as ascending stairs or ramps, are provided for in NFPA 101, Life Safety Code. These recommendations take into consideration the panic that may result when there is no direct access to the outside and no windows to permit fire department rescue or ventilation.

Exits involving upward travel, if more than one-story or 15 feet vertically, must be cut off from main floor areas and provided with outside smoke venting or other means to prevent the exits from becoming charged with smoke. In any fire, some carbon monoxide will tend to collect at the bottom of the stair enclosures. Consequently, there is some life-safety advantage in locating the exit from the stair enclosure above the lowest level of the enclosure. Where downward exiting occurs, provisions must be made to prevent downward travel past the exit door.

d. Elevators. Elevators are not exits but, in practice, they may be used to get people out and firefighters in. A normal person can go down stairs for about five stories before the legs begin to feel the strain and slowing occurs. If people are deep in the ground, they may have to depend on mechanical systems to evacuate them from the place of refuge.

Elevator lobbies could serve as refuge areas if both the lobbies and the hoistways are pressurized to prevent the entry of smoke. If 0.06 inches (1.52 mm) of water column is the pressurization to be expected from a fire in a sprinklered building, the lobby would have to be pressurized to about 0.08 inches (2.03 mm) of water column. Further, all elevators should be programmed to return to the grade-level exit floor. This operation is activated by the building smoke detection system.

Section 13. WATERPROOFING

1. SOURCES OF MOISTURE. The basic sources of moisture are:

- o Surface runoff
- o Water table
- o Temporary water pressure
- o Vapor pressure
- o Capillary draw

Surface runoff must be handled by proper site planning and landscape design. Since building below the water table is a very poor condition for construction, the first objective in site selection and planning should be to avoid the water table. A structure that must be built partially below the water table will require either extensive drainage in order to permanently depress the water table, or it must be completely waterproof and designed to withstand the pressures in both walls and floor. Temporary water pressure against below-grade walls is caused by excessive rains which cannot be adequately handled by surface or subsurface drainage systems. Although the frequency of occurrence and the magnitude of these temporary pressures can be reduced by site planning and drainage techniques, they are unlikely to be completely eliminated.

The final two sources of moisture, vapor pressure and capillary draw, are handled primarily by placing the waterproofing skin against the structure of the building. Reducing the moisture content of the soil through drainage can also help minimize these problems. Depending on the relative humidity and temperature conditions of the soil and the interior air of the building, differential vapor pressures can cause water vapor to flow from the soil into the building. When this occurs, the waterproofing on the outside of the wall can alleviate the moisture problem. Condensation problems caused by high humidity and low surface temperatures on the inside of the space are not a result of inadequate waterproofing. Capillary draw, another moisture transport mechanism, which can occur through concrete walls in contact with moist soil, is usually handled by an impervious waterproofing system. It can also be eliminated by creating an air gap between the earth and the wall.

2. EVALUATION OF WATERPROOFING SYSTEMS. Product evaluation should be based on at least the following criteria:

- o durability and stability underground
- o capability to withstand movement and cracking
- o capability to minimize leaks and facilitate repair
- o appropriate use
- o application procedure
- o relative costs

a. <u>Dampproofing Products</u>. This group of products includes concrete admixtures, polyethylene sheets, and a variety of surface treatments such as epoxy and acrylic paints, simple asphalt and pitch coatings, and cement pargeting (back plastering) that can be applied to concrete and masonry walls. It is important to distinguish between these relatively inexpensive, easily applied products and other systems that use some of the same materials in different ways to yield products with different characteristics and, in some cases, improved effectivenss (see the sections below on cementitious materials, built-up membranes, and modified bitumens). Although commonly used for foundations and basements, these products should not be considered complete and adequate waterproofing systems. The term dampproofing actually connotes the prevention of dampness but not full protection from water that may enter the structure.

(1) Stability and Durability. The effectiveness of two of the most commonly used dampproofing products, asphalt and pitch coatings, is questionable when they are used underground for two reasons. First, asphalt emulsions can re-emulisfy in the presence of ground water over a long period of time and eventually become ineffective. Second, the quality of both asphalt and pitch has deteriorated in recent years. Improvements in the oil refining process have resulted in poorer chemical properties in the asphalt by-product. Similarly, as health standards have caused some of the more volatile substances to be removed from coal tar pitch, a more brittle, less effective product has resulted. It should be pointed out, however, that a great variety of asphalt and pitch products, with a potential difference in properties, is available. Naturally occurring asphalts, for example, would not suffer from most of the disadvantages mentioned above.

Because polyethylene is manufactured with a consistent quality, it can serve as a good vapor barrier if it is carefully applied and seams are overlapped. Although clear polyethylene is subject to degradation if it is exposed to the ultraviolet rays of the sun, this is not a problem in underground applications. Polyethylene is totally inadequate as a waterproofing system, however. It is not puncture resistant and it is difficult, if not impossible, to make watertight seams in the field.

Epoxy and acrylic coatings can be tough and adhere well to concrete surfaces. Using concrete admixtures and pargeting with a dense cement plaster can greatly reduce the permeability of concrete. Although some of these dampproofing products can be effective in preventing water from entering the structure and are also relatively stable underground, they nonetheless do not represent a long-term solution to waterproofing because of the drawbacks discussed below.

(2) Ability to Withstand Movement and Cracking. Dampproofing products may be integral with the concrete structure (admixtures), very brittle (cement pargeting), or applied in a relatively thin layer bonded to the surface (epoxy paints, asphalt, and pitch coatings). In none of these cases does the product have any real ability to respond to movement in the structure or to bridge cracks. Although asphalt and pitch have some softness and flexibility in exposed locations in summertime, they are usually more brittle and subject to cracking when used at consistently low temperatures underground. Since cracks and movement are virtually inevitable in concrete structures, this is the greatest shortcoming of these coatings. Polyethylene, on the other hand, has some ability to bridge cracks and adjust to movement without failing because it is a sheet rather than a coating. For underground applications, clear polyethylene is superior to black polyethylene as it is not exposed to sunlight.

(3) Ability to Minimize Leaks and Facilitate Repair. Dampproofing products are incapable of resealing any punctures, tears, or breaks in the

coating. In a soft and flexible state, asphalt, and pitch have a limited capability to reseal. But, in the presence of cooler, below-grade temperatures, these materials can become brittle. Most dampproofing products are integral with the concrete or bonded to the surface. Because any leaks that do occur cannot travel extensively under the coating (as they could with a sheet or membrane), they are easier to locate. Water can, however, travel through the structural cracks and voids, which can be extensive in block walls. The problem of localizing a leak is considerably greater with polyethylene. Water can travel easily behind the loose-laid sheet. If the sheet is bonded to the surface, the tendency for water to travel is reduced.

(4) Appropriate Use. These products are not recommended for any underground application where a complete waterproofing job is desired. They are appropriate only as a vapor barrier for secondary spaces in which some moisture and dampness can be tolerated.

(5) Application Considerations. The various coatings can be applied by unskilled labor and are either sprayed, troweled, or brushed on. Specifications for temperature, humidity, and surface conditions vary with the products. Polyethylene sheets should be overlapped generously to act as an effective vapor barrier.

(6) Relative Costs. Most of these dampproofing products have relatively low labor and material costs compared with the costs of thorough waterproofing systems.

b. Built-Up Asphalt and Pitch Membranes. This type of membrane, consisting of layers of hot-mopped asphalt or pitch alternated with felt or fabric reinforcing which is bonded to the structure, is a very familiar and commonly used product on conventional roofs. This system has been used with some success in various underground applications, but its long-term durabilty below-grade, as well as its capability to perform in the same manner as on conventional roofs are questionable. Long-term performance is more critical underground because the waterproofing is not easily accessible for repair and eventual replacement.

(1) Stability and Durability. The long-term stability of many asphalt and pitch products underground is not reliable. Although the basic built-up membrane is relatively impervious to water and water vapor when first installed, constant exposure to water can cause deterioration of the asphalt. As for the fabric reinforcing, organic felts will eventually rot with constant exposure to water, whereas glass-reinforced fabrics will last much longer.

(2) Capability to Withstand Movement and Cracking. Built-up membranes have mechanical strength that simple asphalt or pitch coatings do not have. Nevertheless, because membranes are relatively brittle and inflexible at cool below-grade temperatures, they cannot absorb movement or bridge cracks in the concrete structure.

(3) Capability to Minimize Leaks and Facilitate Repair. Built-up membranes used on the roofs of above-grade structures are noted for some capability to reseal punctures, primarily because they become soft and

flexible when heated by the sun. In the cooler below-grade environment, however, they remain too brittle and inflexible to have very good resealing properties. If a leak does occur, the built-up membrane does not always adhere to the structural surface as well as other waterproofing products that are chemically bonded. Thus, water that penetrates the membrane may travel behind it to some degree, making leaks difficult to locate. Any repairs to the membrane must be made from the outside.

(4) Appropriate Use. Horizontal surfaces that have a slight slope for drainage are the most typical and best applications for built-up membranes. Although they can be applied to vertical surfaces, such application is more difficult, particularly when the membranes are applied hot. Because these membranes are not continuous flat sheets or roll goods, but rather are built up from smaller pieces in layers, they are better suited to more complex forms than are other sheet goods. The waterproofing can be formed on curving surfaces and around penetrations as an integral flashing.

(5) Application Considerations. The basic work of mopping on the hot asphalt can be done by relatively unskilled labor. However, experienced supervision and proper equipment are necessary to ensure proper installation. The conditions required for successful application are not as stringent as for many other products. The surface must be relatively clean, dry, and smooth, but some leeway for irregularities exists. The membranes can be applied in a wider range of temperature and humidity conditions as well. Because the membrane may be relatively soft after application, it should be carefully protected from punctures or damage during the backfilling process.

(6) Relative Costs. Material and labor costs for built-up membranes are moderate compared with costs for other waterproofing products. More competitive bidding may be possible, due to the greater number of qualified applicators and the familiarity of built-up membrane.

c. Cementitious Materials. The various cementitious waterproofing materials, which are sprayed, brushed, or troweled onto concrete surfaces, consist of Portland cement and certain organic or inorganic additives. When the mixture is placed on a concrete surface, it comes in contact with moisture and unhydrated cement, causing the formation of crystals in the voids of the concrete. The size of the crystals is such that water molecules cannot pass through, while air and water vapor can. Thus, the concrete can cure properly while maintaining a waterproof surface.

(1) Durability and Stabilty. Most cementitious materials are stable underground and compatible with most soil chemicals and conditions. These alkaline materials (pH of approximately 9.0 to 9.5) will resist pH levels between 3.5 and 11--a range that includes most soil conditions. Products that use sodium-based additives to react with the cement in the concrete may be less desirable because sodium is water soluable and could leach out of the concrete over time. In general, however, cementitious products themselves have a long life span.

(2) Capability to Withstand Movement and Cracking. The major disadvantage of cementitious waterproofing materials is that they have very

little capability to bridge any cracks in the concrete caused by settling, thermal expansion, or other movement. Although very small hairline cracks can be bridged, any larger cracks, which are common in most concrete structures, represent a break in the waterproofing system. Buildings that incorporate precast elements, cold joints, or masonry walls are generally more likely to crack than are monolithically poured concrete structures. Post-tensioned structures in particular resist cracking because the concrete is held in constant compression. Thus, a post-tensioned structure represents the best possibility for successful waterproofing with a cementitious product.

(3) Capability to Minimize Leaks and Facilitate Repair. As stated above, any leaks that occur with cementitious waterproofings are likely to result from cracks in the concrete structure. Since the waterproofing is integrated within the concrete, water cannot travel behind the waterproof layer as it can when an attached membrane material is used. It can, however, travel within and along any cracks, thereby increasing the difficulty of finding the major source of the leak. Cementitious materials have no resealing capabilities, but, if a leaky structure requires repair, these materials can be applied from the inside of the structure, even against hydrostatic pressure. Presumably, application could be done after major cracks were carefully resealed from the inside, though, this might merely cause the entry point of water to shift to a new location by migration of water along unsealed portions of the crack.

(4) Appropriate Use. Cementitious products are limited to use on concrete and masonry surfaces. Basically, a product that can be sprayed or troweled on and that actually penetrates into the concrete appears to have certain advantages. It can be applied to complex or curving forms, for example, as well as vertical or horizontal surfaces. Unfortunately, this apparently wide range of applications must be carefully limited to situations in which cracks are unlikely to appear.

(5) Application Considerations. Unlike most waterproofing systems, cementitious materials can be applied by relatively unskilled or moderately skilled labor. Another advantage is that they cannot be easily damaged during the backfilling process. The major application concerns are the condition of the concrete surface, the moisture content of the concrete, and the temperature. Although the concrete surface should not have any major defects (honeycombing, rock pockets, or faulty construction joints), small irregularities are acceptable. In fact, a surface that is extremely smooth may be undesirable because the waterproofing material must be able to penetrate into the open capillary system of the concrete. Surfaces must also be clean and free from concrete form oil, curing compounds that seal the concrete pores, or other foreign matter that could inhibit penetration. Light sandblasting or waterblasting may be necessary. Modified cementitious materials that can be applied on concrete block walls are available. Additional surface preparation is usually required for block walls to cover cracks and render a more uniform surface.

Because moisture is required to form the crystals, newly poured concrete that is damp throughout is the ideal application condition. Under other conditions some prewatering may be required. It is important that the surface be moist, but if it is wet, it will dilute the penetrating material. Cementitious materials should not be applied at temperatures below 40°F (4°C). Proper curing depends on the temperature and the reaction of the materials with water. Most cementitious waterproofing materials require air for curing and must be sprayed regularly over a period of two to three days with a misty spray. Therefore, the materials should not be covered immediately by polyethylene. During curing, however, they must be protected from excessive wind, sun, rain, and frost.

(6) Relative Costs. Because a number of cementitious materials are available, costs can vary widely. In general, material costs can be considered moderate and labor costs low to moderate in comparison with other waterproofing systems.

d. Liquid-Applied Waterproofing Systems. This group of products includes a variety of urethanes, elastomers, rubbers, and other synthetic compounds which are applied in a liquid form. They typically cure in 48 to 72 hours to form a single, seamless, membranelike coating which is bonded to the surface of the structure. They can be applied in thicknesses of 15 to 100 mils (0.38 to 2.54 mm). But, 60 mils (1.5 mm) is generally the minimum thickness desirable for underground application. These products are applied in one or more coats by spray, trowel, roller, or brush. The liquid-applied systems included in this category are:

- polyurethane
- polyisobutylene (butyl)
- polychloroprene (neoprene)
- chloronated polyethylene (hypalon)
- polyvinyl chloride
- polysulfide
- silicone
- acrylic latex

Because of the large number of products and the almost infinite variations possible with some of them, it is somewhat difficult to generalize about their characteristics. This discussion will emphasize the polyurethanes because they are the most common and the most likely to be used underground. The basic properties of the rubbers--butyl, neoprene, and hypalon--are discussed in greater detail in the subparagraph below on vulcanized sheets, which is their more common application. Although the specific chemical properties of these liquid-applied materials can vary considerably, they can be evaluated as a group because most of the general characteristics--the ability to withstand movement, and design and application considerations--are quite similar.

(1) Stability and Durability. The properties that affect the stability and durability of products within this classification of waterproofing systems can vary widely; hence, these characteristics must be carefully examined. Some products can be prepared to suit a particular application, so that they are especially resistant to certain chemicals, to sunlight, or to abrasion, for example. With respect to underground applications, only the polyurethanes have a reasonable history of use. Although a discussion of the chemical makeup of these products is beyond the scope of this evaluation, some basic points can be made. One-component polyurethane systems are relatively inexpensive and easy to apply but can

become brittle and lose adhesion. They are not recommended for use underground where performance is critical. Two-component systems, although more costly, perform much better. Products based on esters and ethers were developed to resist abrasion above grade; they are not recommended for underground waterproofing because they will become brittle over time.

One characteristic of liquid-applied membranes that would appear to be an advantage over factory-produced membrane sheets is that they are seamless (seams are usually the weak point of loose-laid membranes and roll goods). Attaining the same levels of stability and durabilty characteristic of factory-produced materials in a liquid that is cured in the field may, however, be a significant problem. The rather exacting conditions required to produce a plastic or synthetic rubber are difficult to duplicate in the field, because of the variations in surface conditions, temperature, humidity, and manner of application.

The application procedure can affect the durability of the liquid-applied systems. For example, the use of spray-on applications can seriously weaken the membrane by entrapping air bubbles. Thicker products that can be troweled on are generally more desirable. Some liquids are intended to be applied in a "self-leveling" form. Since it is nearly impossible to build a completely flat, smooth surface, the self-leveled membrane will be thicker in some areas than others. Sections of the membrane that are too thick can blister, adhere poorly, and cure improperly. Polyurethanes cure slowly: when the time required for curing is incompatible with construction schedules, additives are sometimes included to speed up the curing process. The drawback to this technique is that the decrease in curing time results in a product that may be more brittle.

Acrylic latex-based products are not recommended below-grade because the ground water can cause the latex to re-emulsify and migrate from the wall. Polysulfides are composed of synthetic rubber that is quite impermeable to water and resistant to chemical attack. They are costly and excessively soft. Polysulfides are not currently manufactured in the United States. Liquid-applied waterproofing systems based on silicone are relatively new and have little history of application. Because of certain drawbacks, which include poor adhesion to concrete, the silicones seem better suited to above-grade, rather than below-grade, applications.

(2) Capability to Withstand Movement and Cracking. Liquid-applied waterproofing systems are fully bonded to the structure. Therefore, any strain caused by movement and cracking must be taken up by the material lying immediately over the crack. Although these materials have some of the properties exhibited by factory-made membranes--toughness, tensile and shear strength--the capacity of relatively thin, fully bonded materials to bridge cracks is limited. Thus, these products are not recommended for precast concrete roofs, which have a great potential for movement and cracking. Rather they may be more suited to reinforced and post-tensioned poured slabs, in which cracks are minimized.

(3) Capability to Minimize Leaks and Facilitate Repair. Like most membrane and sheet materials, liquid-applied systems have no capability to reseal punctures or tears. They do have two advantages over sheet goods in minimizing leaks in that they are seamless and are fully bonded to the

structure. Provided that the material has good adhesion, leaks cannot travel under the membrane. The source of the leak is easier to locate and therefore easier to repair, but, if a leak is accompanied by loss of adhesion in the membrane, the repair can be more difficult.

(4) Appropriate Use. Liquid-applied systems will bond to a wide range of structural and insulation materials if the product is correctly formulated to do so. The limitations on the appropriate type of surface are based not on the capability of the products to bond, but rather on their incapability to bridge large cracks that may occur during the life of the structure. Direct application is, therefore, not advisable on precast roof decks, masonry surfaces, wood decks, or other surfaces that have a great potential for movement and cracking. These products are best applied to reinforced and post-tensioned poured concrete slabs. In general, the liquid-applied systems are best suited to horizontal surfaces that have a slight slope, although the self-leveling products require a completely flat surface. Some liquid products are formulated to be applied to vertical surfaces, usually in several coats.

Compared with flat sheet membranes and roll goods, materials with good adhesion that can be troweled or sprayed on are more suitable for designs that incorporate complex geometries and curving surfaces. Some question remains, however, as to whether or not the controlled, uniform thickness required for proper curing can be easily achieved on a complicated shape.

(5) Application Considerations. Although the actual brushing or troweling on of these liquid-applied materials does not require highly skilled labor, it is very important to use experienced applicators in order to ensure proper surface preparation, control of the application, and adequate curing time. Some products require a primer or masonry conditioner, which can raise labor and material costs. If the materials are not applied in an even thickness, problems can result. Bubbles can form as gases are released during curing. Adhesion can be lost in areas where the membrane is too thick. Improper application can result in irregularities in the coating, especially in corners and around vent pipes. For best results, two coats are often desirable. On vertical surfaces several coats may be required, along with embedment of the fibers into the first coat. Many of these products release toxic fumes during application and curing. Respirators and good ventilation are required.

The most critical aspects related to the application of these materials are the preparation of the surface, the air temperature, and the humidity conditions. More than any other type of waterproofing product, the liquid-applied systems require a very clean, smooth, dry surface. All oils must be removed, voids filled, and any imperfections in the surface smoothed. The concrete structure should be allowed to cure the full 28 days before the waterproofing is applied, to ensure that most of the moisture is gone from the surface. Some manufacturers do not recommend that their products be applied in situations where moisture must evaporate from the concrete surface underneath, because it will cause the waterproofing to blister and lose adhesion. These situations include lightweight concrete decks, which release large quantities of moisture, and concrete over steel decking, where the moisture cannot escape downward from the slab.

Under the best temperature and humidity conditions, curing of a typical membrane may take 48 to 72 hours, depending on the specific product. Curing of waterproofing that is applied in temperatures below 40°F (4°C) would take several weeks, whereas application above 80°F (27°C) may result in too rapid curing, which can cause brittleness in the membrane. Similarly, if the relative humidity is below 30 percent, the membrane would cure too slowly; above 85 percent humidity, it cures too fast.

After application, the material should be inspected to make sure no voids or bubbles are left in the membrane. Insulation should not be placed over the waterproofing too soon since some volatile substances given off during curing can attack polystyrene insulation. Like most waterproofing products, the fully cured liquid-applied material should be protected from damage during backfilling.

(6) Relative Costs. Labor costs are moderate for the application of this group of products. Material costs can vary considerably because of the many types of liquid-applied systems. Generally, material costs range from moderate to high for the best-quality products compared with costs of other waterproofing systems.

e. Modified Bitumens. These materials, often referred to as rubberized asphalt, consist of asphalt combined with a small amount of synthetic rubber, applied to a polyethylene sheet. In some cases, a second polyethylene sheet is placed between two layers of the rubberized asphalt. The material comes in rolls ranging from 3 feet (914 mm) to 4 feet (1219 mm) wide. The strips of rubberized asphalt adhere to the structural surface and are overlapped to adhere to each other.

(1) Durability and Stability. Since modified bitumens come in factory-produced rolls, they have uniform thickness. The quality of the asphalt itself can vary, however. Rubberized asphalt has good resistance to most chemicals found in the soil. Polyethylene also has good stability in underground conditions, where it is not exposed to the ultraviolet rays of the sun. Generally, these materials will not rot or mildew.

Deterioration of asphalt in contact with ground water is reduced considerably with this system because the polyethylene prevents moisture from coming in contact with the asphalt. Further, the polyethylene acts as a good vapor barrier. The addition of rubber to the asphalt gives the product tensile strength and stability. The rubber makes the asphalt softer and may reduce its tendency to deteriorate with time. The rubberized asphalt products can last a relatively long time if they are carefully installed in an appropriate situation.

(2) Capability to Withstand Movement and Cracks. The tensile strength of the polyethylene and the rubber in modified bitumen materials makes them effective in bridging over cracks up to 1/4 inch (6.35 mm) wide. The softness and flexibility of the rubberized asphalt allow for some movement to occur without stressing the product to the point of failure. The capability of the material to bridge cracks without leaking depends on very good adhesion at the seams, as movement usually creates extra stress in these areas. It is best not to place seams directly over points where cracking is likely to occur, such as at cold joints or other structural connections.

(3) Capability to Minimize Leaks and Facilitate Repair. Modified bitumens are intended to be used with a primer that helps bond the product to the structure. Along the overlapping seams, the rubberized asphalt bonds well to the polyethylene. Ideally, the material will be completely bonded to the structure, thus preventing any water that penetrates the membrane from migrating. This means that the source of a leak will be easier to find. Completely bonding the material to the structure may be almost impossible under most field conditions, however. Because loss of adhesion can occur for a variety of reasons, modified bitumen products must be very carefully applied (see "Application Considerations" below).

(4) Appropriate Use. Modified bitumen products are versatile in that they can be used on concrete, masonry, or wood surfaces. They are well suited to applications on vertical surfaces where there is no continuous head of water. They must be used with more discretion on horizontal surfaces, however. Because these products have numerous overlapping seams, it is not advisable to use them on flat horizontal surfaces where they could be exposed to ponding. If a modified bitumen product is to be used on a horizontal surface, the surface should be sloped slightly to provide drainage and the seams must be overlapped in a manner similar to shingles on a conventional roof.

(5) Application Considerations. Successful application of modified bitumens depends on great care in the preparation of the surface and application only under the proper temperature and humidity conditions. Experienced, skilled labor is usually required for successful application. A smooth, clean, dry surface is necessary for good adhesion of the product to the surface. Mechanical grinding may be required on concrete surfaces, but slight irregularities are acceptable. The waterproofing material should be applied only when the surface temperature is above 40°F (4°C), because colder temperatures reduce the quality of both the bonding and the seams. Space heaters should never be used to warm the surface of the rubberized asphalt because they add moisture, which can cause condensation that may loosen the bond.

Modified bitumens are incompatible with pitch and certain solvents and sealants. Because membranes are combustible, they should not be exposed to flames, sparks, or temperatures over 100°F (37.8°C). Modified bitumens, like most elastomeric materials, have a high degree of memory, such as a tendency to return to the original sheet or roll configuration. Although wrinkles or voids created during application are rolled out, the material will tend to return to the wrinkled state.

During the backfilling process, the waterproofing products should be protected from damage. Insulation can serve this purpose. On uninsulated roofs a layer of sand can be used, whereas some form of protection boards is necessary on uninsulated walls. The backfilling operation should occur relatively soon after the waterproofing is installed so that the polyethylene is not exposed to ultraviolet rays from the sun.

(6) Relative Costs. Compared with costs of other waterproofing products, the cost of modified bitumen products can be considered moderate or average.

f. Vulcanized and Plastic Sheets. This classification of waterproofing materials includes various natural and synethetic rubber compounds and plastics that are formed into sheet membranes by vulcanization or other processes. The generic names and chemical compositions of the six major types of sheet membranes are:

- o isobutylene isoprene (butyl)
- o ethylene propylene diene monomer (EPDM)
- o polychloroprene (neoprene)
- o chlorosulfonated polyethylene (Hypalon)
- o chloronated polyethylene (CPE)
- o polyvinyl chloride (PVC)

Most of these materials are available in roll stock or sheets in sizes up to 50 feet (15 240 mm) wide and 200 feet (60 960 mm) long, depending on the product. Flexible sheets of PVC are available in sizes up to 80 feet (24 384 mm) wide and 700 feet (213 361 mm) long. Thickness ranges from 1/32 inch (0.79 mm) to 1/8 inch (3.18 mm) in the vulcanized products and typically between 10 mils (0.25 mm) and 45 mils (1.14 mm) for CPE and PVC. They can be seamed at the site, using special cements or solvents, or in the factory to form a single membrane that will cover the entire structure. The membranes can be loose laid or partially or fully bonded to the structure. Some are used for above-grade conventional applications as well as below-grade. Some of the products, EPDM and neoprene in particular, are used as flashing materials and are available on rolls as narrow as 12 inches (305 mm) for this purpose. Most of the generic types of sheets are also available in a liquid form that has the same basic chemical composition. Although most of the characteristics of stability and durability are similar for these liquids, the other criteria--such as the ability to withstand movement or facilitate repair and the application considerations--are quite different. For this reason the liquid forms of these products are discussed in the section on liquid-applied systems.

(1) Stability and Durability. With the possible exception of PVC sheets, the stability of this group is quite good. In addition, the products in this group have among the longest life spans of all the waterproofing products. The high quality control in the manufacture of the sheets results in very consistent products. Generally, these membranes are moderately tough, puncture resistant and resistant to most chemicals. Soft and flexible, most of these products can be elastic in temperatures ranging from 40°F (4°C) to 200°F (93°C). The vulcanization process helps prevent the stress cracking that can occur when sheet membranes are used on sudden, sharp bends.

Although the membranes themselves have excellent characteristics for underground applications, the presence of seams is always a concern. The vulcanization process gives butyl, EPDM, and neoprene great strength and resistance to permanent deformation under long-term loading. Seams that are made in the field with cold-applied solvents or cements, unfortunately, do not have these same inherent characteristics. They can be sufficient, however, if they are not located in areas with great potential for movement or stress from other forces. For best results, the number of seams should be minimized and application should be done with extreme care by professionals. The sheets can be seamed in the factory into one custom-fit membrane. Although this process guarantees a good bond, the resulting rather large, heavy membrane may be difficult to work with.

The six major types of membranes, although similar in many of their general characteristics, differ in some ways. The key characteristics of each type are briefly discussed below.

Butyl membranes are lightly vulcanized, resulting in high strength, flexibility, and softness. They can be reinforced with nylon and have a high resistance to heat and ozone. Although they are resistant to bacteria, fungi, and most soil chemicals, they should not be exposed to acids, oils, or solvents. The very low permeability of butyl rubber to gas makes it a good vapor barrier.

EPDM is a synthetic rubber that is quite similar to butyl in most respects. It is even more resistant to weathering, chemicals, and the ultraviolet rays of the sun. Like butyl rubber, it can be reinforced with nylon. The sulfur included in its composition provides EPDM with high strength.

Neoprene is a synthetic rubber that has good resistance to chemicals, oils, solvents, high temperatures, and abrasion. It is more sensitive to degradation caused by exposure to the sun than are the other vulcanized membranes. It is also more vapor permeable than the other vulcanized membranes. Generally, neoprene membranes are not used in underground applications as often as butyl membranes or EPDM. Neoprene is commonly used for flashings because it can be formed into complex shapes in the field when heat is applied to it. Rolls of neoprene flashing material are available in cured or uncured form.

Hypalon is distinguished from the other sheet membranes in a number of ways. Its chemical composition gives it some unique characteristics. Because Hypalon is highly resistant to the ultraviolet rays of the sun, ozone, and high tempeatures, it is suitable for exposure above grade. Unlike the other membranes, it can be manufactured in a variety of colors. Perhaps the most undesirable characteristic of Hypalon is its relatively high rate of water absorption if it is constantly exposed to water. Thus, it is generally not recommended for use underground.

CPE, which is also quite durable and stable underground, is available as a 20-mil (0.51-mm) sheet laminated to a polyester backing that can be fully bonded to the structure. Seams can be made on site by welding the sheets with solvents, cements, or adhesives, and in the factory by means of an electrothermal process. If properly done, the seams can have the same characteristics as the sheet itself. This is one of the assets of CPE that distinguishes it from some of the other sheet membranes.

PVC is a well-known plastic. As a raw material, it is hard and brittle. As a sheet material, it is flexible, strong, and resists tears and punctures. It is also resistant to ultraviolet degration and soil chemicals. A drawback of PVC sheets is that shrinkage can occur. They can also become brittle as plasticizers leach out of the material over time. Some products have additives that slow this process.

(2) Capability to Withstand Movement and Cracking. Basically, vulcanized and plastic sheet membranes have excellent properties for bridging any cracks that occur in the structure. They are flexible under a wide range of conditions and have great tensile and sheet strength. Their capability to bridge cracks is affected by the manner in which the membrane is bonded to the structure. Total bonding of the membrane reduces the flexibility of the material over cracks and concentrates the stresses in the small portion of membrane lying directly over the crack. A loosely laid or partially bonded membrane allows for this stress to be dissipated over a greater area, thus reducing the strain on the material. At points where movement is expected, such as at an expansion joint, it is desirable to leave some extra material to take up any stress that occurs.

(3) Capability to Minimize Leaks and Facilitate Repair. A vulcanized or plastic membrane has no capability to reseal itself once punctured. Membranes can be repaired from the outside of the structure with patches that are bonded in the same manner as seams with cold-applied cements. The major drawback of sheet membranes when loose laid is the difficulty of locating leaks, because water can travel behind the membrane and enter the structure at a point remote from the original source. This is one of the main reasons why membranes are bonded to the structure. Completely adhered sheets can localize leaks, but, this technique is costly and reduces the capability of the membrane to bridge cracks. A compromise solution that is often used is a partial bonding of the membrane in a regular grid pattern so that any leaks will be localized in one section of the grid.

(4) Appropriate Use. Sheet membranes can be applied over both precast and cast-in-place concrete surfaces, as well as on masonry and wood. These large, heavy membranes are best suited to horizontal surfaces, which can be completely flat. No slope is required because the membranes can hold standing water indefinitely. Although the material can easily resist water on vertical surfaces as well, application is quite difficult because of the tendency of the heavy sheets to stretch from their own weight, especially in the heat.

Application of a large, flat membrane over complex shapes is difficult. Flat surfaces and simple shapes are the best applications for sheet membranes. Minimizing projections and penetrations through the membrane simplifies the application by reducing the number of seams, flashings, and other field-bonded details, which are always the potential weak points of the system. Of course, if projections and penetrations are required, they can be waterproofed by using flashing materials and specially formed boots, corners, and other accessories.

(5) Application Considerations. The application of vulcanized or plastic sheet membranes--particularly the seaming and bonding--is quite exacting work and requires experienced, skilled professionals. If the membrane is factory seamed, installation requires experience because the material is heavy and difficult to adjust. Clean, dry, smooth surfaces that are free of oil and grease are required if the membrane is to be bonded. Because the membranes are tough, the surface can be somewhat irregular without causing damage, although sharp edges and foreign objects should be removed. The membrane materials remain flexible over a wide range of

temperatures, but applying vulcanized membranes in extremely cold or hot temperatures is inadvisable. Heat can cause the membrane to expand considerably; when placed in the cooler underground environment, it will contract, causing stresses in the membrane and at the seams.

Before bonding or seaming takes place, vulcanized membrane sheets should be laid out and allowed to relax and return to their original size. If they are stretched during application, greater stresses will result. One advantage of the solvents and cements used with these products is that they are cold-applied--no hot mastic is required. (The adhesives can be moderately toxic.) After installation but before backfilling, the system should be water-tested for leaks. Field inspection and water-testing are advised to ensure watertight seams. Pinhole punctures can sometimes occur in the manufacturing process, however. Although the materials are relatively tough, insulation, or other protective materials should be used to protect them from punctures by sharp objects during backfilling.

(6) Relative Costs. The cost of vulcanized and plastic membranes is higher than that of any other category of waterproofing materials. This high cost is largely attributable to the cost of a very high quality durable material, although skilled labor is also required. The cost can be affected by the complexity of the job and by the number of seams and flashings required. The cost of labor for a membrane that is fully bonded to the structure is significantly higher than for a loose-laid application.

g. Bentonite Clay Products. Bentonite (montmorillonite) clay is used in several forms to provide waterproofing on underground structures. This highly plastic clay, which is mined in the western United States, has the unique property of swelling from 10 to 20 times its original size when saturated with water. As it dries, it returns to its original volume. This process of expanding and contracting can continue indefinitely without wearing out the material. The bentonite material is applied in a thin layer confined between the structure and the soil. As the clay material comes in contact with water, expansion of the material is restrained, resulting in a gel- or paste-like barrier characterized by a high density and impermeability.

The many types and grades of bentonite clays have different characteristics. The major types of these products use specific clays and are available in the following forms:

- Raw bentonite
- Bentonite mixed with asphalt
- Bentonite contained in cardboard panels
- Bentonite mixed with binding agent in a trowel- and spray-grade product
- Bentonite mixed with polymers for caulking joints

The materials most commonly used on underground buildings are the spray-on and trowel-on mixtures and the cardboard panels.

(1) Durability and Stability. The fact that bentonite is inorganic and will not deteriorate means that bentonite-based products in general are characterized by long-term stability and flexibility. They should not be used in highly salinated soils, however, because salt diminishes the

swelling action of the bentonite clay. Another concern is that, if bentonite is allowed to dry out completely and is then saturated with water, there will be a slight delay before the clay is activated and expands to seal all leaks. Thus, it may not be advisable to use bentonite in a hot, arid climate subject to sudden downpours. Although bentonite products function effectively when exposed to a continuous head of water, they should not be exposed to running water that could cause the clay to wash off the surface.

Raw bentonite in its dry, granular form is often used for well casings and for sealing the bottoms of reservoirs. Although it can be used to waterproof buildings, this use is generally not recommended because the bentonite does not adhere as well with other applications, its application is difficult to control, and it will not work on vertical surfaces. It is a mistake to mix raw bentonite with water so that it can be applied in a paste-like form. After expanding during application, it will later dry out and shrink to its original size, and thus allow water to leak through because the amount of material is inadequate. Bentonite mixed with asphalt is also unsatisfactory and is not recommended. Although this product has been successfully applied, the asphalt tends to coat the clay particles, thus reducing their activity.

Application of bentonite contained in the voids of cardboard panels results in a very consistent thickness of the raw material. The panels must be very carefully handled during transport and application to prevent damage. The biodegradable cardboard is intended to deteriorate as a result of the bacteria in the soil. Problems can occur if the backfill soil does not contain sufficient organic matter to cause this deterioration. For example, water may penetrate past the bentonite and run behind the cardboard and along the seams between panels.

The effectiveness of the trowel-on and spray-on mixtures of bentonite is dependent on good quality control in product formulation and application. Trowel-on applications do not require the high level of skill and the exacting tolerances needed for applying other waterproofing products. Two types of spray-on products are currently available. The first, which is similar to the trowel-on product, is applied in a paste-like form and must be covered with polyethylene to help cure and protect it. It is partially activated and remains in a gel-like state that can be maintained at 50 percent relative humidity. A serious disadvantage of certain gel formulations is their tendency to separate upon moisture activation, causing leakage of the binder through cracks in the structure and possible further moisture leakage. The second type of spray bentonite is a relatively new product that dries immediately on application. A polyethylene cover sheet is recommended on the second type of spray to contain the bentonite, even though it is not needed for curing.

(2) Capability to Withstand Movement and Cracks. Bentonite clay products have excellent ability to respond to movement and cracks in the structure that occur after installation. Because it can expand to many times its original volume, bentonite can bridge cracks up to 1/4 inches (6 mm) wide and fill voids created by movement, provided that these cracks or voids do not provide a large enough path to transmit bentonite particles and wash them through the crack to an interior void. Extra protection can

be applied to cold joints or other points where cracking is anticipated by means of bentonite-based caulking or tubes of bentonite that decompose in a manner similar to the cardboard panels.

(3) Capability to Minimize Leaks and Facilitate Repair. The same expansion capability that allows bentonite to bridge cracks also permits it to reseal any punctures or holes that may be present in the waterproofing. Another important advantage is that if any leaks should occur, they will enter the structure near the source of the leak, thereby making it easier to locate. This benefit results from the full adhesion possible with the trowel-on and spray-on products; the adhesion prevents water from traveling away from the source of the leak, as it can when a loose-laid membrane is used. Water can travel somewhat behind cardboard panels unless they are embedded in a gel that is applied to the surface.

If a leak does occur and is located from the inside of the structure, it can also be repaired from the inside by injecting bentonite through a small hole in the area of the leak. This type of repair has considerable cost advantages over having to excavate in order to locate and repair a leak from the outside. One drawback of bentonite is that, if waterproofing problems occur, options for future repair or replacement by other products may be limited because the bentonite is difficult and messy to remove.

(4) Appropriate Use. Generally, bentonite products, particularly in the spray-on and trowel-on forms, are quite versatile in comparison with most other waterproofing systems. They can be applied to virtually all types of surfaces, including poured or precast concrete, masonry, and wood (provided that it is pressure-treated). Both vertical and horizontal surfaces can be waterproofed with these products; there are no special requirements for a sloping surface or fast-draining soil. As mentioned above, these bentonite-clay products can perform effectively against a constant head of water. Trowel-on and spray-on systems are well suited to complex geometries. Bentonite can be applied to curving forms, complex penetrations, and on very rough, irregular surfaces such as stone, blocks, or corrugated metal.

The cardboard panels containing bentonite have more limitations than do the spray-on and trowel-on products. They are not recommended for use on block walls unless they are first covered with a cement plaster coat. Generally, there is more concern over their effectiveness on horizontal (as opposed to vertical) surfaces.

Problems with the cardboard not deteriorating as expected seem to occur more often on horizontal surfaces. If the panels are used on horizontal surfaces, a slope is desirable. Finally, the large, flat panels are not well suited for situations involving complex geometries, curving forms, irregular surfaces, or numerous penetrations.

(5) Application Considerations. Application of each type of bentonite product requires a different level of skill and experience. The trowel-on product can be applied by relatively unskilled labor. The cardboard panels require a moderate level of skill and experience in order to avoid pitfalls. The spray-on application requires an experienced

applicator who has invested in the proper equipment for this application. Because bentonite is relatively new as a waterproofing product, the number of qualified applicators is limited.

Very little surface preparation is required for the spray-on and trowel-on products. Rough, irregular, surfaces such as those of spalled or honeycombed concrete are acceptable. The cardboard panels require a smoother surface unless they are applied with a gel. One of the major advantages of bentonite products is that they can be applied in virtually any temperature and humidity. In addition, they are nontoxic.

A critical element in the successful application of the trowel-on and spray-on products is that an even thickness of material must be achieved. For the spray-on product, a 3/16 to 3/8 inch (4.76 to 9.53 mm) thickness is acceptable, depending on which system is used. The trowel-on product requires a thickness of 3/16 inch (4.76 mm), with greater thicknesses at corners and construction joints. The thickness of product remaining after evaporation of solvents and binders should be sought from the manufacturer and tested before full application. The application of these products can be interrupted without leaving any seams or joints that could weaken the material. Because bentonite should not be sprayed over 8 feet (2438 mm) high without using scaffolding, it may be necessary to backfill before spraying higher; thus, application of spray-on waterproofing products may require more than one stage. In the second stage of the application process, it is important not to leave spillage of bentonite on top of the first layer of backfill. This would interfere with the free-draining of the soil.

Polyethylene and insulation are usually placed over spray-on or trowel-on bentonite. Because the bentonite will reseal around any punctures, nails can be used to hold insulation in place. One of the most important application concerns is to prevent the bentonite from getting wet before backfill takes place. Thus, backfilling must be scheduled as soon as possible after the material has been applied and has had time to cure. Curing time for both the trowel-on product and the most common type of spray-on product ranges from 4 to 30 hours depending on temperature and humidity conditions. The polyethylene keeps the material from completely drying out during the curing process. The material is too soft to walk on without damaging it before it is cured; however, a newer type of spray-on product dries immediately and can be walked on or backfilled immediately. Backfilling can also be done immediately when cardboard panels filled with bentonite are used.

During the backfilling process, it is important to protect the bentonite from damage. One concern with bentonite applications is that settling of the earth can drag the insulation down the wall and scrape the waterproofing layer off. Because nails can exacerbate this problem they are not recommended unless absolutely necessary. Although the polyethylene helps reduce friction from settling, it is most important that the backfill be compacted to minimize settling.

(6) Relative Costs. The cost of the bentonite products is low to moderate in comparison with the other general types of waterproofing systems. One of the key factors influencing costs with the spray-on product is simply the availability of a qualified applicator.

3. WATERPROOFING SELECTION PROCEDURE. The flow charts below (Figures 20 through 26) summarize the criteria in this section as well as the sections dealing with drainage, structural design and roof or wall details. The flow charts are not intended to be fully self-explanatory or exhaustive in its coverage of all considerations. It is provided to give a framework and process to the selection of a waterproofing system for known building and construction parameters.

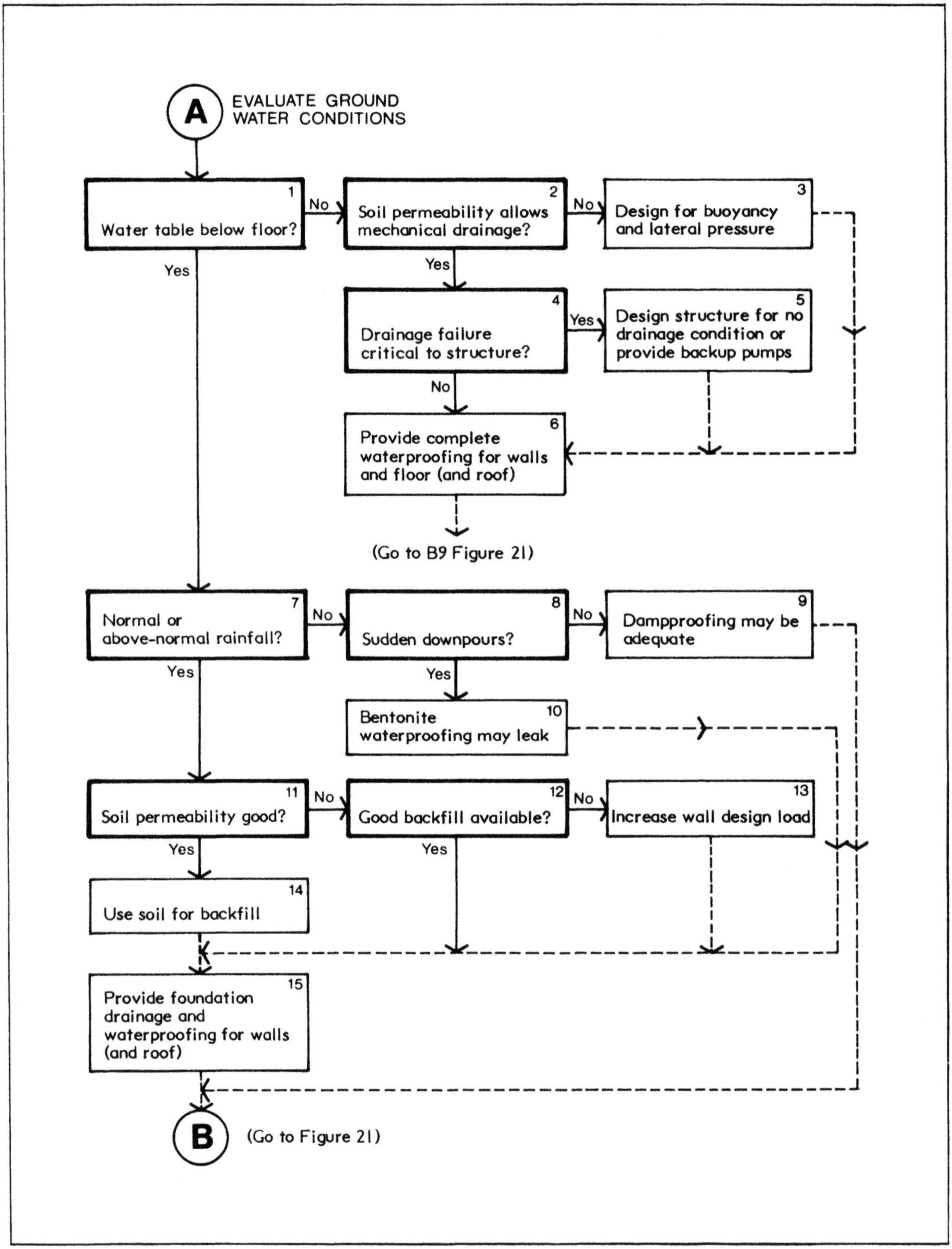

FIGURE 20
Waterproofing Flowchart--Ground-Water Conditions

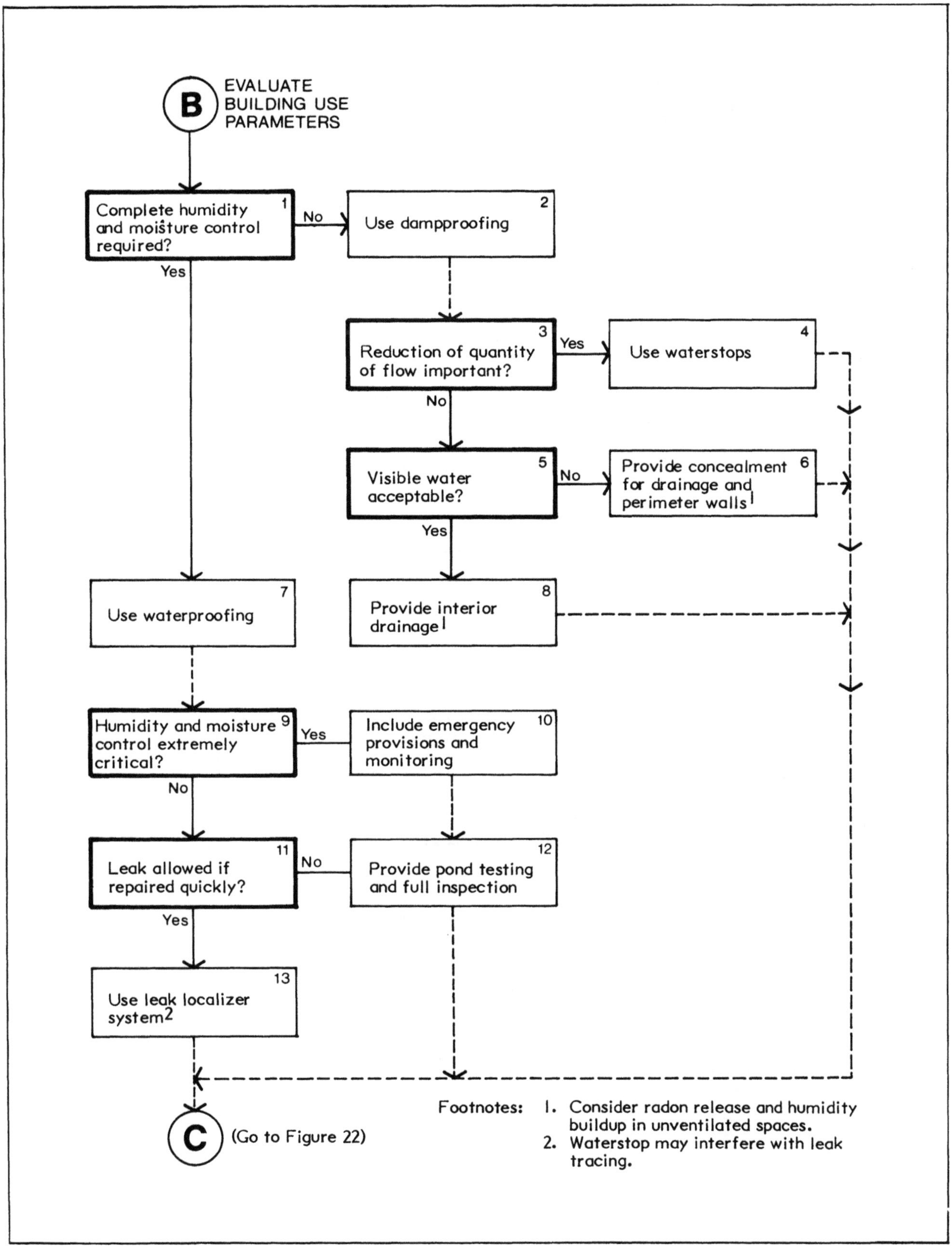

FIGURE 21
Waterproofing Flowchart--Building Use Parameters

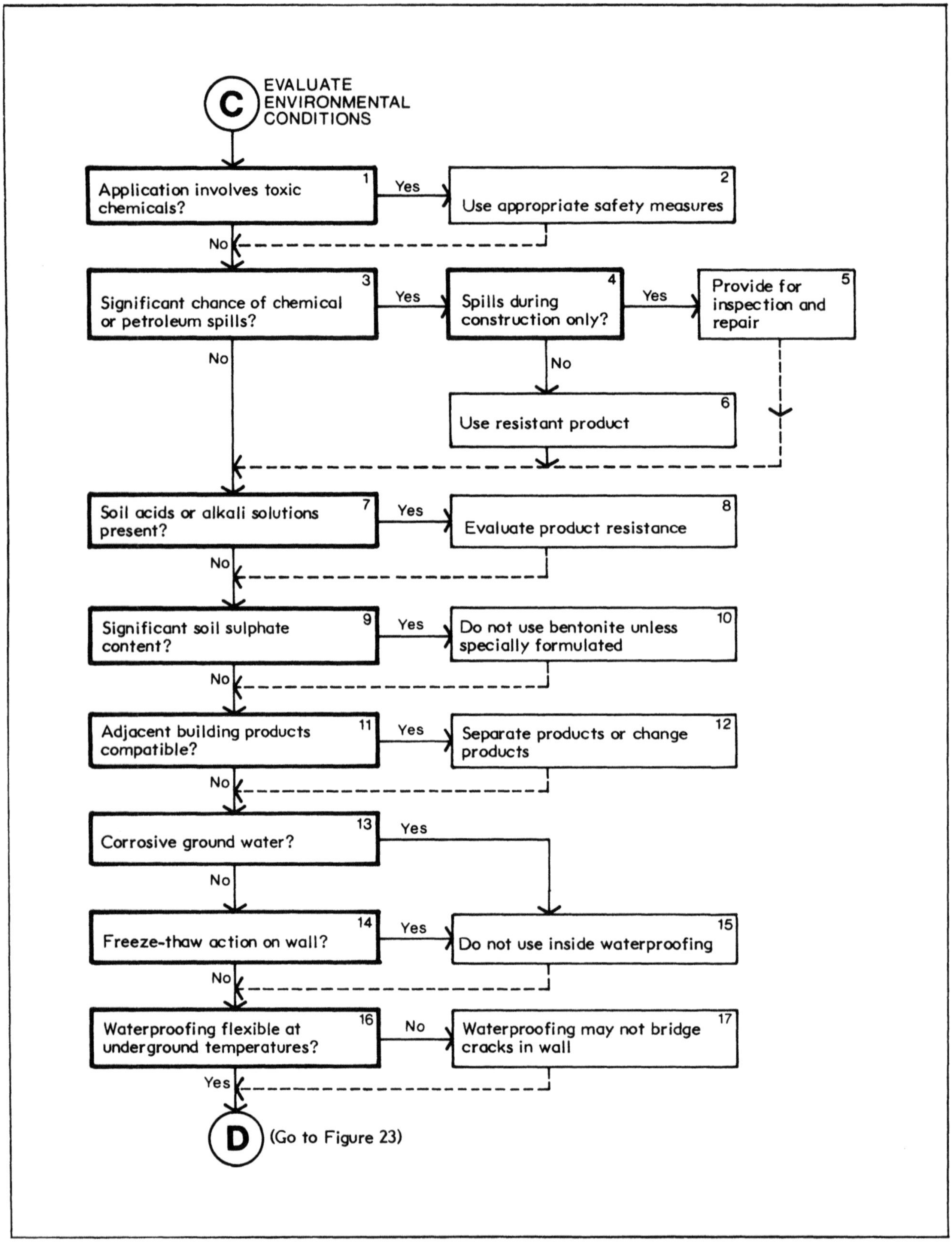

FIGURE 22
Waterproofing Flowchart--Environmental Conditions

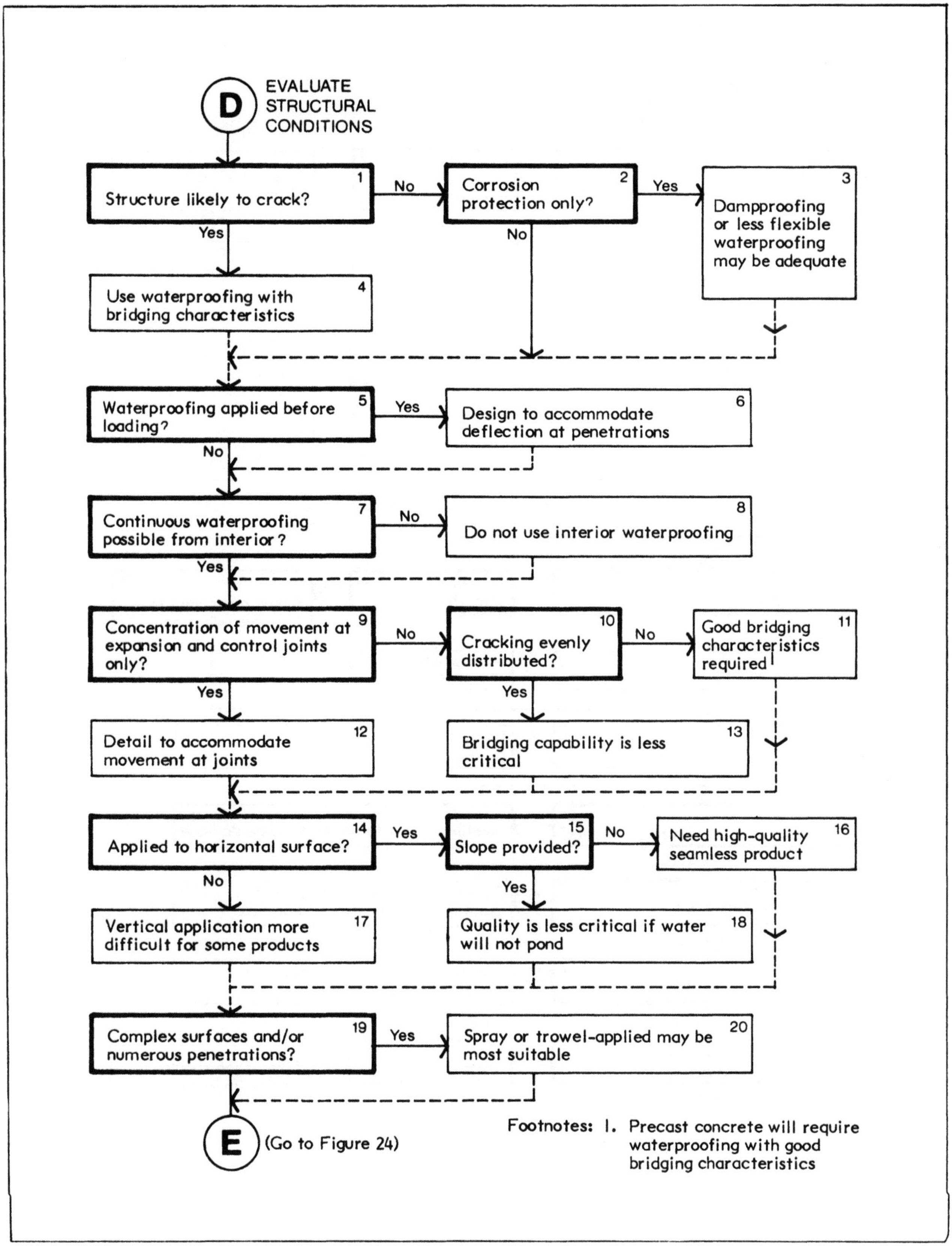

FIGURE 23
Waterproofing Flowchart--Structural Conditions

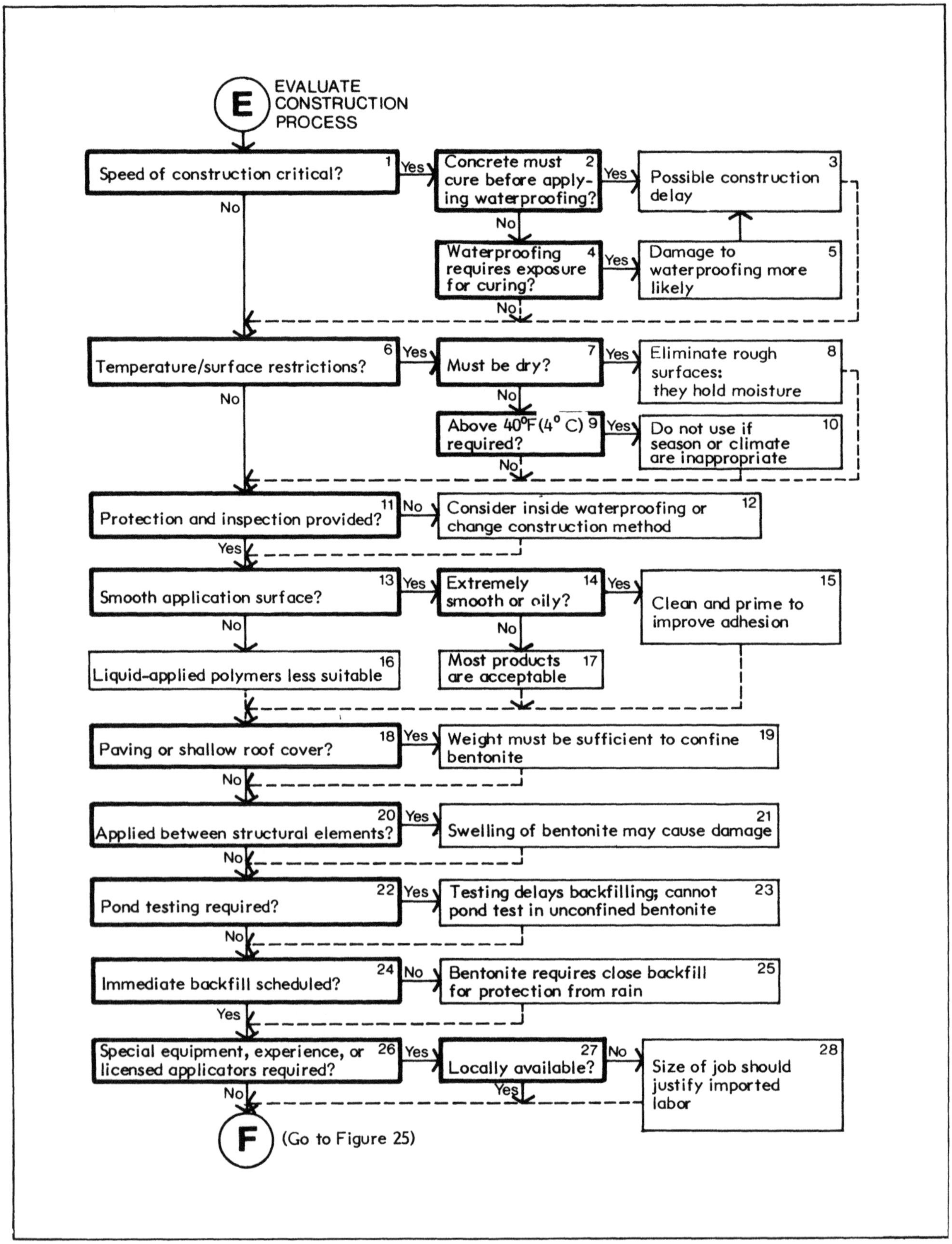

FIGURE 24
Waterproofing Flowchart--Construction Process

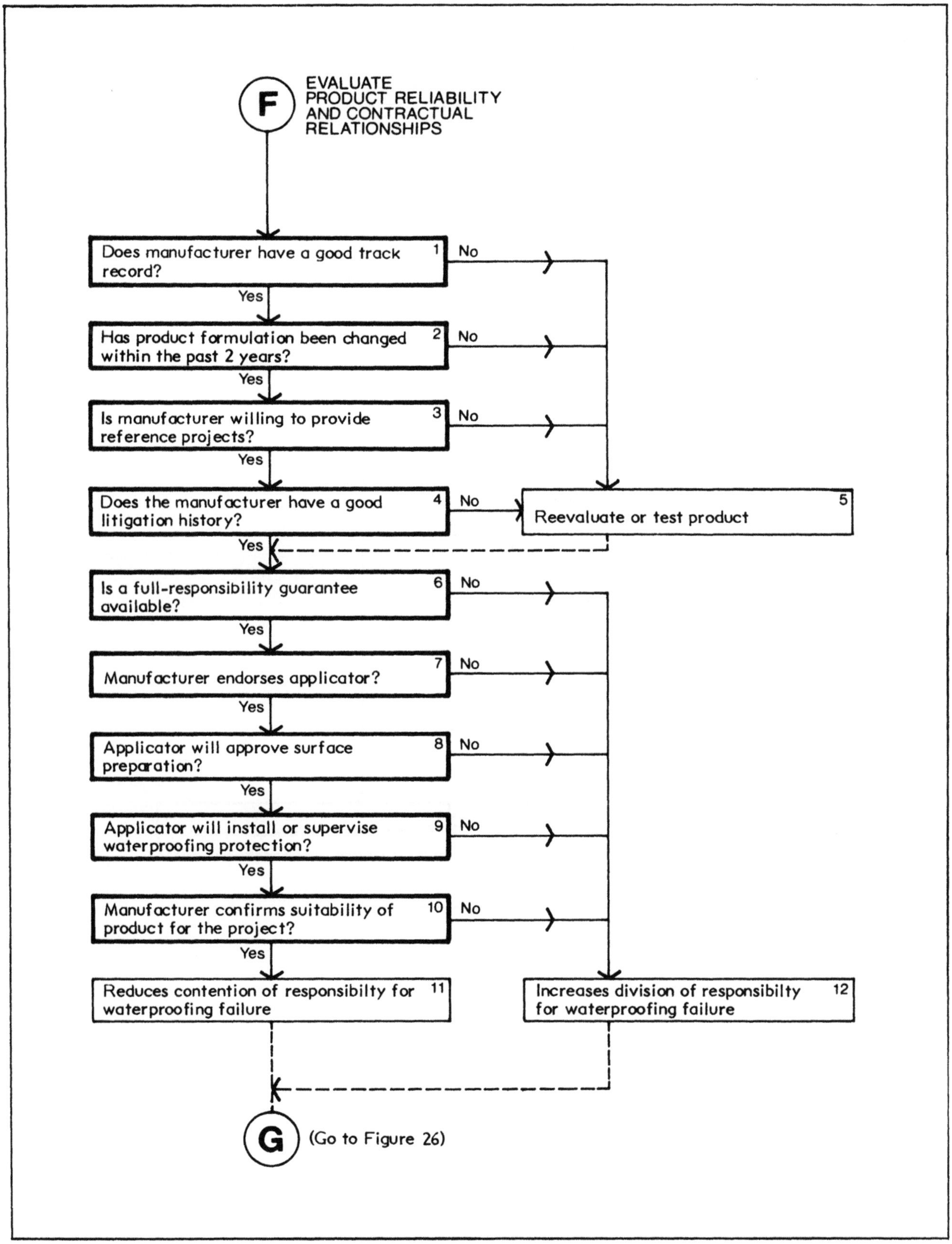

FIGURE 25
Waterproofing Flowchart--Product Reliability and Contractual Relationships

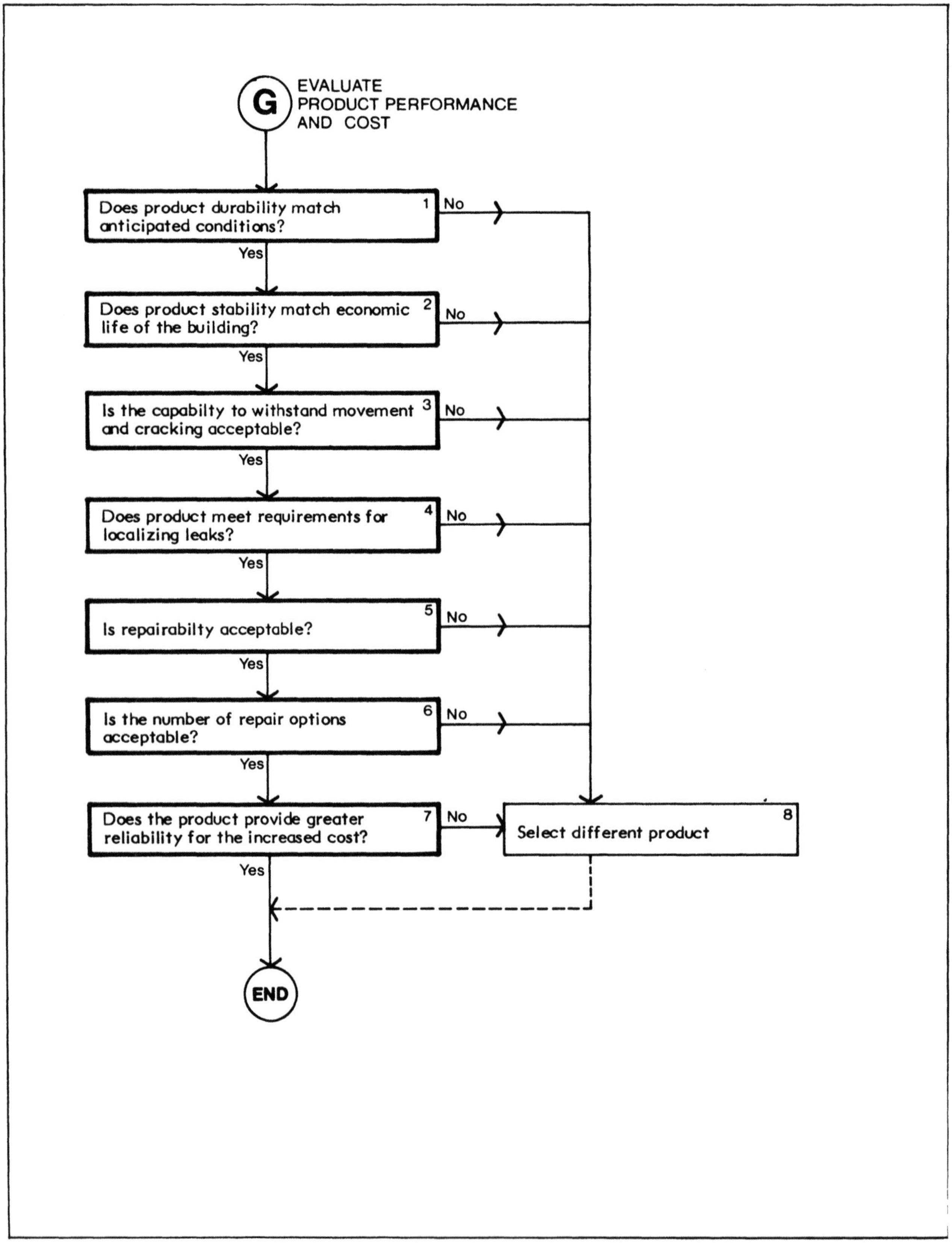

FIGURE 26
Waterproofing Flowchart--Product Performance and Cost

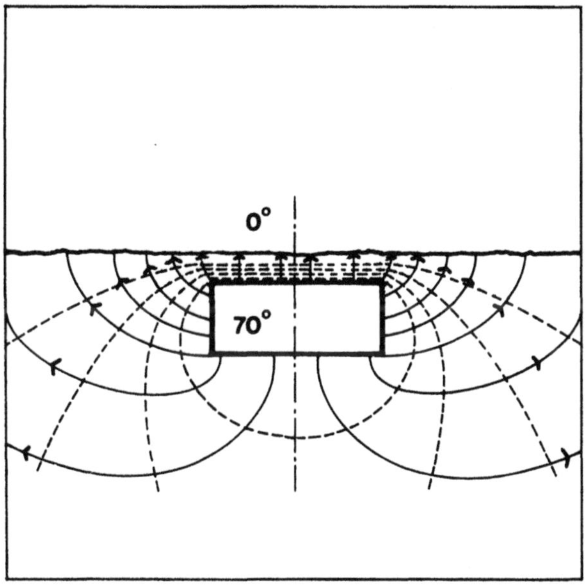

FIGURE 27
Heat Flow from Buried Structure
in Winter

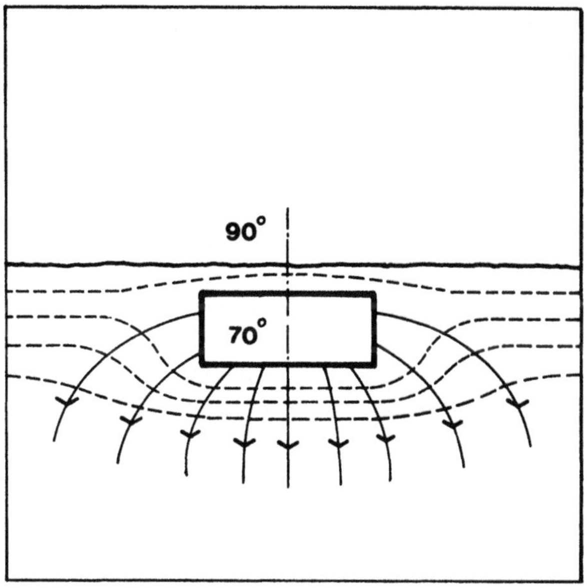

FIGURE 28
Heat Flow from Buried Structure
in Summer

1. EFFECT OF INSULATION ON THERMAL PERFORMANCE.

a. Reduction of Heat Flow. Figure 27 schematically illustrates the heat flow through the ground from an uninsulated buried structure for nearly steady-state mid-winter conditions. The solid lines indicate the direction of heat flow. It can be observed that the highest heat flux is experienced through the roof and the upper portions of the walls. Introducing insulation into this picture will modify both the direction and the magnitude of the heat flux. Clearly, the portions of the building shell nearest the surface require the most insulation. However, before non-uniform insulation strategies are chosen, particular care must be taken to ensure that thermal short circuits around or through the insulation--via relatively high conductivity paths in the walls or soil--are eliminated.

Figure 28 schematically illustrates a hypothetical heat flow pattern for summer conditions. In this case, the ground surface is warmer than the temperature within the building and heat will flow into the building through the roof and out of the building to the cooler ground at greater depths. Insulating the roof under these conditions would reduce unwanted heat flow into the building but insulating the lower portions of the structure would retard desirable heat flow out of the structure.

b. Thermal Stability. Placement of insulation outside a high-mass, earth-sheltered building structure, such as concrete, will enhance the thermal stability of the building since the mass inside the insulation envelope will be available to moderate room air temperature swings (for example, those due to passive solar gain).

Earth-sheltered structures have a strong capacity to maintain very even temperatures despite high levels of passive solar gain in the winter. Furthermore, when all sources of heat are removed from the building, the air temperature will fall slowly because of the heat stored in the walls and earth around the building. Even temperatures result from the typical use of heavy materials for the structure. These can absorb heat during the day and release the heat at night--but daily temperature fluctuations are not significantly affected by the earth outside the structural mass. Hence, interior insulation which isolates the structural mass of the building from the interior environment will increase the magnitude of daily temperature fluctuations, whereas exterior insulation will not.

2. AMOUNT OF INSULATION.

a. Climate. Insulation schemes for northern climates will not be appropriate or cost effective for cooling-dominated climates. Since a primary benefit of insulation is in reducing winter heat losses, insulation requirements can be reduced or possibly eliminated in areas where winters are less severe. Cooling conditions are usually enhanced by using no insulation on the surfaces in contact with cooler ground. Hence, a balance must be struck between the heating and cooling needs for a particular area.

Although insulation requirements must be evaluated on an individual basis, Table 3 may be used as a general guide to cost-effective supplementary insulation values for different areas of the country for small envelope-dominated buildings. (No detailed cost-benefit analyses have been carried out to determine more precise figures.) Large buildings with high internal heat gains may not require these levels of insulation even in cold climates but before reducing insulation levels a useful purpose should be sought for the excess heat.

Where cooling is the dominant concern, thermal performance is enhanced the most by eliminating insulation where appropriate rather than reducing insulation levels.

b. Internal Heat. Large-scale buildings--office buildings as well as those involving process heat--usually have a year-round cooling requirement in interior zones of the building. The amount of insulation to use in such a building must be considered in conjuction with (1) the amount of energy used for cooling it, (2) the HVAC zoning plan for the building (that is, exterior zones cannot be too cold), (3) the possible use of excess energy in other buildings, and (4) the probable future reduction of energy release within the building.

c. Ground Moisture. The thermal conductivity of most soils can increase greatly as soil conditions change from dry to very moist. Although the heat loss will undoubtedly increase when the ground adjacent to a building is wet, the effect on the building may be reduced by the following conditions.

(1) Conductivity Versus Heat Flow Path Length. As the ground thermal conductivity increases, the zone of warmed earth around the building will spread further out. For deep-ground heat loss, this will mean a higher conductivity but a longer heat flow path thus reducing the impact of the increased conductivity on actual steady-state heat loss.

TABLE 3
Suggested Resistances for Insulation

Climatic Region		Suggested Range of Supplemental Insulation[1]		
Heating Degree Days, Base 65°F	Cooling Degree Days, Base 65°F	Roofs and Upper Wall[2]	Lower Wall[3]	Remote Floor Areas[4]
8,000 - 11,000	0 - 500	R20 - R40	R5 - R20	none - R5
5,000 - 8,000	500 - 1,500	R20 - R30	R3 - R10	none - R5
2,000 - 5,000	1,500 - 2,500	R10 - R20	none - R5	none
less than 2,000	more than 2,000	R10 - R20	none	

Footnotes:

This table should only be used as a general guide for envelope-dominated buildings.

1. Section 14 discusses distribution of insulation and exceptions to the norms presented in the table. Values are expressed as thermal resistance in ft^2 h °F/Btu.
2. Earth covered roofs with 12 inches to 30 inches (300 mm - 760 mm) of cover and walls within 8 feet (2440 mm) of the ground surface. In cooling-dominated regions, upper wall insulation will depend on adjacent ground temperatures.
3. Earth-covered wall surfaces further than 8 feet (2440 mm) from the ground surface.
4. Floor areas more than 10 feet (3050 mm) from the ground surface which are not used as a solar storage areas or for heat distribution.

1 Btu/h ft^2 °F = 5.6783 W/m^2K

(2) Portions of the structure close to the ground surface will usually have substantial insulation in addition to the ground cover. In this case, the earth provides only a small component of the overall steady state R-value but functions primarily as a thermal mass and temperature moderator. The thermal diffusivity of the soil (a combination of specific heat and conductivity which measures the speed at which heat will propagate through the material) is not as greatly affected by an increase in moisture content.

Seasonal changes in ground moisture content can provide an energy advantage in some climates. In northern climates, ground freezing and winter snow cover combined with heat loss from a building will tend to dry out the ground adjacent to the building, lowering the heat loss. In summer, when the ground surface is open to percolation and subject to higher rainfall, an increase in ground moisture content will promote summer cooling. Other climates, for example, where heaviest rainfalls occur in winter and dry conditions exist in the summer, may reverse this situation. Where very moist ground or the ground-water table is adjacent to a building in the heating season, consideration should be given to increasing insulation levels in those areas of the building affected (bearing in mind the loss of cooling capacity in the summer this will cause). Flowing ground-water adjacent to an underground structure should be given particular attention.

d. Ease of Adding Insulation. After analyzing the basic insulation requirements of the structure for cold climates, it may be worthwhile in regions with severe winters to consider adding extra insulation. When the insulation is added outside the structure, there is no structural limit to the amount of insulation that can be added. Because there should be only a small increase in the labor cost of placing additional insulation, higher insulation levels may be more cost effective despite the diminishing rate-of-return on higher insulation levels.

3. DISTRIBUTION OF INSULATION AROUND THE BUILDING ENVELOPE.

a. Optimum Distribution for Minimizing Heat Loss. To minimize total heat loss from a structure under steady-state heat loss conditions, insulation should be distributed so that the rate of heat loss through all areas of the structure is the same. The lowest heat loss from an uninsulated structure (in a uniform material in mid-winter) occurs at the center of the buried floor of the structure. The highest heat loss occurs at the upper portions of the building. Consequently, insulation should be added to the upper portions of the building so that heat flow from all surfaces is approximately equal.

The resulting optimum insulation pattern would be similar to the solid line in Figure 29. The optimum insulation pattern is also affected by each corner of the structure because an area of wall near or at the corner has a larger zone of ground adjacent to it through which to lose heat. The same situation will also occur as a three-dimensional effect at the ends of the building--an effect that creates an inherent inaccuracy in any two-dimensional heat loss model. It is reasonable to assume such a steady-state heat flow pattern in a cold climate in mid-winter but not under more variable or warmer conditions. In addition, stopping insulation abruptly will radically alter heat flow paths in both the wall and the soil. This can be seen in Figure 30 (based on Reference 8, Effect on Insulation Placement, by Meixel). Hence, the insulation pattern in Figure 29 should be used as an aid to understanding rather than a fixed requirement

Line 'B' in Figure 29 represents a guide to the relative placement of insulation around an underground structure to minimize the total heat loss per unit of insulation. These guidelines do not determine the amount of insulation to be used, but only the suggested distribution. Lower levels of insulation would be optimized by subtracting a constant thickness of insulation around the structure from the previous distribution (line 'A'). Higher levels would be optimized by adding a constant thickness of insulation (line 'C').

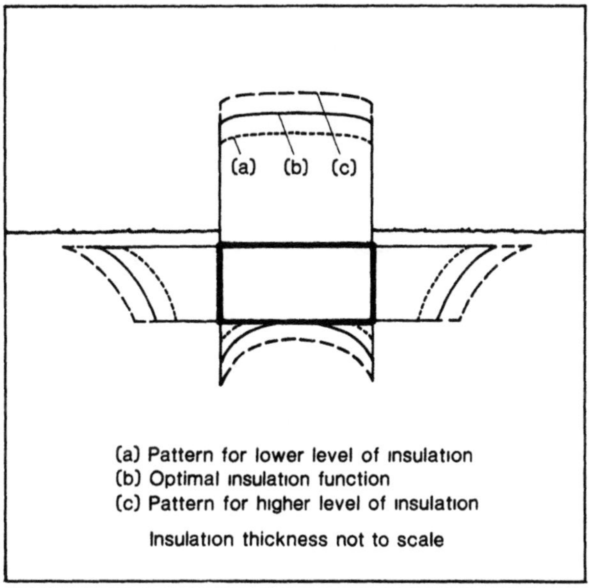

FIGURE 29
Optimum Insulation Pattern for a Buried Structure
(Heating Conditions Only)

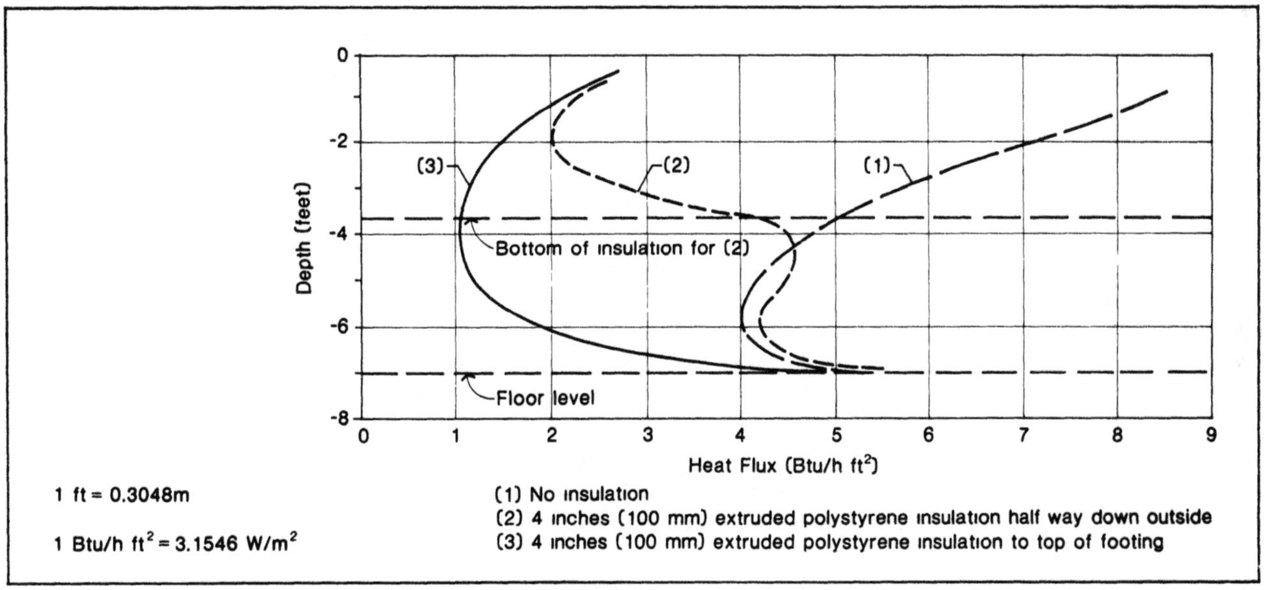

FIGURE 30
Simulated Below-Grade Heat Flux
(February, Minneapolis)

Because the continuous distribution shown in Figure 29 cannot be adhered to exactly when using rigid sheets of insulation, a practical approximation should be made after the optimum cooling placement for insulation has been considered. See Figure 31 for an example.

The two-dimensional section shown through the building does not illustrate all the high heat loss areas that should receive thicker

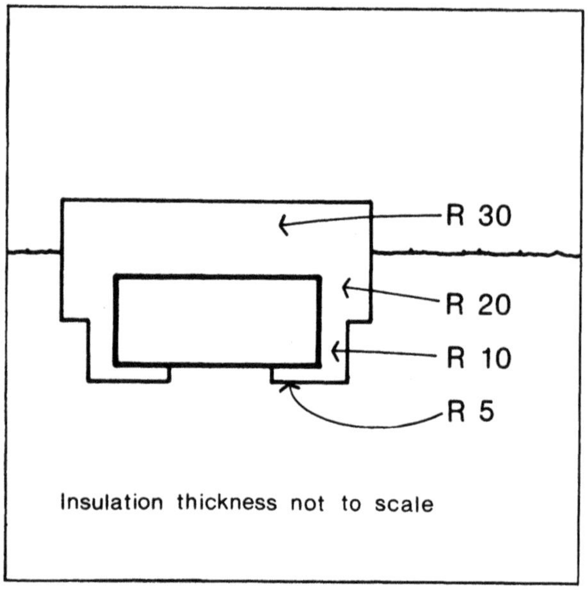

FIGURE 31
Optimum Insulation Placement
(Heating Conditions Only)

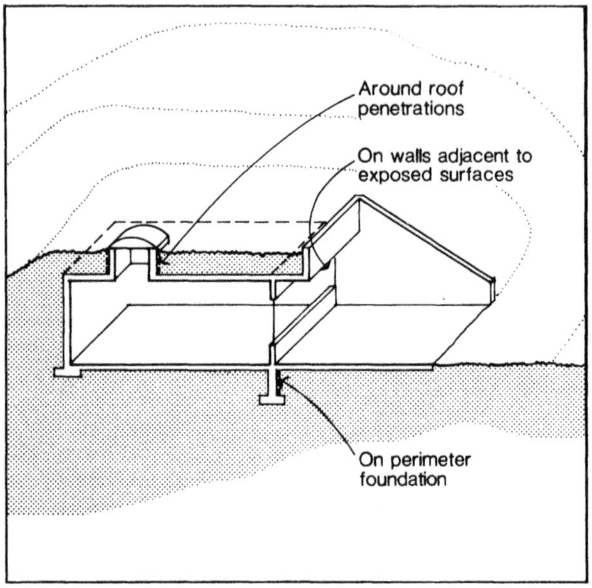

FIGURE 32
Areas Requiring Extra Insulation

insulation. Figure 32 indicates other surface areas of an underground structure requiring extra insulation because they are also close to the ground surface--for example, walls close to exterior retaining walls and sections of floors close to a walkout level.

b. Optimum Distribution For Maximizing Cooling. When cooling requirements are the most important, it is desirable to maximize the heat flow into the ground to aid in cooling the interior of the structure. When the ground temperature is below 78°F (26°C), insulation will only retard heat flow out of the structure and hence will detract from the maximum cooling effect. Eliminating insulation will also slightly lower the interior surface temperature of walls in contact with the ground. This can increase human comfort by the radiation of heat from the body to the cooler wall surfaces. In very hot climates, (where the roof and the upper wall surface may be warmer than 78°F (26°C)), insulation of those areas will help retard additional heat flow into the structure.

c. Compromises. When designing for both heating and cooling requirements, some sacrifices in optimum insulation placement for each condition will be necessary. The obvious compromise in climates with balanced heating and cooling requirements is to reduce or leave off the insulation in those areas where the heat loss in winter is smallest--for example, floor areas remote from the ground surface and the lower areas of walls more than 7 to 10 feet (about 2 to 3 m) from the ground surface.

Climates with severe winters and little cooling requirement can be optimized for the winter condition. Predominantly overheated climates will only require insulation for portions of the structure which will be in contact with ground temperatures above 78°F (26°C).

d. Insulation as a Protection Board for the Waterproofing. When applied to the outside of the structure, waterproofing usually requires protection from damage during backfilling. Although from a thermal point of view it might be most economical to stop off the insulation completely partway down the wall, a more effective overall compromise (for one- and two-story structures) may be obtained by carrying a small thickness of insulation down to the footing to serve as the protection board. The exception to this procedure would be in cooling-dominated climates where insulation is actually detrimental to annual performance. Under these conditions, a protection board with high conduction properties should be used.

e. Passive Solar Gain Areas. When a floor is being used for direct passive solar gain, insulation can be placed below the slab to maintain a higher slab temperature and thus improve re-radiation (see reference 7).

f. Insulation as a Protection Against Cold Floors and Condensation. It is desirable in earth-sheltered structures to reduce the sensation of cold floors and walls during the heating season. Insulation outside the structure does not significantly raise the surface temperature inside the structure. The main reason for the common perception of cold floors is that materials such as concrete or tile can rapidly conduct the heat away from a hand or foot, whereas a wood flooring or carpet locally insulates the part of the body in contact with it. Thus, the sensation of warmer floors or walls is best obtained by using appropriate interior finishes rather than by exterior insulation. After the building has been in operation for a few months, the interior surface temperature will be primarily controlled by the interior air temperature adjacent to the surface, together with any radiation falling on the surface. In the case of no radiation, an uninsulated floor or wall remote from the ground surface will usually have a surface temperature within 4°F (2°C) of a relatively stable air temperature. A small amount of insulation outside the wall will probably not raise the surface temperature by more than 1°F (0.5°C).

Much the same argument applies to the prevention of condensation. Interior surface temperatures (except passive solar areas) will probably be raised only 1°F (0.5°C) or less by adding exterior insulation. Compared with the uninsulated wall, this added insulation will only eliminate condensation problems when the dewpoint lies in the 1°F (0.5°C) range between the two potential surface temperatures. This small improvement is not enough to guarantee a greatly superior performance to the insulated case. If condensation is a real problem in a particular climate, a small amount of interior insulation or a vaporbarrier on the warm side of an air gap will ensure that the surface on which condensation could occur will be essentially at the interior air temperature. Otherwise, provisions for occasional dehumidification can be made.

Abruptly stopping insulation can cause worse condensation problems than if the wall were not insulated at all. The insulation keeps the ground temperatures outside it cool--an effect that spills around the edges of the insulation, causing the section of wall immediately adjacent to the insulation to be cooler than the insulated section and the rest of the uninsulated section (see Reference 9, Heat Loss Through Basement Walls and Floors, by Houghten).

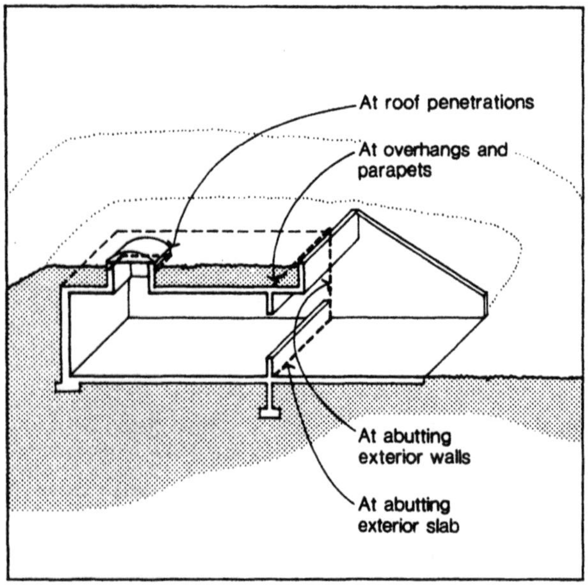

FIGURE 33
Thermal Break Locations

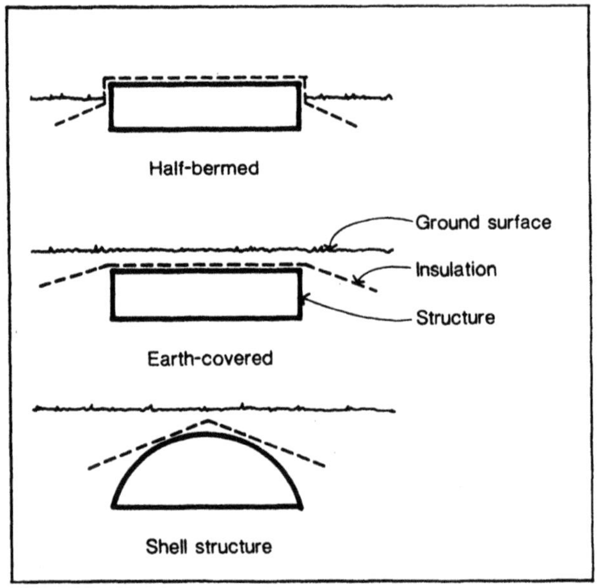

FIGURE 34
Insulation Within Soil Mass and Detached from Building

g. Insulation for Thermal Breaks. Neglecting thermal breaks can cause significant heat loss in an otherwise well-insulated building envelope. A thermal break refers to the insertion of insulation in parts of the building where a high conductivity material is in contact with both the inside air and the outside conditions. Although these sections usually have a relatively small area, they can begin to dominate the heat loss as the overall insulation of the building is improved.

Thermal breaks can be provided by inserting small thicknesses of insulation at appropriate points within the structure or by wrapping the exposed structural element with insulation. An example of thermal break locations is shown in Figure 33.

4. LOCATION OF INSULATION. It is recommended that insulation be placed only on the outside of the structure and the waterproofing envelope. This insulation must be extruded polystyrene or meet the requirements of paragraph 5 below.

Insulation may also be placed within the soil mass as shown in Figure 34. The advantage of this location is that it increases the cooling benefit if the soil temperature is lower than the indoor temperature. However, winter heat loss for this location can be up to twice as high as for an equivalent amount of insulation placed directly on the wall. This location should, therefore, be considered only in climates with cool soil temperatures and for exterior earth-contact zones which have dominant cooling requirements. Backfilling will also be more difficult with this arrangement. When cooling is not the major consideration, placement directly adjacent to the structure should be the most practical solution.

TABLE 4
Influence of Moisture Boundary Conditions on the 100-Day Moisture Gain of Various Insulations

Insulation	100 day moisture gain (percent of dry weight) Condition A	Condition B
Fiberboard, 1 inch (25 mm)	5.5	140.0
Perlite board, 1 inch (25 mm)	2.0	90.0
Cellular glass, 1 inch (25 mm)	0.7	11.0
Expanded bead polystyrene, 1 inch (25 mm)	0.2	1000.0
Extruded polystyrene, 1 inch (25 mm)	2.5	5.5
Urethane with asphaltic skins, 1 inch (25 mm)	4.0	160.0
Composite of 1 inch (25 mm) urethane with asphaltic skins and 3/4 inch (19 mm) perlite board below	45.0	340.0

Moisture Boundary Conditions

Condition	Cold side vapor seal	Dew point on warm side	Vapor pressure difference across sample (in Hg)
A	No	No (70% RH there)	0.66
B	No	Yes	1.03

Footnotes:

1. The relatively low vapor permeability of the skins causes the dew point to occur within the sample.

5. INSULATION TYPE. A few research reports (see Reference 10, Laboratory and Field Investigations of Moisture Absorption, by Dechow and Reference 11, Moisture Gain and its Thermal Consequences for Cannon Roof Insulation, by Tobiasson) compare the performance of the major types of exterior insulation under various moisture exposure conditions. When used without waterproofing or a vapor-retardant facing, extruded polystyrene clearly outperforms expanded polystyrene (beadboard) and extruded polyurethane in reducing moisture absorption and retention of its insulating capabilities. Table 4 shows the absorption of moisture by various insulation products under differing moisture boundary conditions.

Of the insulation types shown on Table 4, only two have an acceptable resistance to moisture absorption. These are cellular glass and extruded polystyrene.

All below-grade insulation must meet the moisture absorption resistance of these two products as measured under condition 'B' in Table 4.

Insulation which is used above grade or on the interior of a building must meet the fire requirements of NAVFAC DM-8 and DOD 4270.1M.

6. INSTALLATION.

a. Method of Adhering Insulation. The method of fixing the insulation to the wall or waterproofing membrane should not cause damage to the membrane if settlement of the backfill adjacent to the building occurs and causes drag forces. To help avoid these problems, the backfill should be well compacted in several layers. Refer to NAVFAC DM-7 Series for soil compaction criteria.

b. Backfilling. Backfilling procedures on the roof of the structure over the insulation should not involve pushing earth into place, as this can shift that insulation and create uninsulated gaps. Earth or gravel should be tipped just ahead of the previously covered portion and the new material leveled-off in place.

c. Prior to Backfilling. A layer of polyethylene outside the insulation can prevent dirt from being worked between the insulation boards during backfilling and can help the insulation remain dry. The polyethylene should be laid loosely at the edges of the building to allow drainage if any water should penetrate the sheet (see subparagraph above).

d. Thermal Short Circuits. Care must be taken to avoid thermal short-circuit effects in transitions from above- to below-grade insulation methods. Refer to Section 15, Roof and Wall Details, for examples of thermal breaks.

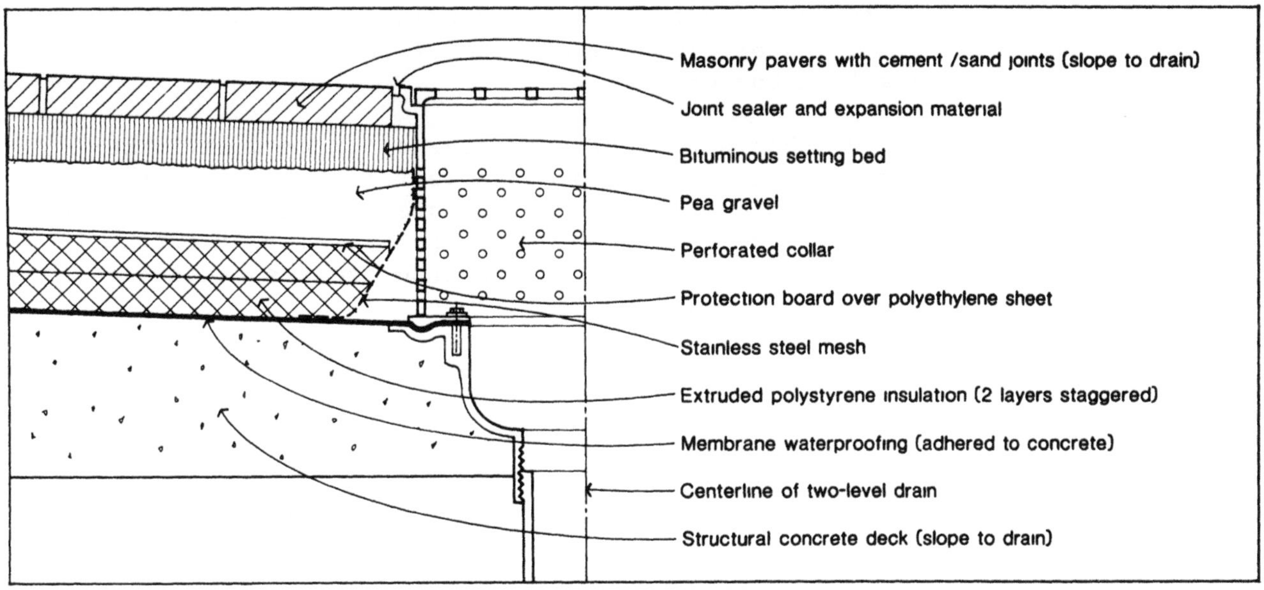

FIGURE 35
Drain at Brick-Paved Roof--Bituminous Setting Bed

1. GENERAL. As necessary, each detail in this section will be discussed in terms of cost effectiveness, energy efficiency, constructability, quality control, resistance to failure or deterioration, repairability, maintenance implications, fire resistance, and warranties. All details must be evaluated for (1) suitability to climate and (2) suitability to the type of functions to be housed. Freeze-thaw, which is an important consideration in northern climates, is an example of the former. Live loads, vapor transmission and interior finishes are examples of the latter. As necessary, the discussion of the recommended details will include climatic and functional suitability.

2. RECOMMENDED DETAILS. The details are presented in the following order: (1) roof construction (including control joints and expansion joints), (2) edge conditions for roofs (including conditions at abutting walls), (3) wall construction at grade level, (4) wall construction at retaining walls, and (5) wall/floor construction at footings. All of the details involve waterproofing and most of the details involve insulation. The installation methods for different kinds of waterproofing and insulation are discussed in sections 13 and 14.

a. Area Drain at Paved Roof--Pedestrain Traffic--Minimal Freeze-Thaw Cycles. Refer to Figure 35. The intent of this detail is to provide an economical plaza surface of brick pavers with a minimum total depth between the high point of the paving and the bottom of the structure.

This detail applies to all concrete deck systems. However, a separate concrete topping above the structural deck should be avoided since this may allow horizontal water migration in the event of a leak. This would make locating the actual source more difficult. The concrete surface must have a positive slope to the internal drains or the roof edges. The slope should not be less than one percent. A greater slope is recommended in the vicinity of the drain so that any water backing up from the drain (because of partial freeze-up or clogging combined with a heavy rain) will not extend too far back from the drain.

The waterproofing should be adhered to the concrete deck. The finished surface of the concrete must be acceptable for this. (The finish required will depend on the waterproofing.) The waterproofing is terminated in a clamp ring at the drain (some manufacturers may call for an extra ply at this critical area). The insulation, in this case, serves as the protection board for the waterproofing.

The drain must be accessible for cleanout. The drain shown is an all-level drain with a perforated collar and removable drain cover. A stainless steel mesh is placed around the lower portion of the drain to prevent adjacent construction material from washing into the drain. Washout of materials is just as critical to the stability of the paving construction as it is to the clogging of the drain. The drain should have a removable internal strainer or filter.

In climates having frequent freeze-thaw cycles, it is advisable to cut back the insulation thickness around the drain so that heat from the space below can prevent ice from forming inside the drain. The insulation must not absorb water. However, in order to preserve the thermal value of the insulation, a layer of polyethylene sheet can be placed over the insulation. Most of the water entering the paving construction will then be channeled away above the insulation. The effectiveness of this depends on the proper direction of overlap for the polyethylene sheet, sealing of the edges of the polyethylene, the slope of roof, and the effectiveness of the drainage course. The insulation should be placed in two layers with joints staggered.

The protection boards serve to protect the polyethylene sheet, should it be used. The protection board as well as the polyethylene sheet are optional. In terms of maintenance and potential future damage, especially where the roof cover is soil, the protection board is an added precaution for stakes and other objects that may be driven into the roof cover.

The gravel layer serves three purposes; first, to provide a primary drainage layer for water (especially if the polyethylene sheet is installed below it); secondly, to provide a base for the bituminous; and, thirdly, to provide protection to the insulation from the hot bituminous. The gravel layer may be omitted if the following is done: (1) the thickness of the bituminous is increased--use a 1-inch (25-mm) sand asphalt leveling bed over a 2-1/2-inch (63.5-mm) bituminous hot-mix underlayment, and (2) an asphalt protection board is placed over the insulation.

The brick pavers are set in the bituminous immediately after leveling, while the bituminous is still hot. Alternatively, the bituminous may be rolled and covered with a 1/4-inch (6.35-mm) neoprene setting bed. The main advantage of bituminous over concrete in this case is the economy of construction and the reduction of cracking in the brick pavers. A

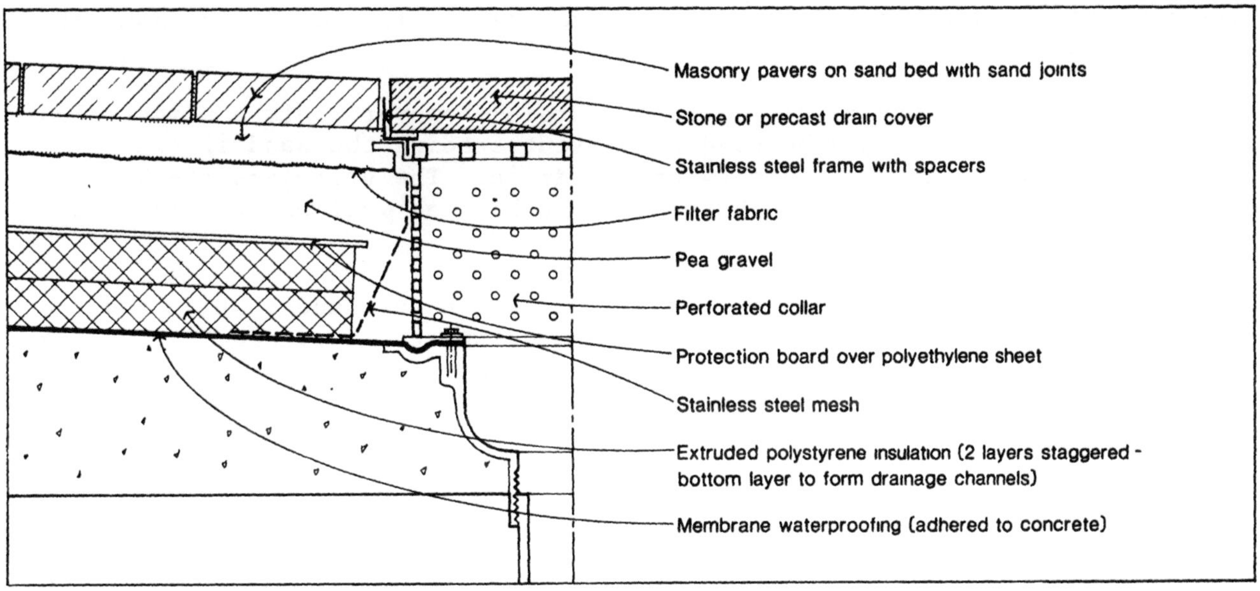

FIGURE 36
Drain at Brick-Paved Roof--Sand Setting Bed

concrete base with pavers grouted in place would also create a thicker roof section (if the drainage course were also required) resulting in a higher dead load on the structure (however, paving equipment will be required for the bituminous), and would require control joints aligned with the control joints in the pavers. The disadvantage of bituminous is that it has less strength to resist stresses from frost and thermal movement. In extremely warm conditions, the bituminous will tend to soften and will not support heavy concentrated loads. Consequently, this detail is suited to pedestrian traffic and light maintenance vehicles only.

The paving should be sloped to the drains (1 percent slope minimum) in order to drain as much of the water as possible before it soaks into the paving construction. The joint at the drain should allow for lateral movement of the pavers (use expansion material) and should be caulked. If bituminous is used as a base, the pavers should be at least 2 inches (50 mm) thick. The joints are filled with a cement-sand dry mix. The pavers may be as thin as 3/4 inch (19 mm) if concrete is used with a 1/4-inch (6.35-mm) minimum mortar bed. In this case, the joints in the pavers should be mortared.

b. <u>Drain at Paved Roof--Frequent Freeze-Thaw Cycles--Pedestrian Traffic</u>. Refer to Figure 36. This detail serves the same function as the previous detail, the major difference being that is is better suited to harsh northern climates. The major disadvantages of this type of construction are that (1) it is less capable of supporting traffic than the previous detail, (2) snow removal by light equipment may cause more damage because the bricks are not adhered or grouted, (3) the slope of the concrete slab becomes more critical since more water will reach the membrane/insulation layer, and (4) the joints are a potential place for weeds to grow (however, this can be remedied by chemical spraying--an additional maintenance consideration). The major advantages over the previous detail are (1) tolerance to freeze-thaw cycles and (2) economy of construction.

The pavers may be sloped to the drain--however, only a slight slope (1 percent maximum) should be used to prevent eroding the sand in the joints, especially in the vicinity of the drain. The stainless steel frame at the drain serves to hold back the sand cushion as well as the adjacent pavers.

A filter fabric must be installed between the sand cushion and the drainage layer to prevent the washout of sand into the drainage layer. The drainage layer will carry more water than the one in the previous detail. Consequently, in this case, the polyethylene sheet above the insulation becomes more useful.

Criteria for the other components of this detail are the same as for the previous detail. The remaining commentary for this detail will discuss the location of the drainage layer. The main purpose of the drainage layer is to prevent or reduce freeze-thaw action in the paving and to reduce the amount of water in the paving. The gravel in this location also serves to hold down the insulation during construction. Consider what would happen if the insulation were located above the gravel drainage layer. First, the thermal mass above the insulation will be decreased and, simultaneously, the interior warmer temperature gradient will be brought closer to the surface. In a cold climate, the warmer substrate will cause thawing to occur to a greater depth. The freeze-thaw cycles in the paving will be increased by a decrease in the thermal mass. If snow is left on the paving to thaw, the paving will become saturated with melt-water. Removal of this water is improved by eliminating as much capillary rise as possible. The gravel provides a capillary break. Free flow of water will take place in the lower portion of the gravel layer and in the insulation layer in the joints between the insulation and under the insulation. The thickness of the gravel drainage layer can be reduced if the flow in the insulation is improved. This can be done by cutting drainage channels in the underside of the insulation. This should be done by the supplier before the insulation is delivered to the job site.

If heat loss is a concern due to the water in the insulation layer, polyethylene sheet can be placed over the insulation. The edges should be overlapped and sealed. Further, if puncturing of the polyethylene is a concern, a thin layer of dimensionally stable protection board can be placed over the insulation. The polyethylene will restrict most of the water flow to the gravel layer and will decrease the rate of drainage within the insulation so that the water does not remove as much heat on its way to the drains.

c. <u>Area Drain at Concrete Paved Roof--Light Vehicular Traffic</u>. Refer to Figure 37. In this condition, the drainage layer is omitted. The amount of water reaching the level of the waterproofing is small (this water will tend to collect at the membrane waterproofing and drain between the joints in the insulation). The concrete (which is reinforced) is divided into panels to allow for expansion and contraction and to minimize freeze thaw action. The joints are formed with expansion material and sealed. (The sealant is necessary to reduce water penetration.)

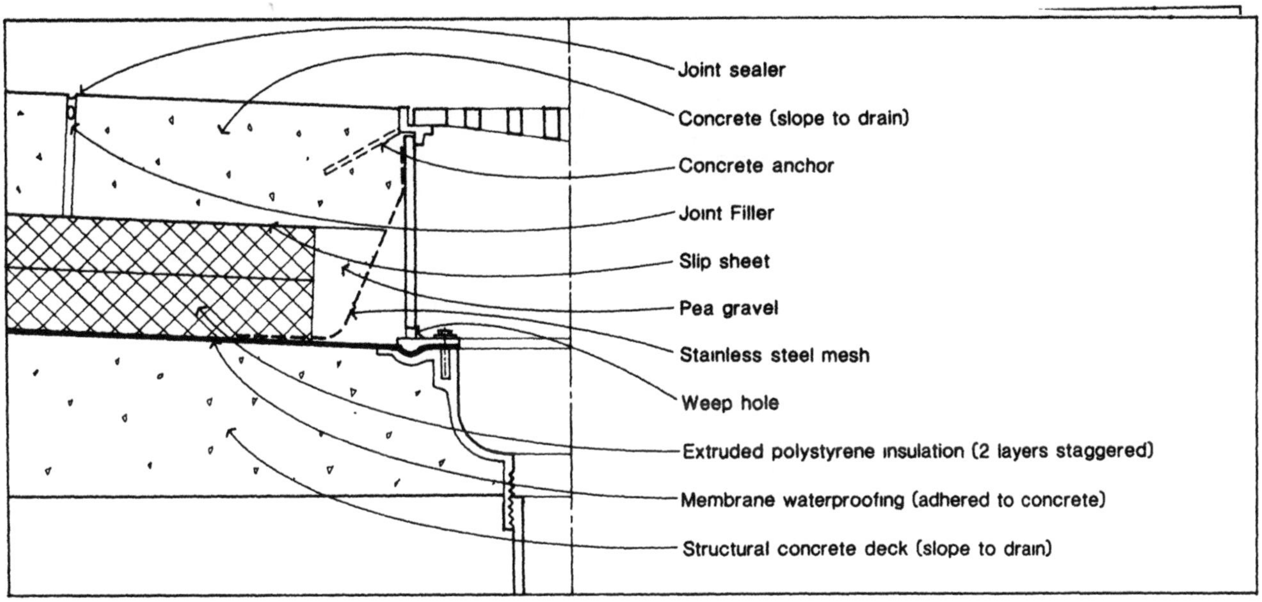

FIGURE 37
Drain at Concrete-Paved Roof

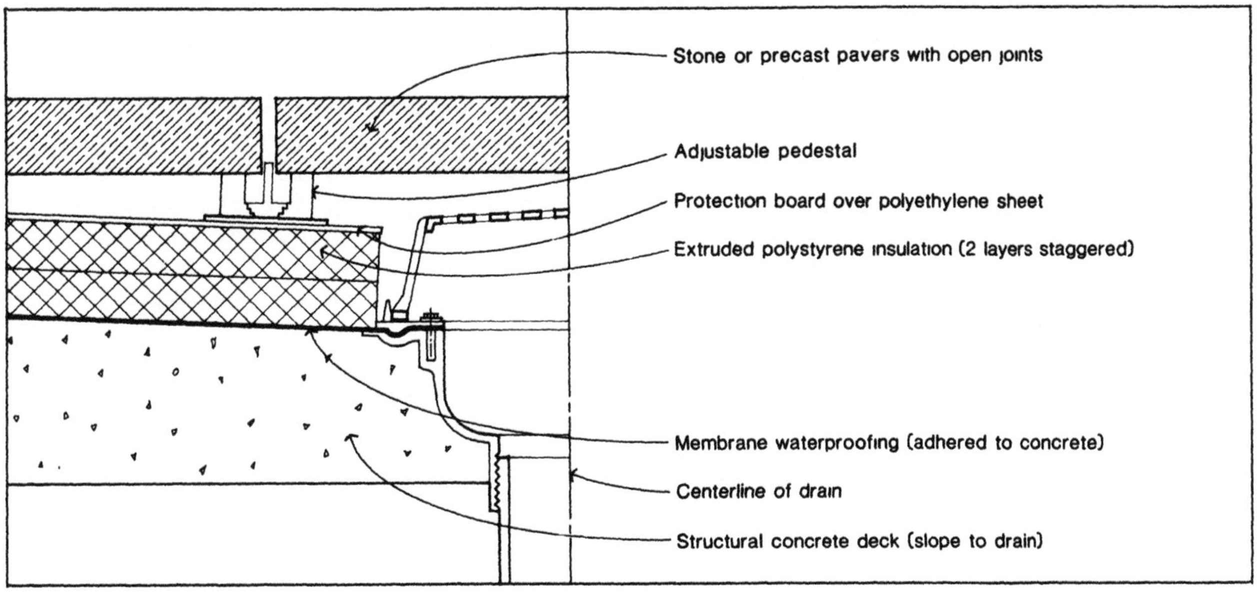

FIGURE 38
Drain at Paved Roof--on Pedestals

Both the paving and concrete deck must slope to the drain. The slope of the concrete deck, however, should be increased adjacent to the drain to help evacuate back-up water that has entered from the top of the drain. The drain itself must withstand lateral stresses and must have a removable drain cover for cleanout and inspection. A removable internal strainer or filter is also necessary.

d. Drain at Paved Roof--Frequent Freeze-Thaw Cycles--Pedestrian Traffic. Refer to Figure 38. This type of construction solves the

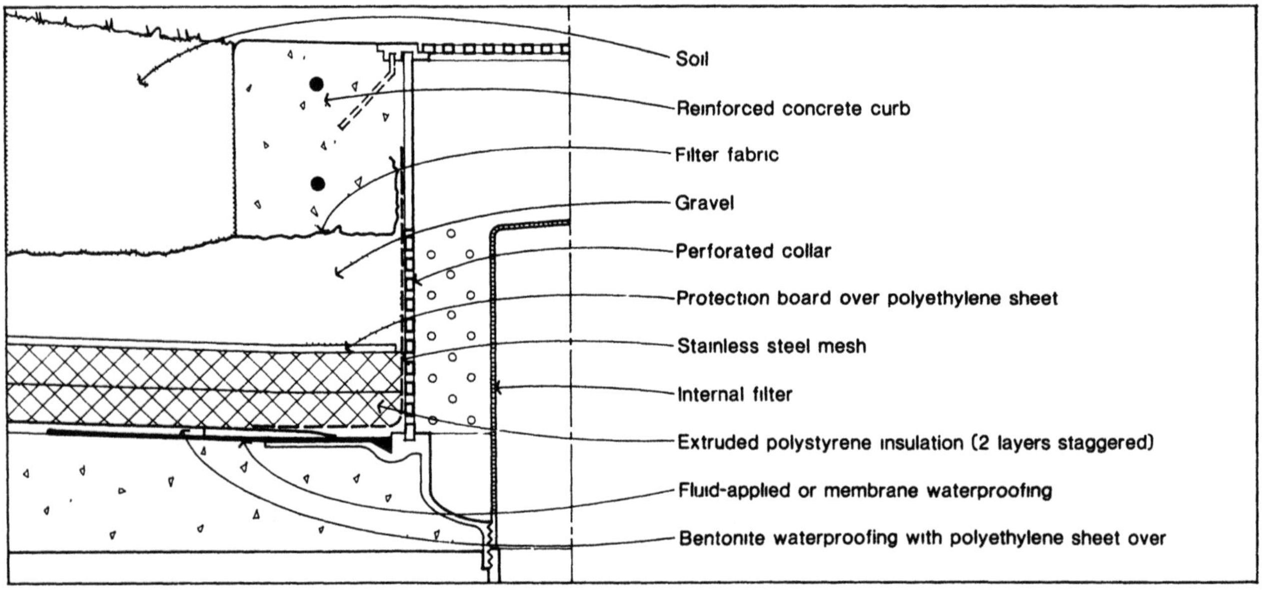

FIGURE 39
Drain at Earth-Covered Roof

freeze-thaw problem by setting the pavers on pedestals. The pedestal shown is a manufactured product and is adjustable over a range of heights. However, other pedestals such as concrete blocks or concrete may be used (consult a local contractor on the relative material/labor cost). The pavers are typically 2 to 4 feet in width (about 600 mm to 1200 mm). The size affects labor costs as well as the thickness required for the paver). The weight of the pavers should be sufficient to counter buoyancy of the insulation in the event that the roof becomes flooded. Normally, the thickness of the insulation will be considerably less than the thickness that would be required to produce buoyancy problems.

The insulation must tolerate water without losing its thermal resistance. Cellular glass or extruded polystyrene (staggered in two layers) is recommended for this. If extruded polystyrene is used, the insulation must be covered with a durable protection board to protect the insulation from fire as well as objects which may be poked through the joints in the pavers.

Two important operational advantages of this system are that (1) it allows easy access for repairs and maintenance, and (2) it provides ventilation of the construction which promotes evaporation of water as well as cooling (by reducing heat build up). Consequently, the same detail with a pea-gravel setting bed in lieu of the pedestals would not be as desirable functionally (although it might be less expensive initially). An additional advantage, aesthetically, is that the pavers can be installed level rather than sloped.

e. <u>Area Drain at Earth-Covered Roof</u>. Refer to Figure 39.

(1) Component Parts of an Earth-Covered Roof. The roof construction shown Figure 39 has six principal parts which must be provided for all earth-covered roofs. These consist of the following:

(a) Soil Layer. The soil thickness and mixture must be adequate to support the selected vegetation. An irrigation system may also be required depending on the type of vegetation and the thickness of the soil layer (see Section 8, Planting and Irrigation).

(b) Soil Separator. A soil separator must be provided. This may consist of a manufactured filter fabric, an improvised filter mat (such as fiberglass insulation). Both have the function of preventing smaller particles from entering the bottom drainage layer. Filter fabrics, however, have the additional advantage of discouraging plant roots from penetrating into the drainage layer.

(c) Drainage Layer. Sometimes referred to as a percolation layer, the main function of the drainage layer is to allow water to migrate to the drain (or roof edge). In some instances this layer may be placed adjacent to the waterproofing with an intervening layer of staggered protection board. In most situations, however, it is economical (and equally effective) to install the insulation directly above the waterproofing system rather than introducing a layer of protection board under the gravel, which, in turn, must be carefully leveled to receive the insulation boards. In either location, pea-gravel is recommended for ease of placement and leveling.

(d) Insulation. The insulation must be capable of withstanding submersion in water without substantial loss of its thermal resistance. Extruded polystyrene is suitable for this. The insulation should be installed in at least two layers with staggered joints. This will improve the thermal integrity at joints as well as prevent materials from reaching the waterproofing layer through the joints in the insulation. A polyethylene sheet may be added above the insulation layer to help confine water to the drainage layer. The edges of the polyethylene must be overlapped and sealed to be effective. Also, a protection board layer may be added as a construction consideration to help protect the polyethylene and to provide additional protection from stakes and other objects driven into the soil.

(e) Waterproofing. The type of waterproofing chosen depends on the construction methods, timing, type of structural deck and cost considerations (see Section 13, Waterproofing). If bentonite waterproofing is used, a layer of polyethylene over the bentonite is recommended. This prevents the bentonite from eroding in concentrated flow areas such as at the joints in the insulation, for example. The polyethylene also serves as a rain cover for the bentonite before placement of the gravel and soil layers. The edges of the polyethylene must be overlapped and sealed to be effective. If bentonite is used for the waterproofing, it may also be used to seal the polyethylene.

(f) Structural Deck. In most instances, the structure will be cast-in-place concrete, however, other systems including precast concrete, concrete fill over metal deck, and wood plank may be used, depending on other design criteria. In all cases, the surface of the deck must slope to drains or roof edges and have a finish suitable for the selected waterproofing.

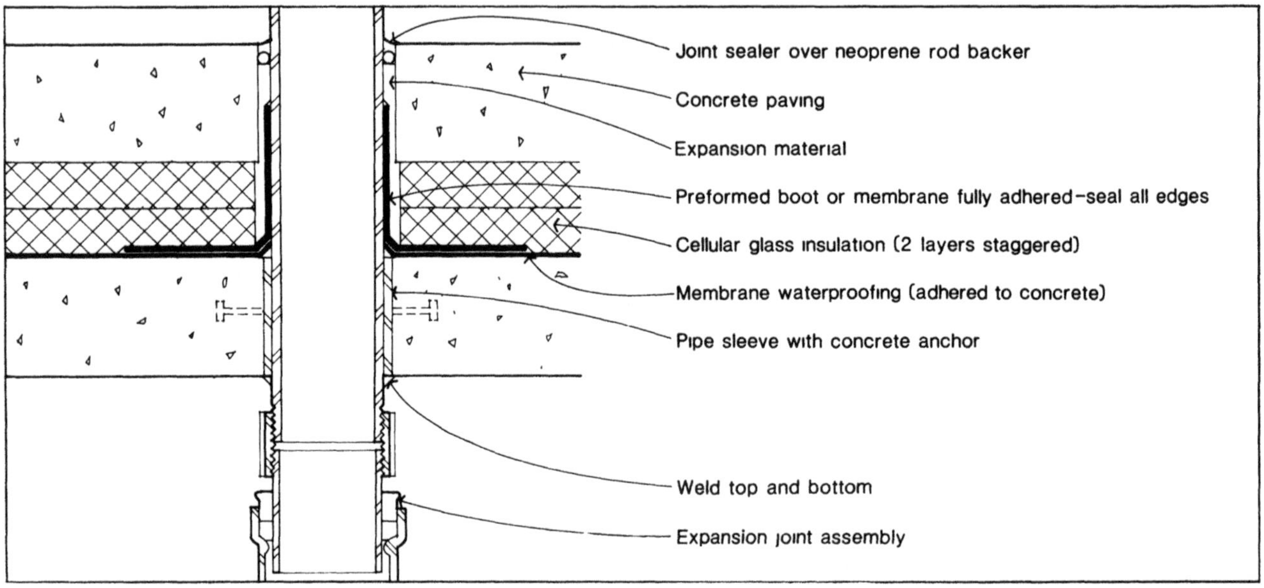

FIGURE 40
Pipe Penetration at Paved Roof

(2) Interior Roof Drains. The drain shown in Figure 39 is an all-level drain with a perforated collar, removable drain cover, and removable internal filter. The perforated collar must be wrapped in a stainless steel mesh to prevent the erosion of the gravel layer around the collar. Bentonite waterproofing, if used, must terminate at least 1 foot (300 mm) back from the drain (refer to the manufacturer's technical notes) and overlap a layer of membrane waterproofing adhered to the concrete deck and drain flange. If membrane waterproofing is used, this should extend into a clamp ring as well. The drainage pan and flange detail shown in Figure 39 are for fluid applied waterproofing. Using a drainage pan, as shown, reduces the potential for the bentonite to erode around the drain, leading to clogging of the drain. An increased rate of slope around the drain will reduce the extent of backup water and thus help prevent erosion of the bentonite near this condition.

In Figure 39, a concrete collar is provided around the top of the drain. This collar helps hold back grass and lawn clippings and reduces soil erosion where other types of planting may be used. With a smaller tributary drainage area and a soil surface that is graded flat, the concrete collar may be omitted. In this case, a solid drain cover may be used. Drain collars should be wrapped in filter fabric and the collar should not be perforated at the soil layer. If it is possible for the roof to flood, because of a continuous parapet for example, the drains must have a grate or strainer at the top. In addition, should flooding be possible, overflow drains would have to be provided. Overflow scuppers are not recommended. If an overflow (either drain or scupper) is not provided, then the structure and the parapet must be designed to hold the water.

f. Pipe Penetration at Paved Roof. Refer to Figure 40. This detail illustrates a pipe penetration through a roof paved in concrete. However, this type of solution will work equally well for other types of paving

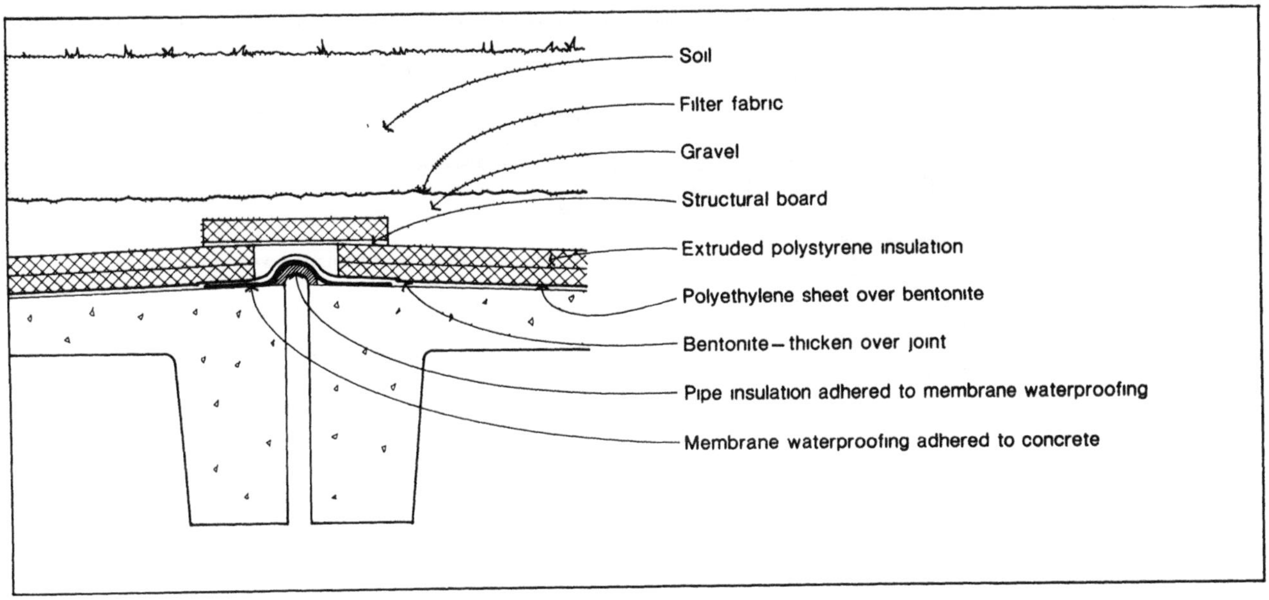

FIGURE 41
Expansion Joint at Earth-Covered Roof

situations as well as earth cover. The basic strategy in this solution is to eliminate flexible connections above the roof and to restrict all movement below the level of the structural deck. This reduces the vulnerability of the connection above the roof construction and also allows a neater-looking appearance. This particular detail uses a pipe sleeve with concrete anchors. After casting the assembly in the concrete, the steel pipe is then welded to the top and bottom edges of the pipe sleeve. An alternate solution that would work equally well would be to weld the concrete anchors to the section of vent pipe and to cast this into the concrete directly. The end of the pipe is threaded so that proper connections can be made using manufactured expansion joint assemblies.

The waterproofing illustrated is membrane waterproofing. Normally the manufacturer will call for additional plies to be used at this critical point. The vertical legs of the membrane waterproofing must be terminated so that it can be caulked with a compatible waterproofing material. Different manufacturers will have different recommended methods for terminating the waterproofing, however, and reference should be made to the technical information for the product selected. Expansion material should be used around the pipe to reduce stresses on the pipe. The joint at the surface should be sealed. If the roof is earth-covered, the soil should be sloped away from the pipe. Sloping the concrete away from the pipe is also advisable.

g. Expansion Joint at Earth-Covered Roof. Figure 41, illustrates a condition at an expansion joint at an earth-covered roof. Because of the continuity of the insulation, the plastic air stop below the expansion joint is not required. However, a similar type of construction for the waterproofing will be required. In the detail, bentonite waterproofing is the main waterproofing for the structure. The bentonite must overlap the membrane waterproofing at the expansion joints. The membrane waterproofing

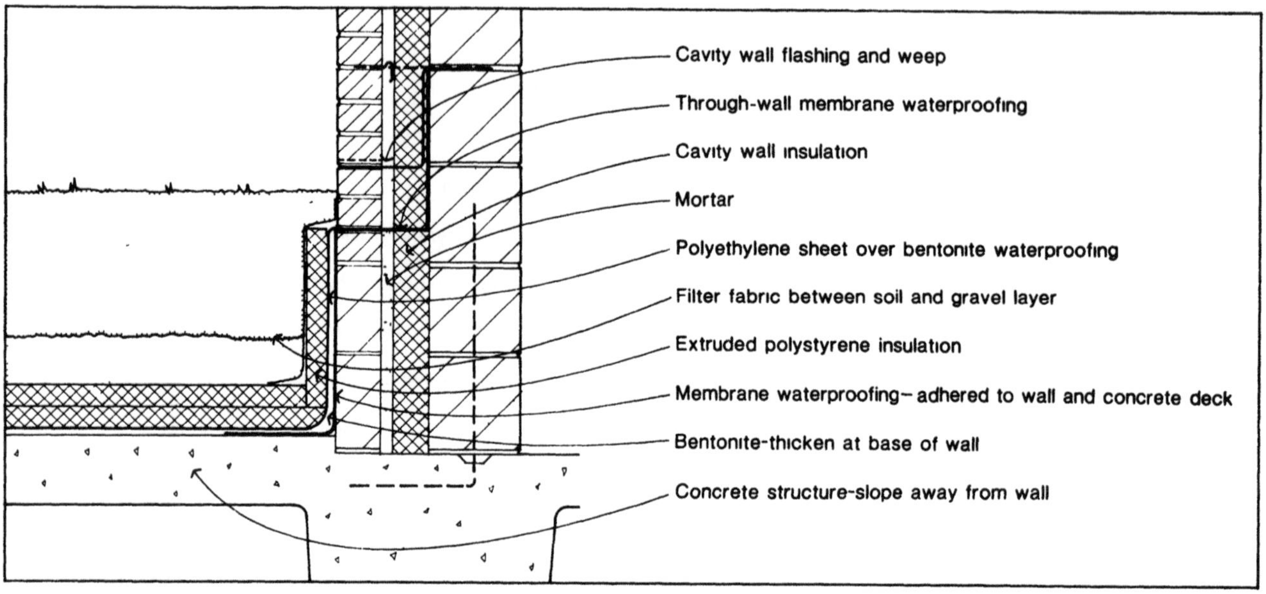

FIGURE 42
Earth-Covered Roof at Masonry Cavity Wall

must be thoroughly adhered to the concrete surface and extend out from the expansion joint for a width of at least 2 feet (600 mm) on either side of the expansion joint. The membrane waterproofing may be formed over a prefabricated expansion cover, neoprene pipe insulation, or large-diameter form rod stock. This material must be secured in place by adhering the top to the underside of the membrane waterproofing. If rod stock is used, it can be sandwiched between two layers of waterproofing to hold it in place. The bentonite is then carried over the completed assembly. Increasing the thickness of the bentonite would be advisable in this area. The polyethylene sheet, which is normally placed over bentonite is imperative in this case. Not only does the polyethylene sheet prevent erosion of the bentonite in the open area where the water would tend to flow in channels, but it also helps confine the bentonite. This is necessary to reduce the swelling of the bentonite in an unconfined space. The insulation is held away from the expansion assembly either by notching the insulation or by stacking the insulation as shown in the detail. This eliminates pressure on the expansion joint assembly which might cause collapse of the domed shape. In this case, to ensure that the weight of the earth-cover and gravel layers above will not collapse through the insulation, a structural board (of treated wood) is layed under the top layer of insulation. It would be advisable to increase the rate of slope of the concrete deck away from the expansion joint.

h. Earth-Covered Roof at Masonry Cavity Wall. Refer to Figure 42. For additional commentary on the roof construction shown in this detail refer to Figure 39 above. The concrete deck should slope away from the wall. The additional thickness required for the concrete slab at this point will depend on the size of the area to be drained and slope of the concrete surface. Bentonite waterproofing is the main waterproofing used in this case. However, at this condition, membrane waterproofing must be used so that the bentonite waterproofing overlaps the membrane waterproofing at the

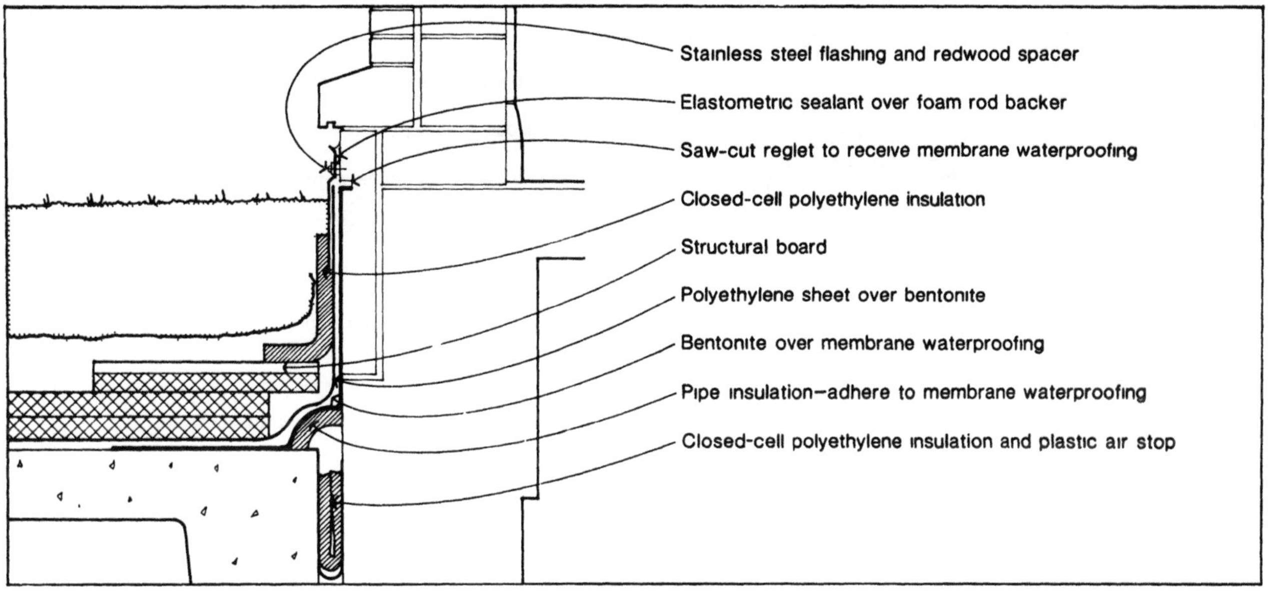

FIGURE 43
Expansion Joint at Earth-Covered Roof and Existing Structure

base of the membrane waterproofing and so that the integrity of the waterproofing system can be continued above grade. This is accomplished by extending the membrane waterproofing through the wall construction at a point just below grade and then by extending the waterproofing up at least 1 foot (300 mm). The membrane waterproofing must be adhered to the masonry along the upper portion. The bottom edge of the through-wall waterproofing should be left unadhered until a separate ply can be installed at the wall and slab. The through-wall membrane is then overlapped and adhered to this second ply. To ensure a positive overlap of the bentonite and the membrane waterproofing, the bentonite is continued part way up the wall over the membrane waterproofing. The bentonite is thickened at the cove at the base of the wall and at the edge of the membrane waterproofing at the slab. The top of the bentonite should be flashed to prevent washout of the bentonite adjacent to the ground surface. Polyethylene sheet applied to the wall with an appropriate mastic will serve this purpose.

As with conventional building construction, the cavity wall should not extend below grade. The cavity space below the cavity wall flashing should be grouted solid. The cavity wall flashing is extended up and terminated in the mortar joint of the concrete block backup according to criteria for conventional cavity wall construction. The insulation (which should be extruded polystyrene insulation) serves two functions: (1) to protect the waterproofing and (2) to develop a thermal overlap with the insulation in the wall. This overlap does not constitute a complete thermal break. However, in this case, the overlap provides a cost-effective way of reducing the heat transmission at this point.

i. <u>Expansion Joint at Earth-Covered Roof and Existing Structure</u>. Refer to Figure 43. This expansion joint occurs near the foundation wall of an existing building. The expansion joint construction is similar to the construction for an expansion joint within the typical roof area. Refer to

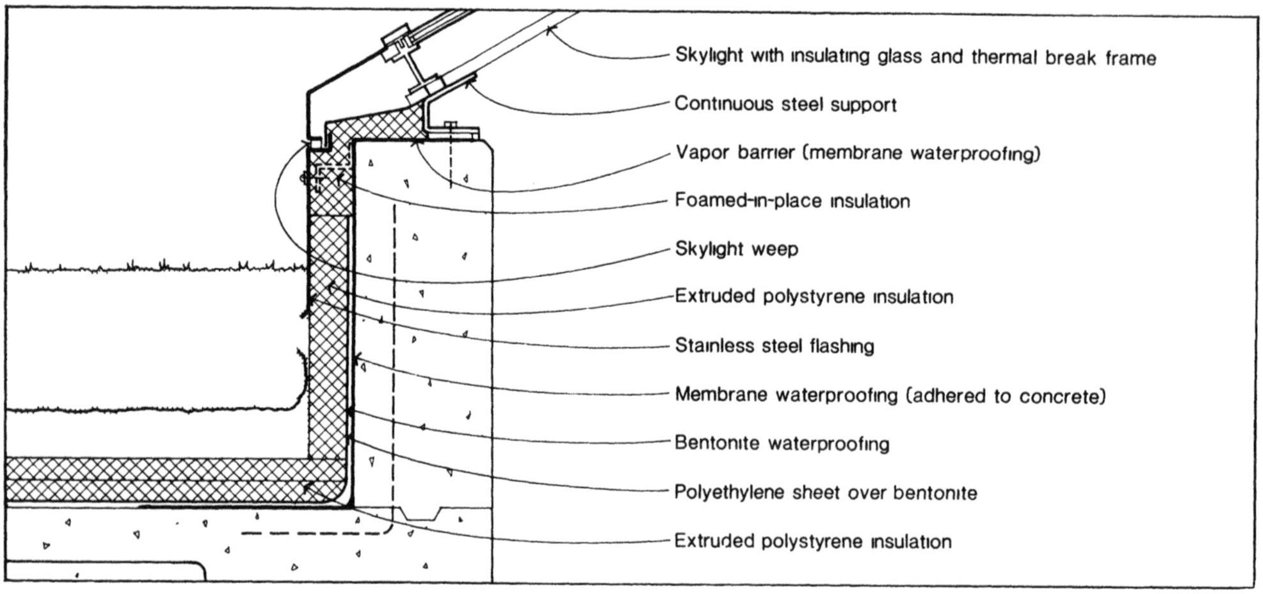

FIGURE 44
Earth-Covered Roof at Skylight with Concrete Curb Exposed to Interior

Figure 41 for criteria on this type of construction. The membrane waterproofing which is adhered to the concrete deck and to the cleaned and primed surface of the existing wall is carried up to a point above grade level and returned into a saw-cut reglet and caulked. The bentonite waterproofing, which is the principal waterproofing for the roof in this case, is overlapped and carried up the wall over the membrane waterproofing. The polyethylene sheet, which is normally installed over the bentonite on the roof, is also carried up the wall over the bentonite. The upper edge of this assembly is flashed with stainless steel flashing and continuous redwood spacers and anchored into the existing stone with stainless steel screws and neoprene washers. The top of the flashing is formed such that a positive joint may be formed and sealed with elastomeric sealant over a foam rod backer. The flashing is removable for maintenance and repair.

An additional layer of extruded polyestyrene is cantilevered over the bottom two layers with a layer of treated wood to ensure that the soil and gravel materials do not fall into the space above the expansion joint. To complete the closure, a layer of closed-cell polyethylene insulation is placed over the structural board and carried up the wall as high as possible without extending above-grade. This insulation serves two purposes. First, the insulation provides protection to the waterproofing and, secondly, the insulation (which is compressible) will take up the movement between the soil and the existing building. The compressible insulation will not completely eliminate the development of cracks between the soil and the building, however. Consequently, it is imperative that the waterproofing and, especially the bentonite, be thoroughly flashed to shed water away from these critical areas. In addition, it would be advisable to increase the slope of the concrete deck adjacent to the expansion joint.

j. <u>Earth-Covered Roof at Skylight with Concrete Curb Exposed to Interior</u>. Refer to Figure 44. This detail illustrates the lower

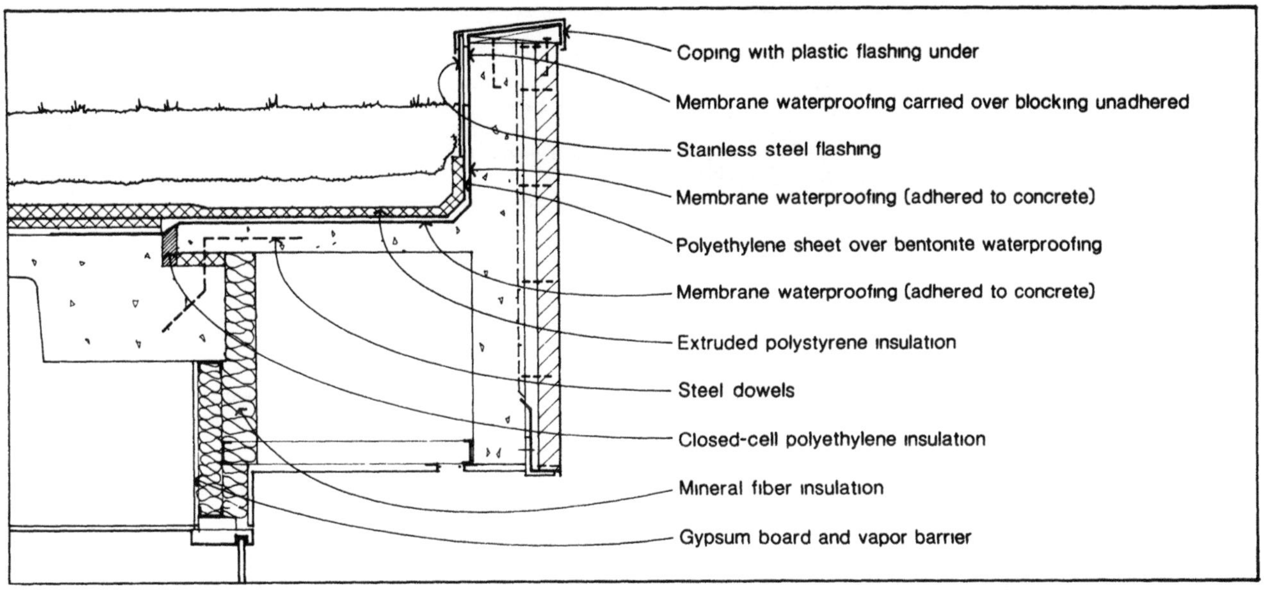

FIGURE 45
Parapet at Earth-Covered Roof with Thermal Break

termination of a skylight at an earth-sheltered roof. The primary aesthetic constraint was that the concrete be exposed on the interior. Bentonite, which is the primary waterproofing in this case, is shown overlapping membrane waterproofing which is adhesively applied to the concrete curb below the skylight. This detail, however, would be similar for a condition where membrane waterproofing was the primary waterproofing system. The insulation which is continued up the concrete wall serves as protection for the waterproofing as well as completing the thermal continuity of the insulation. It is important to carry the insulation as far as possible, not only to improve the thermal efficiency, but also to help prevent condensation on the backside of the membrane waterproofing (which acts as a vapor barrier). By detailing the membrane waterproofing as a vapor barrier, it can be terminated in a caulked joint between the concrete and a continuous steel support angle for the skylight. This configuration, in effect, extends the vapor barrier all the way to the first purlin on the skylight. The metal flashing which covers the above-grade insulation is positioned so that it accommodates the weep at the skylight as well as performing a flashing function.

k. Parapet at Earth-Covered Roof With Thermal Break. Refer to Figure 45. Rather than extending the insulation envelope around the parapet and soffit, a positive thermal break is provided within the concrete construction. The connection at the thermal break is held in place by steel dowels, which permit torsional movement of the parapet (the parapet spans between bays). This movement will be minimal. However, the movement must be allowed for in the detailing of the waterproofing. In this case, the slope of the concrete is away from the parapet. The joint is designed so that water will pass freely over it and onto the main roof area.

The life-cycle economy of this detail must be evaluated on an individual basis. In many climates, especially mild climates, the energy

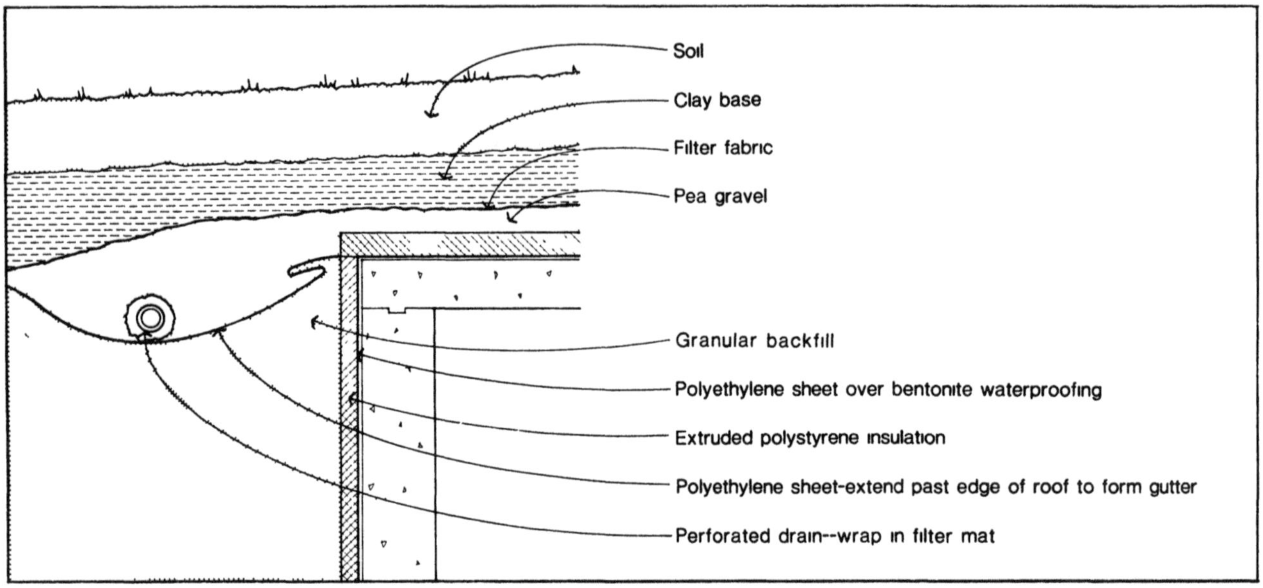

FIGURE 46
Backwall Drainage at Earth-Covered Roof

savings attributable to the thermal break construction will not justify the additional cost.

l. Back Wall Drainage at Earth-Covered Roof. Refer to Figure 46. This condition occurs at the back wall of a structure which is located below a watershed. A drainage swale should be provided at the base of the slope so that the surface water is carried away from the building. However, in addition to this, a clay base may be used below the soil layers to provide additional protection. This construction will tend to increase the total thickness of the soil and drainage layers required for the roof construction. This particular detail represents a condition where the concrete is sloping to the roof edge and the water draining over the edge. This method of drainage increases the vulnerability of the roof edge. To compensate for this, the polyethylene sheet, which was installed directly above the bentonite waterproofing, is extended out beyond the roof edge and downward, leaving a small tuck in the polyethylene to allow for potential settlement of the backfill. The polyethylene is then formed into a gutter. The perforated drain and the gutter should be below the level of the roof. An additional perforated drain will be warranted in many instances near the footing. The advantage of the upper system of drainage is that it reduces the amount of water washing down the exterior wall. This may also be beneficial from an energy efficiency point of view in that the soil adjacent to the wall will be less saturated and therefore will have a lower conductance. If a layer of polyethylene sheet is used over the insulation, this too can be extended out to help the water move away from the vertical wall.

m. Window Sill at Berm. Refer to Figure 47. This detail illustrates a wall that is bermed to within 1 foot (300 mm) of a window sill. The aesthetic requirements called for exposed concrete on the exterior below the window. To meet this requirement and to continue the waterproofing to the

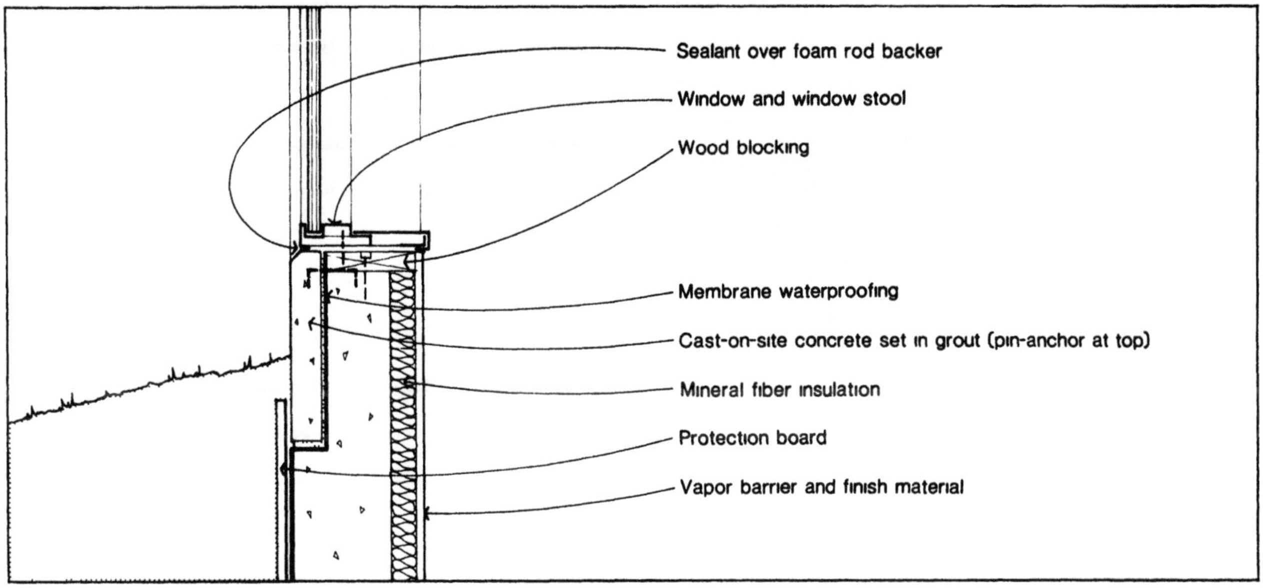

FIGURE 47
Window Sill at Berm

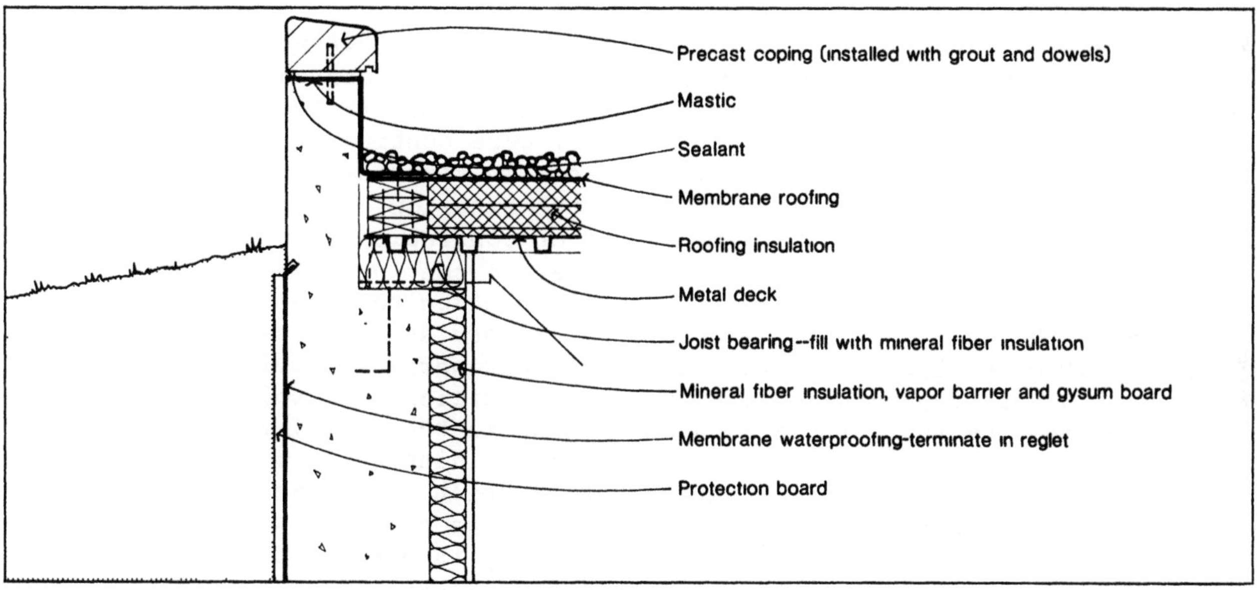

FIGURE 48
Edge of Conventional Roof at Berm

underside of the window sill, a notch is formed in the concrete wall. The waterproofing is adhered to the concrete following this notch. A cast-on-site concrete panel to match the concrete on the remainder of the building is grouted in place with a pin anchor located at the top. The building insulation is located on the interior side of the structure in this case.

n. <u>Edge of Conventional Roof at Berm.</u> Refer to Figure 48. With the type of structure shown in the detail, it is possible to achieve fairly

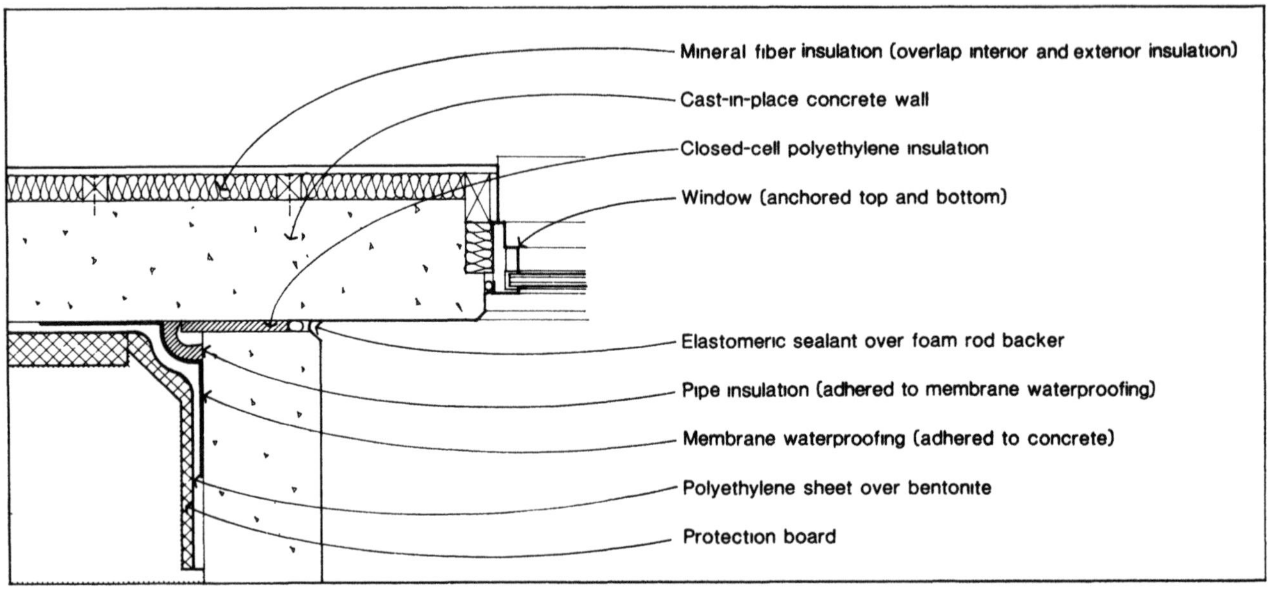

FIGURE 49
Juncture of Retaining Wall and Exposed Concrete Wall (Shown in Plan)

uninterrupted insulation envelope that is also cost effective. This is done by ensuring that the space between the joist bearing plate and the metal deck is filled with an adequate amount of insulation. If the insulation shown on the interior of the structural wall were continued down to the floor, this approach would be disadvantageous in terms of benefiting from thermal mass. To place the insulation on the exterior of the structural wall and still maintain an adequate degree of continuity in the insulation, the insulation shown on the interior wall may be continued down to the level of the finished ceiling. This will allow insulation to be placed on the exterior of the wall overlapping the interior insulation by as much as 4 to 5 feet (1220 mm to 1525 mm). The interior vapor barrier should not be omitted, however, depending on the humidity conditions and soil temperatures. The roof of the structure shown in the detail is designed as a conventional roof (in this case, a loose-laid, single-ply membrane with ballast is used). The membrane waterproofing which is adhered to the concrete wall is terminated in a reglet as high as possible without being exposed above grade. If bentonite waterproofing is used on the wall, the top of the waterproofing should be flashed (see Figure 43).

o. <u>Juncture of Retaining Wall and Exposed Concrete Wall (Shown in Plan)</u>. Refer to Figure 49. This detail (in plan view) shows a building wall with an abutting retaining wall. The aesthetic requirements for the exterior finish of the building, in this case, called for exposed concrete. At typical above-grade conditions, this building must have insulation located on the interior side of the concrete wall. The type of insulation and its protection must conform to the requirements of NAVFAC DM-8. The same type of insulation and interior finish is used along the inside of the structural wall in the condition depicted in the detail. However, the interior insulation (but not necessarily the vapor barrier) may be terminated once a sufficient overlap length (about 3 feet (900 mm)) with the exterior insulation is developed.

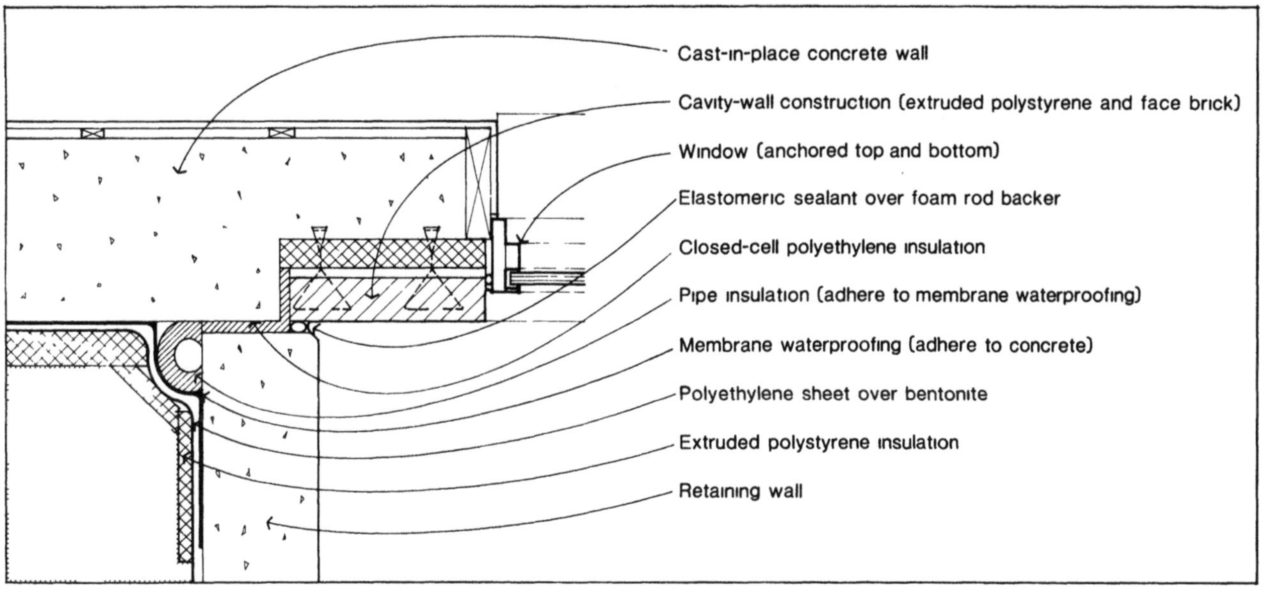

FIGURE 50
Juncture of Retaining Wall and Masonry Cavity Wall (Shown in Plan)

In terms of waterproofing, bentonite is the primary waterproofing system for the building. At this condition, however, membrane waterproofing must be introduced at the joint where movement is expected to occur. This detail is similar to the conditions shown in Figure 41. In this case, the waterproofing is extended onto the retaining wall for a distance of 3 to 4 feet (915 mm to 1220 mm). Rather than tying the retaining wall to the building structure with steel dowels, the retaining wall is free-standing and is designed to withstand the earth and possible hydrostatic pressures without assistance at the end conditions. Consequently, movement can be anticipated at the joint between the building and retaining wall. The joint must therefore be sized and sealed with elastomeric sealant of sufficient width to accommodate this movement. The backfill shown in the detail is undifferentiated. However, the fill against the retaining wall should be gravel to provide drainage of water and therefore, relief from potential hydrostatic pressure. The gravel should be drained at the bottom of the retaining wall with plastic pipes sloping to the exterior side of the wall. Because the waterproofing of the building also depends on the waterproofing of this portion of the retaining wall, this portion of the retaining wall must be treated as if it were a building. The slope of the finished grade must slope away from the retaining wall as well as from the building. If water is allowed to flow towards the retaining wall, for all practical purposes, it is also flowing towards the building.

p. Juncture of Retaining Wall and Masonry Cavity Wall (Shown in Plan). Refer to Figure 50. This detail illustrates the same situation as the previous detail. However, in this case, the architectural requirement for the exterior finish of the building is face brick. On the right hand side of the retaining wall a cavity wall situation exists. It is important that the retaining wall overlap the face brick rather than the face brick overlapping the retaining wall. Otherwise, the movement of the retaining wall, which will certainly occur, will tend to shear off the face brick.

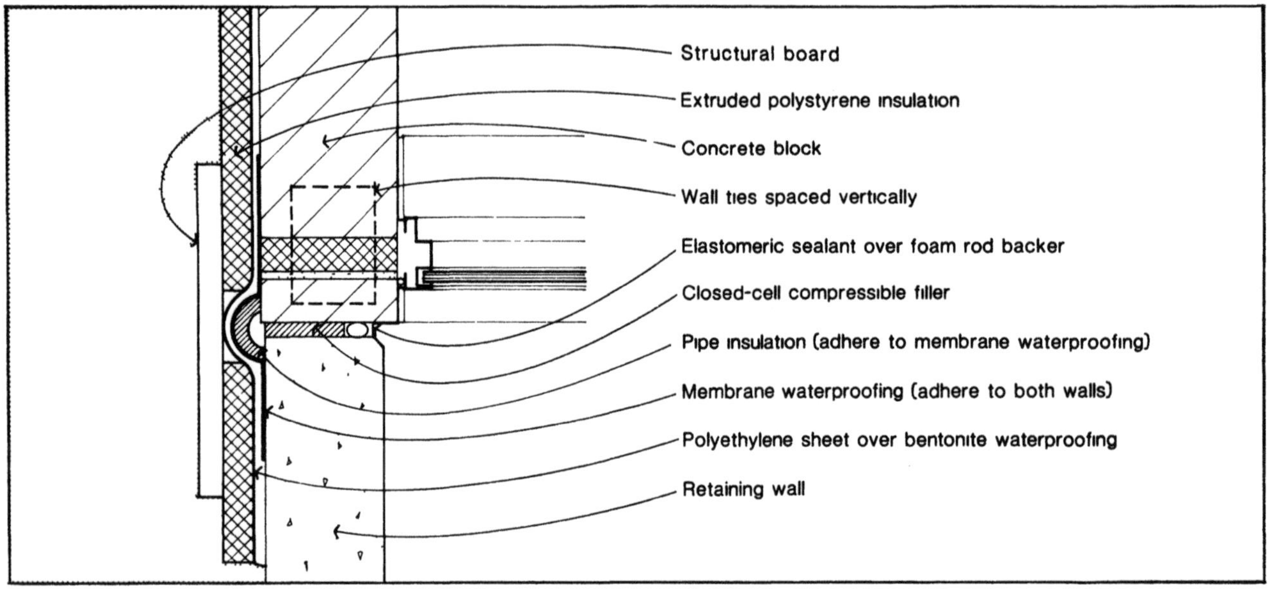

FIGURE 51
Thermal Break and Waterproofing at Retaining Wall (Shown in Plan)

Continuity of the insulation for the building is maintained by inserting a compressible closed-cell polyethylene insulation in the space between the retaining wall and the building.

q. Thermal Break and Waterproofing at Retaining Wall (Shown in Plan). Refer to Figure 51. At above-grade locations, the exterior material used on this building is burnished concrete block. This is used in a cavity wall system with concrete block as a back-up. Concrete block is also used for below-grade conditions. The condition shown in the detail occurs at a corner of the building with the retaining wall parallel to the end wall. The finish material that occurs above and below the window is burnished concrete block, also utilizing a cavity wall system. The insulation shown adjacent to the window jamb is aligned with the insulation that occurs above and below the window (this is the standard location for the insulation in this building relative to the face of the exterior wall). The location of the building insulation, due to the choice of a cavity wall system, easily facilitates maintaining the continuity of the insulation envelope. As in the previous details, the waterproofing is applied to the portion of the retaining wall adjacent to the building. Again, the joint between the retaining wall and the building is treated as an expansion joint. Refer to the commentary for Figure 41 for a description of the waterproofing at this joint.

r. Foundation Near Water Table. Refer to Figure 52. This condition occurs where there is no hydrostatic head. To allow for the possibility of a higher water table, foundation drainage is incorporated into the design. Drainage tile is located at the outside of the footing as well as at calculated intervals below the basement slab. Plastic pipes are cast into the footing to transfer water from the interior side of the footing to the drainage pipe on the exterior side. The dashed line in the detail indicates the draw-down profile of the water table. To assist drainage of water at

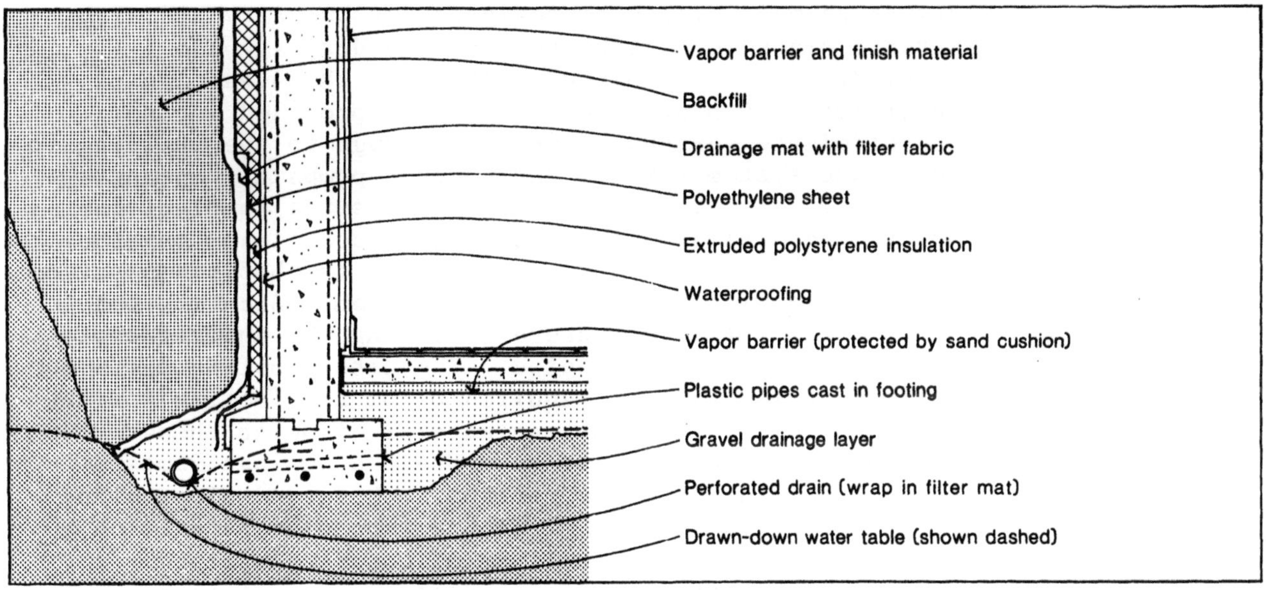

FIGURE 52
Foundation Near Water Table

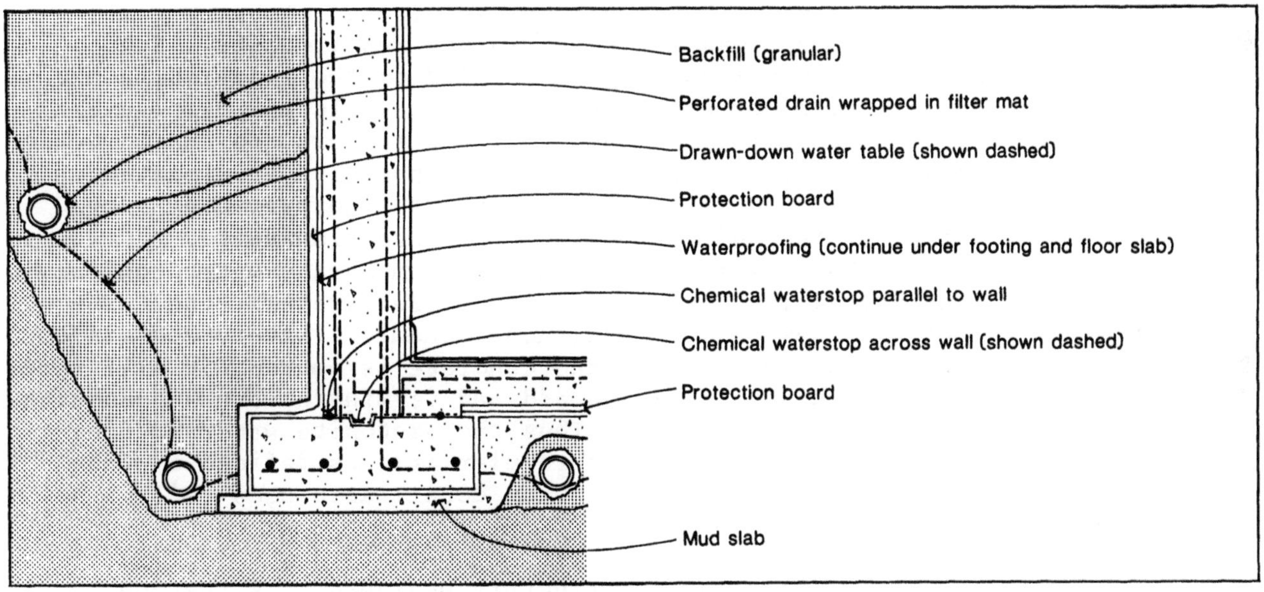

FIGURE 53
Foundation Below Water Table
(Hydrostatic Pressure Relieved in Drainage)

the exterior wall, a drainage mat with a filter fabric is installed to carry water to the drainage tile at the footing. To maintain the thermal efficiency of the insulation, a layer of polyethylene sheet (with overlapped and sealed edges) is placed over the insulation. This also terminates at the drainage tile at the footing.

s. Foundation Below Water Table--Hydrostatic Pressure Relieved by Drainage). Refer to Figure 53. This condition occurs below a water table.

Consequently, the structure must either be designed for buoyancy and slab uplift or the hydrostatic pressure must be relieved by drainage and pumping. In this case, drainage is provided. This has the advantage of reducing the structural cost. A blowout panel must be provided in the structure in order to eliminate serious structural damage to the building in the event that the drainage system fails. Backup sump pumps are essential in this case. The dashed line in the detail illustrates the level of the drawn-down water table. The drainage tiles must be spaced closer together than in the previous detail. To allow for the possibility that the position of the drawn-down water table will not be at the location indicated, but might be slightly higher, the entire structure including the underside of the footing and the floor slab is waterproofed. The waterproofing shown is a membrane waterproofing. A mud slab is required below the footing and the floor slab for the application of the membrane waterproofing at these locations. The membrane waterproofing must be protected against damage during the placement of steel and concrete for the footing and floor slab. In this case, protection board is shown. Often the type of construction will be such that an additional protection layer of concrete will be required, however. Chemical waterstops are an integral part of this system. Chemical waterstops should be provided at critical joints in the structure, such as between the wall and the footing and between the floor slab and the footing. Chemical waterstops should also be used at regular intervals to help confine the leak, should the waterproof membrane fail. This facilitates locating the leak and thus reduces repair costs. Note that, because the membrane waterproofing is applied to the mud slab rather than to the footing or the floor slab directly (this would not be possible), a path for water migration will exist between the structure and the membrane waterproofing. Consequently, the chemical waterstops which are located in the floor slab will not help localize a leak without also providing a grid of chemical waterstops in the joints between the protection board. Bentonite waterproofing has a slight advantage over membrane waterproofing in this regard. The protection board will also be required for the bentonite waterproofing. However, no additional chemical waterstops will be required in the protection layer. Once the concrete is poured, it will tend to fill the gaps between the protection board and terminate at the bentonite waterproofing at the bottom of the gap. Thus the bentonite waterproofing will be in contact with the concrete in a regular grid as defined by the layout of the protection board. The following detail also uses this waterproofing system.

t. <u>Foundation Below Water Table--Structure Designed to Withstand Hydrostatic Pressure</u>. Refer to Figure 54. This condition, which occurs below the water table, utilizes the excavation shoring for one-half of the formwork for the wall. This condition occurs infrequently. However, when land area must be maximized and there is no room (in terms of programmatic and life-cycle requirements) to set the building back from the shoring, then this kind of construction may be necessary. The bentonite waterproofing is applied directly to the wood lagging. This is covered with a fiberglass mesh which, in a vertical position, acts as a rain cover during construction and also provides a degree of protection for the bentonite waterproofing during pouring. The mesh allows the bentonite to come into contact with the concrete through the spaces of the mesh. The most critical point in this construction is the juncture between the waterproofing at the bottom of the shoring and the waterproofing over the mud slab. The bentonite should be thicker in this area.

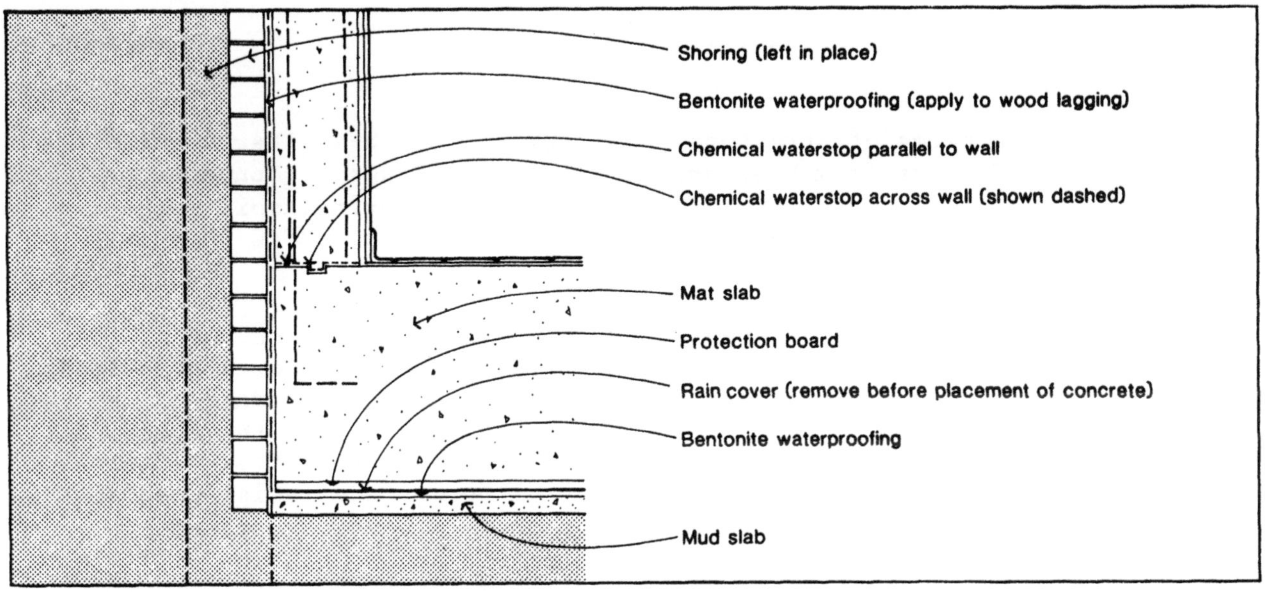

FIGURE 54
Foundation Below Water Table
(Structure Designed to Withstand Hydrostatic Pressure)

Section 16. STRUCTURE

1. GENERAL. In terms of structural engineering, earth-sheltered buildings are not very different from basements of large buildings or conventional retaining structures. Criteria for structural engineering, soil mechanics, and foundation engineering are covered elsewhere in the NAVFAC DM series (see Related Criteria in Section 1).

The primary difference between an earth-sheltered building and a conventional building in terms of structure is the much higher static loads present in most cases on the exterior structural surfaces of an earth-sheltered building. These heavier loads tend to limit the types of structural systems which can be used for below-grade structures. Internal bracing of the building to transfer horizontal forces and resist unequal earth pressures becomes important and, when roofs are earth-covered, vertical supporting elements within the structure and foundation will also carry substantially increased loads. It is thus important that the overall design of the building (smaller bay sizes, for example) allow for an economical structural system.

2. DESIGN LOADS. Soil loads are normally considered dead loads. However the effects of possible later partial excavation of the earth on the roof or adjacent to the walls of an earth-sheltered building must be considered in design. If the structural design depends on balanced lateral earth pressures at opposite walls of the structure, unequal excavation or backfilling may cause structural movement or damage. The likelihood of maintaining a permanent soil load must be considered in designing an earth-sheltered building, particularly if it is a post-tensioned structure with final tensioning designed to balance earth loads.

a. <u>Roof Loads</u>. The most dominant roof load is the presence of soil on the roof. Normal soils can vary in weight from 90 lbs/ft^3 (1442 kg/m^3) to 135 lbs/ft^3 (10 500 kg/m^3). It is possible to design lighter soil mixes by including a lightweight filler material, but, this may be detrimental to plant growth. A fully saturated soil may weigh more than 135 lbs/ft^3 (10 500 kg/m^3). If the roof is well drained, however, the water loading can be considered a live-load condition. The soil density most commonly assumed in design is 120 lbs/ft^3 (9350 kg/m^3). See Section 8, Planting and Irrigation Design, for planting and soil loads.

Live-load allowances for earth-covered roofs usually range from 50 to 100 lbs/ft^2 (about 10 to 20 kg/m^2). The actual value will depend on snow load, any vehicular traffic, and whether allowance must be made for rain-saturated soil. When heavy snow is typical and extensive drifting of snow on the roof can occur or when large vehicles can drive onto the roof, the design must allow for these possibilities. Minimum must be in accordance with NAVFAC DM-2.2, <u>Structural Engineering, Loads</u>, Section 3.

b. <u>Wall Loads</u>. Walls in an earth-sheltered structure must resist lateral earth pressures if they are placed against the earth; if they are exposed, they must resist wind loads. The walls also act as components of the structure, transferring forces from the roof and other walls to the foundation.

Lateral soil pressures will vary considerably, depending on whether the wall is able to deflect away from the load (active pressure), is held rigid (at-rest pressure), or is actually forced towards the soil (passive pressure). Although the actual soil pressure conditions are greatly affected by backfilling procedures and building geometry, at-rest pressures should generally be used in the design of a fairly rigid structure such as a concrete wall, roof and floor system; active pressures can be used for external cantilever retaining walls.

When the lateral pressure exerted by a backfill material is used in design, it is important that other soil materials that would exert a higher pressure on the wall be placed outside the zone in which they could influence the wall. A rule of thumb is to assume that this zone lies within a line drawn at 30 degrees from the vertical, extending out from the base of the wall.

Other loads that must be considered in wall design are:

(1) Saturated Soil Conditions. Water pressure can add significantly to the wall loads. A granular backfill, which provides good drainage in addition to lowering normal lateral pressures, is generally used for backfill whenever it is available. When gravel backfills and drainage tile are used, the wall is normally designed for unsaturated conditions.

(2) Swelling Pressure/Frost Pressure. These pressures must be considered in design whenever soil conditions are potentially hazardous.

(3) Surcharge Loads/Live Loads. The vertical pressure from a surcharge load adjacent to a retaining wall causes a corresponding increase in the horizontal pressure on the wall. If any live loads are considered in retaining wall design, they are usually treated as a surcharge load. A major live-load item will be the surcharge load from any heavy vehicle that may be driven on the surface adjacent to the top of the retaining wall. Firefighting equipment access should be considered in calculating these loads.

(4) Backfilling Pressures. Overcompaction of backfill as a result of using heavy equipment can place a higher load on a retaining wall than the wall might have to withstand in the course of normal service. Backfilling procedures should be specified to avoid this problem. Refer to DM-7 Series for criteria.

c. Seismic Loads. Seismic loads for buried structures are generally less severe than for conventional structures. Below-grade tanks or pipes of moderate size generally do not require special seismic design considerations if applicable non-seismic design criteria are satisfied. Tanks, tunnels, or pipes of critical importance or with large cross sections will require special considerations for seismic design. Issues of particular concern for earth-sheltered buildings include: (a) inertial forces of a heavy earth-covered roof with respect to its attachment to the walls of the structure (This is very important when the roof element is not confined by surrounding earth on all four sides of the building.); and (b) relative movement or settlement of the building foundation. Buried structures are usually relatively rigid and will not accommodate significant relative movement without cracking and structural damage.

In addition to the direct effect of ground motion on the building, other hazards include subsidence and settlement due to consolidation or compaction, landslides or liquefaction. Design of earth-sheltered structures to be built into hillsides in seismic zones should carefully evaluate slope stability under earthquake conditions.

Refer to the following sections of NAVFAC P-355, Seismic Design of Buildings, for specific guidance: Retaining Walls--Chapter 11, Section 6; Buried Structures--Chapter 11, Section 7; Design of Foundations--Chapter 4, Section 8; Overturning--Chapter 4, Section 4b; Other Hazards--Chapter 2, Section 7c (see Criteria Sources).

d. Blast Loads. For bomb- and blast-resistant design, obtain the latest criteria from NAVFAC. See also NAVFAC P-397, Structures to Resist the Effects of Accidental Explosions (see Criteria Sources).

3. CHECKLIST OF LOADING CONDITIONS. The following conditions should be considered in calculating the potential loads on a structure:

- self-weight of structure and finishes
- vertical soil loads on roof
- ground-water pressures/uplift
- loads from landscaping elements
- mechanical equipment on roof or on intermediate floors
- rain saturation of roof and wall backfill
- snow loads
- construction equipment loading on roof and adjacent to walls
- service vehicle access to roof and adjacent to walls
- swelling clay conditions
- frost-heave pressures
- earthquake forces
- wind loads on exposed portions of the structure
- live loads on interior floor structures
- soil/structure interaction (for flexible structures such as thin shells)
- slope stability (for structures on hillsides)
- required resistance to blast loads

Section 17. MECHANICAL AND ELECTRICAL SYSTEMS

1. AIR-HANDLING SYSTEMS. The model building codes referenced by NAVFAC DM-3.3, Heating, Ventilating, Air Conditioning, and Dehumidifying Systems state that ventilation shall be provided either by natural means or by a mechanically-operated ventilation system. Because of the imposed design constraints of large below-grade structures, ducted supply, air and positive return air systems must be utilized to provide ventilation for the building.

It is essential that the designer thoroughly examine the overall energy-efficiency of the system. The designer should: (1) seek to reduce the resistance to air flow; (2) reduce air volumes where possible; (3) minimize the thermal losses. Energy savings are achieved by reducing air volume and resistance to air flow.

In general, the design of ventilation systems for below-grade structures is very similar to that of above-grade structures. The most significant difference in system design between above-grade and below-grade buildings is the placement of outside air louvers. In an earth-sheltered building louvers will tend to be more prominent visually. Further, where the above-grade portion of the building is concentrated in a small area, care must be taken to ensure that the air intake louvers have adequate separation from exhaust louvers and fumes at vehicle access points.

2. VENTILATION REQUIREMENTS. Because most large above-grade buildings are hermetically sealed and mechanically ventilated, the ventilation requirements for below-grade and above-grade structures will be similar. The requirements for ventilation quantities are defined by NAVFAC DM-3.3 (Chapter 4, Ventilation Requirements). Where power ventilation is provided, the ventilation air quantity (including pressurization and infiltration) used for calculating heating and cooling load shall be established as the greater of the exhaust requirement plus 10 percent or 0.125 ft^3/min per square foot of net floor area (0.305 m^3/min per square meter), provided that ventilation air exclusive of infiltration is furnished at 5.0 ft^3/min per person (0.14 m^3/min). Refer to ASHRAE Standard 62-1981 (see Reference 5) for minimum ventilation air requirements.

Adequate ventilation is important to prevent the buildup of indoor air pollutants and to remove excess heat from an occupied underground space. Low uncontrolled infiltration rates make ventilation particularly important in earth sheltered structures. A pollutant of particular concern is radon. Radon is a radioactive gas released in minute quantities by soil and rock materials including the materials used in building construction, such as concrete and building stone. Radon is also absorbed by ground water and then released at a free ground-water surface. Normal ventilation rates in excess of 0.5 air changes per hour are believed to keep radon levels to below permissible standards. In addition, it is desirable to prevent the passage of ground-water or water vapor from the surrounding ground to within the building envelope since this is significant source of radon.

3. WATER SYSTEMS. In multistory buildings, pressure diminishes at upper stories due to associated gravitational and frictional losses, requiring upfeed pumping to augment the system pressure. Such losses will be reduced or eliminated in earth-sheltered buildings since distribution will typically be nearer or below the city main. In remote areas where domestic water is pumped from ground water tables, holding tanks may have to be incorporated into the system to provide adequate reservoir for domestic water and firefighting. Earth-sheltered buildings can benefit by placing tanks on grade.

4. SANITARY SYSTEMS. Where building waste lines are lower than the sanitary sewer, pumping will be required, necessitating (1) a sump pit or receptacle into which waste may flow by gravity; and (2) a sewage ejector (operated by either a motor driven centrifugal pump or by compressed air). Sewage and drainage pumps must be provided with two power sources or must be equipped with dual electric and diesel drives. Two pumps should be provided which alternate in operation (each cycle). Each pump should be sized to take the entire peak flow.

Sump pits can be located in the building, even with little head room, under heavily travelled pavement or in remote areas. Except in areas subject to prolonged below-freezing temperatures, it is recommended that sewage sumps be sited away from the building to prevent spillage or flooding in the event of equipment failure. Otherwise, sumps must be of sufficient size to take sewage flow for a period of time which will allow detection of the problem and installation of portable pumps. Sewage sumps must have proper ventilation to the outside. Sumps located away from the building will require a manhole installation. Refer to DM-3.1, Plumbing Systems, for criteria on sizing of sumps.

5. LIGHTING SOURCE. Source efficacy (lumens per watt) and color are both important considerations in the selection of a light source for earth-sheltered construction. Generally, wherever possible, high efficacy sources should be used. Internal loads due to lights and equipment play a more significant, and sometimes different, role in the heat balance of underground spaces. Spaces which would normally require heating, often have no heating requirements when earth-sheltered. The amount of usable waste heat from lights and equipment is, in effect, decreased. Often cooling will be required where heating would have been necessary previously. Consequently, the lighting system not only affects the electrical load in terms of direct energy usage but also has a more significant impact on the HVAC load for the building. Because of recent developments in high-intensity discharge sources, color-corrected metal halide and high-pressure sodium lamps, the designer can utilize the most energy-efficient lamps while still providing good color rendering characteristics in interior applications. (See NAVFAC DM-4.4 for lighting design requirements.)

6. LIGHTING CONTROLS. In single-story, earth-sheltered buildings, windows and/or skylights can provide effective daylighting. Special care must be taken to provide adequate control of glare and excessive luminance ratios. The artificial lighting system should be designed to compensate for periods of inclement weather or night operation. An analysis should be done to determine the cost effectiveness of a photo-controlled dimming system. In multiple-story, earth-sheltered buildings, standard skylights are ineffective past the second level. To utilize any natural lighting component in lower

levels, the beam shaft method would have to be employed (refer to Section 11, Natural Lighting, Skylights and Passive Solar).

Motorized or manual blinds and louvers are other means of controlling natural lighting into the building. These controls, especially when combined with automatic thermal shuttering, tend to be more important where skylights are used. Skylights make up for the loss of daylighting at the perimeters of earth-sheltered buildings. However, depending on climate and orientation, skylights can be an energy liability. To offset this, controls must be used. Automated building systems are an ideal way to interface lighting and solar gain control with total energy control of the building. If total automated building control is not available, a programmable lighting controller can be used in conjunction with transceivers and contactors to control the lighting automatically, according to the schedule of the building. As in conventional buildings, multiple switching and other energy-saving control methods should be investigated for their cost effectiveness.

7. PSYCHOLOGICAL RESPONSE TO LIGHTING. Lighting can have a significant impact on the human psychological response to an environment. The lighting designer should work with the interior designer during the color selection process so that adequate lighting levels and luminance patterns on the room surfaces can be achieved. Lighting systems that provide nearly 100 percent direct illumination and very little intensity at angles above 50 percent such as downlights, louvers, deep cell parabolics, and refractive lenses help control direct glare and increase visual comfort probability. But these fixtures tend to bring the ceiling down perceptually, since they appear dark when looking across the ceiling and do not give high luminance patterns on the ceiling.

Source color is also very important to psychological response. Sources rich in the red tones, such as incandescent and high-pressure sodium, help to make the space feel more intimate and relaxed. Some studies have shown that this feeling increases productivity and employee comfort. However, in earth-sheltered buildings, the use of sources rich in the warm tones (especially when used with dark-colored surface finishes) tends to reinforce the underground associations of the space. A whiter source, richer in the cooler tones, increases perception of the openness and lightness of the room by visually causing the walls and ceiling to recede.

8. EMERGENCY LIGHTING. The emergency lighting system for an earth-sheltered building must conform to the requirements set forth in the Life Safety Code. Where there is no natural lighting, adequate egress lighting is critical in preventing panic, especially in multiple-story earth-sheltered buildings.

9. ELECTRICAL COOLING ENERGY AND FAN ENERGY. The electrical load for cooling will normally be lower for earth-sheltered buildings. Refer to Section 18, Energy Calculations, for HVAC requirements calculations. For the purpose of life-cycle costing electrical fan energy can be considered to be the same for comparable conventional and earth-sheltered buildings.

10. TRANSFORMER LOCATION. Transformer location is dependent upon the building type and the landscape design. Consideration must be given to the standard advantages and disadvantages of outdoor and indoor locations, including construction cost, maintenance, security, noise, and visual appearance.

Frequently, an exterior location will be more visually prominent, especially if the building is substantially buried. On the other hand, if the transformer is located within the building, provision of louvers for ventilation will restrict the extent of earth-contact. Also, an indoor location will be more expensive. Refer to NEC Code 1981, Article 450-2 for additional location requirements.

If the equipment room is on a below-grade lower level in a multistory building, there may be additional costs for primary cable and ventilation. If equipment space is made available on an upper level, vibration, and noise may be a problem. Flexible conduit connections and vibration-dampening mounting methods may be necessary.

11. SERVICE LOCATION. If an earth-sheltered building is fed by means of a unit substation located outdoors, the advantages and disadvantages will be similar to those of the transformer location outdoors. If the service entrance equipment is located indoors, as it is in most commercial buildings, special consideration must be given to the length of the secondary conductors, especially in multi-story, earth-sheltered buildings. For a deep earth-sheltered building, if the service entrance equipment is located on a lower level, additional cost may be incurred for running the secondary conductors down the entire depth of the building and then distributing upward as opposed to locating the service entrance equipment at near ground level and distributing in only one direction. In addition, the service disconnecting means, or service switch, must be accessible to firefighters. For additional criteria on service entrance equipment selection and transformer selection, refer to NAVFAC DM-4.2, Electrical Power Distribution Systems.

Power must be taken off ahead of the service disconnecting means if waste fire water pumps are required. (See Section 12, paragraph 6, Drainage.)

12. EMERGENCY POWER. An emergency power system--whether it is a second means of utilization voltage, emergency generators, or an Uninterruptible Power Supply (UPS) system, is normally dependent on the building occupancy and the nature of the work being done in the building. Earth-sheltered buildings tend to have more requirements for emergency power than conventional buildings. The most typical of these are emergency power for sump pumps and fire protection systems. For design criteria, refer to NAVFAC DM-4.4, Electrical Utilization Systems.

a. Auxiliary Power Source. If the choice is made to utilize an auxiliary method of feeding the service entrance switchgear, such as a utility-supplied power source or Navy-generated power source, there would be very little difference between the earth-sheltered building requirements and the conventional building requirements. Additional costs may be incurred depending on the location of the transformer and service entrance equipment as discussed above in Section 3.2.

b. Emergency Generators. The considerations for the location of the emergency generator are similar to those for the high-voltage transformer. If the emergency generator is located on the roof of the building, there may be additional structural requirements due to the weight and vibration of the generator. If the generator is to be located indoors, close proximity to the switchgear reduces the chance of voltage drop. Provisions for combustion air, adequate ventilation, and exhaust from the generator, will tend to conflict with maximizing the extent of earth-contact, however.

c. Uninterruptible Power Supply. For requirements of UPS systems and critical loads, refer to NAVFAC DM-4.4, Electrical Utilization Systems, NAVFAC DM-4.1, Preliminary Design Considerations, and the National Electrical Code, NFPA 70, Article 701, (see Criteria Sources). The UPS system will normally be located indoors, adjacent to equipment that must be supplied with uninterruptible power. Special provisions may be needed for the structural support of the batteries unless the UPS system is located on the lower level. However, in a deep underground location, costs will tend to be higher for ventilation of the batteries. Refer to DM-4.1 for "Legally Required Standby Systems."

13. FIRE ALARM. Early fire detection warning systems are critical in most earth-sheltered buildings due to their limited means of egress. For requirement criteria, refer to NAVFAC DM-8, Fire Protection Engineering.

14. COMMUNICATIONS. Refer to NAVFAC DM-4.7 for criteria and design methods for telephone and wired communication systems. Service entrances for telephone and communications will tend to be most economical if located on a floor closest to grade (noise and vibration will not be a concern in this case).

Section 18. ENERGY CALCULATIONS

1. GENERAL. This section presents an overview of heat transfer principles as they relate to earth-sheltered buildings. Heat flux data is included here and in Appendix A to aid the system designer in sizing mechanical systems and computing the total annual space conditioning requirements of earth-sheltered buildings. Sample load calculation procedures are presented

2. OVERVIEW OF HEAT TRANSFER PRINCIPLES IN BELOW-GRADE STRUCTURES. Heat transfer processes within the ground are extremely complex for the following reasons: (1) thermal conductivity of the soil is not easily established with precision, (2) thermal inertia of the earth tends to delay and dampen change in outdoor weather conditions, (3) heat flow through below-grade walls does not occur between parallel planes as in above-grade walls, but involves thermal movement in a three-dimensional environment, and (4) moisture migration through the soil is difficult to quantify. These interrelated factors make a precise solution difficult. However, all of these must be considered to obtain reasonably-accurate heat transfer predictions.

a. Thermal Conductivity of Soils. Conductivity values for soils range from about 0.30 to 1.30 Btu/h ft °F (0.52 to 2.25 W/m °C) depending on factors such as soil type, density, and moisture content. Fine-grained soils such as clay tend to have lower values than granular materials. Increasing the moisture content will increase the conductivity. In general, the soil surrounding an earth-sheltered building will tend to have conductivities between 0.70 and 0.90 Btu/h ft °F (1.21 and 1.56 W/m °C). Refer to Appendix D for a listing of soil thermal conductivity factors for various soil types and moisture contents.

New, simplified methods are being developed for determining on-site thermal conductivity of soils by use of "Conductivity Probes." See Reference 12, "The Measurement of Apparent Thermal Conductivity of Soils in the Field and Laboratory Using Thermal Conductivity Probes," by the Department of Energy.

b. Temperature Regime Around An Underground Structure. In the simple condition of steady-state heat flow between two parallel surfaces at different temperatures (as for a wall above grade) the paths of heat flow are parallel lines at right angles to the surfaces. When the two surfaces are not parallel but at an angle to one another (as for a below-grade wall) the paths of heat flow are curvilinear. Heat flow from below-grade surfaces is three dimensional in nature, especially at building corners. However, the heat flow can be represented with reasonable accuracy in two dimensions when the surface in question is substantially removed from the corners of the building.

The two-dimensional heat loss around an uninsulated basement in Minneapolis for typical February and August conditions is illustrated in Figures 55 and 56. Heat flow is perpendicular to the isotherms. Heat loss in February is greatest near the ground surface where the isotherms are closely spaced.

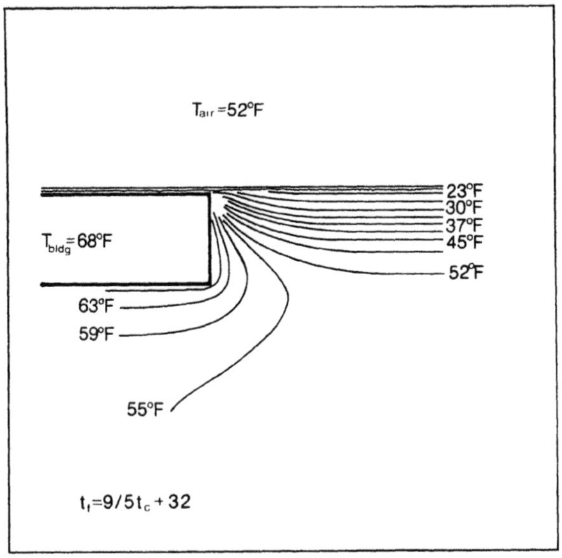

FIGURE 55
Isotherm Distribution Around an Uninsulated Underground Building for Typical February Conditions in Minneapolis

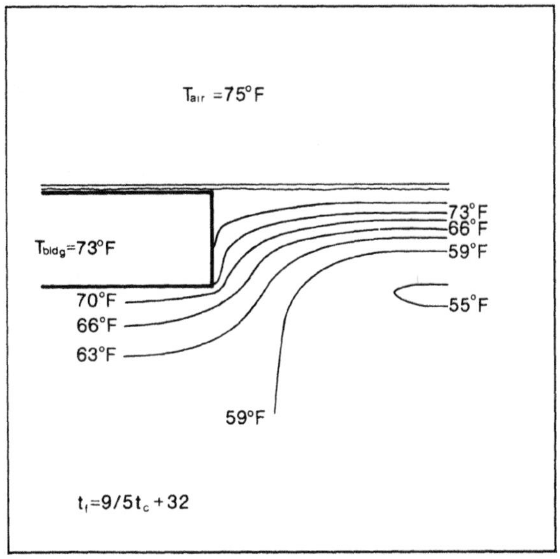

FIGURE 56
Isotherm Distribution Around an Uninsulated Underground Building for Typical August Conditions in Minneapolis

c. Simplified Heat Conduction for Undisturbed Earth. The earth temperature oscillation near the surface varies cyclically with the annual weather cycle. The variation is almost sinusoidal and is reflected below grade by an amplitude that decreases with increasing soil depth until at about 30 to 50 feet (9 to 15 m), the temperature remains essentially constant throughout the year at a value called the annual average earth temperature.

Usually the mathematical treatment of earth temperatures for undisturbed earth starts with the assumptions that: (1) the earth is a homogeneous heat-conducting medium of a semi-infinite solid system, that is, the thermal diffusivity is constant throughout the temperature domain; (2) the temperature of the surface exposed to the atmosphere varies periodically with time. For methods or predicting ground temperatures for undisturbed soils, refer to "Regional Analysis of Ground and Above-Ground Climate," in Underground Space, by Kenneth Labs (see Reference 1).

Because of the thermal diffusivity, subsoil temperature variations lag more and more in time behind surface temperature as depth increases. At depths of 10 to 15 feet (3 to 5 m), this time lag is measured in months. Figure 57, which illustrates this time lag, is a plot of temperature values versus depth for various times of year in Ottawa, Canada. The coldest ground temperature at the 6 foot (2 m) depth occurs in April--a thermal lag of approximately 2 months from the coldest air temperature. The soil temperature lag in effect shifts the cold temperature pulses of the winter into the summer months where cooling benefits can be derived. The same principle also applies to the heating season; warm temperature pulses received from the previous summer months help moderate the buildings thermal environment.

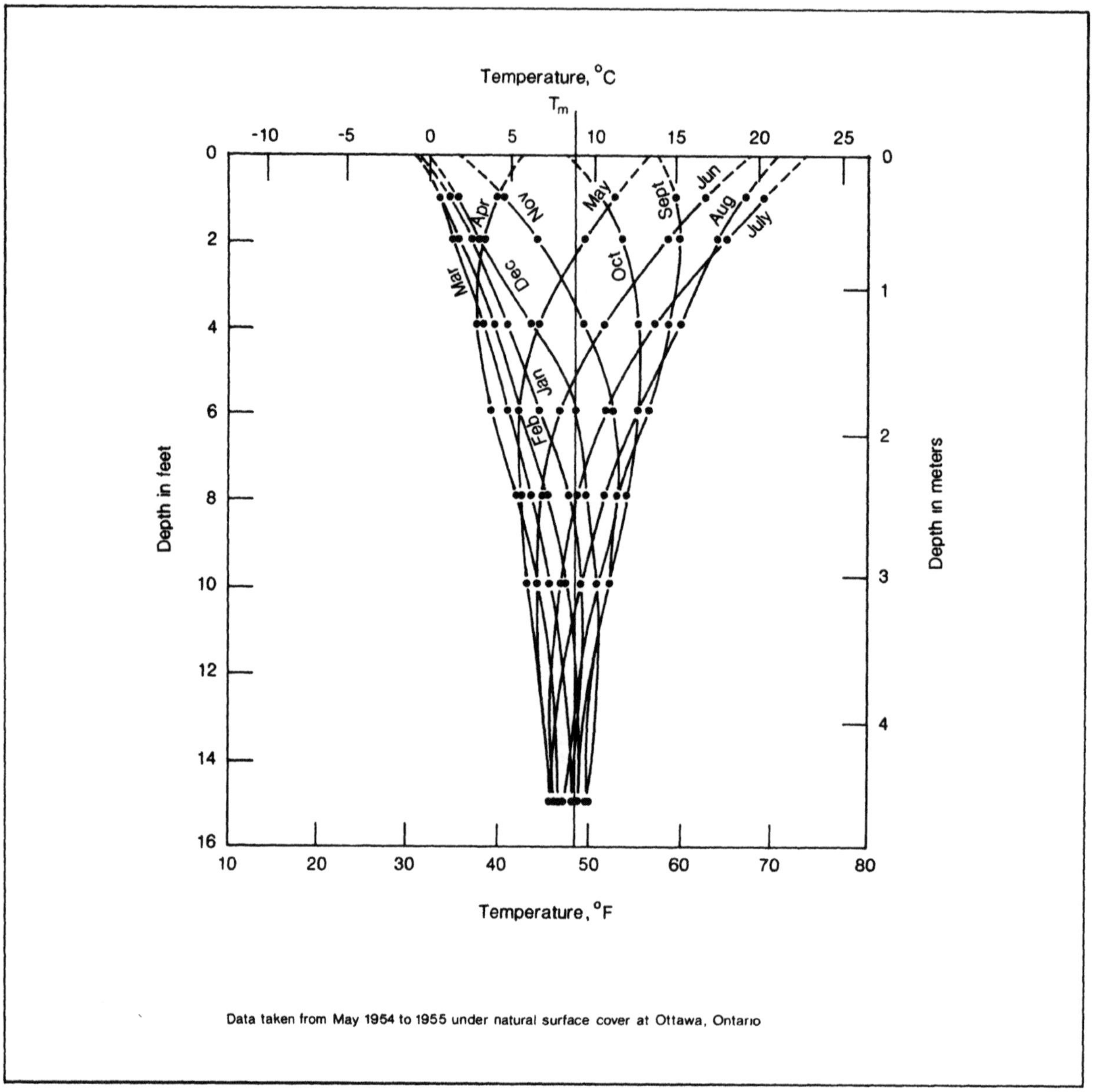

FIGURE 57
Monthly Average Ground Temperatures (Ottawa, Ontario)

3. CALCULATION PROCEDURES. Several limitations are inherent in hand calculation procedures. These are discussed in the following subsections. A transient, two-dimensional finite-difference computer program for predicting heat flow in below-grade structures is also presented.

a. Existing Earth-Contact Methods. The 1981 ASHRAE Handbook of Fundamentals, referenced by NAVFAC DM-3.3, Heating, Ventilating, Air-Conditioning and Dehumidifying Systems (see Reference 13, ASHRAE Handbook of Fundamentals), recommends the use of an approximate graphical procedure developed by Boileau and Latta (see Reference 14, Calculation of Basement Heat Losses, by Boileau) for estimating the winter heat loss from

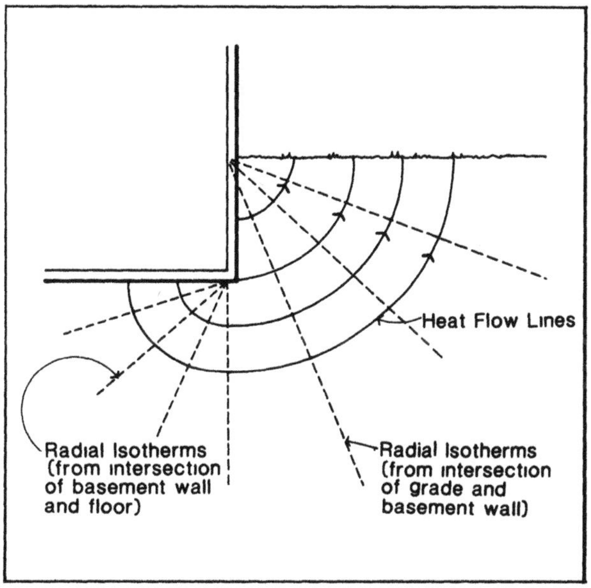

FIGURE 58
Radial Isotherms and Circular Heat Flow Lines
Assumed by ASHRAE/Boileau and Latta

below-grade basement walls and floors. Boileau and Latta suggest that this procedure is a reasonable approximation only for the nearly steady-state mid-winter conditions. The bases for the ASHRAE/Boileau-Latta procedure are the assumptions that: (1) heat flows along circular paths perpendicular to radial isotherms from the basement wall to the surface of the ground (see Figure 58), (2) the soil behaves as a low-mass insulator with relatively low resistance, and (3) the earth is a homogeneous heat conducting medium.

Figures 59 and 60 present heat flux data for both an uninsulated and an insulated basement wall in Minneapolis for a typical February day. Two sets of data are plotted: one from the ASHRAE/Boileau-Latta calculation procedure, and the other from a transient two-dimensional finite-difference computer analysis.

Figure 59 demonstrates that good agreement is realized between the two methods when no insulation is utilized. This is graphically apparent in Figure 61. The isotherms (generated by the two-dimensional finite-difference model) conform to the Boileau-Latta method in that they are radial in nature and emanate from the point where the ground surface meets the basement wall. The graphic method predicts a slightly smaller wall heat loss than does the computer model. As explained in detail by Meixel, Shipp, and Bligh (see Reference 15, The Impact of Insulation Placement, by Meixel), this is largely due to the fact that the graphic procedure does not take into account the vertical gradient in the building's wall temperatures and the resultant heat flux. This error increases dramatically when insulation is introduced. Figure 62 is a plot of the computer-predicted profile around the basement wall when R-21.0 insulation is applied to the exterior. The shape of the isotherms is altered dramatically from the no-insulation case. In this case, the isotherms depart from the Boileau-Latta model where the isotherms radiate from the intersection of the ground surface and basement wall. In particular, the

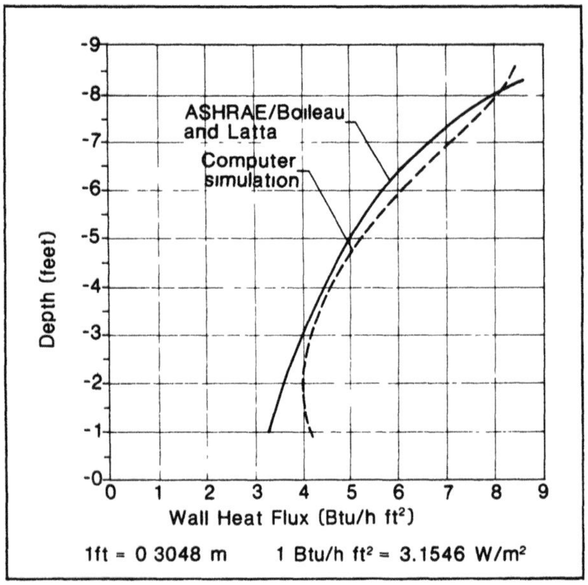

FIGURE 59
Comparison of ASHRAE and Computer-Predicted Heat Flux for an Uninsulated Basement Wall

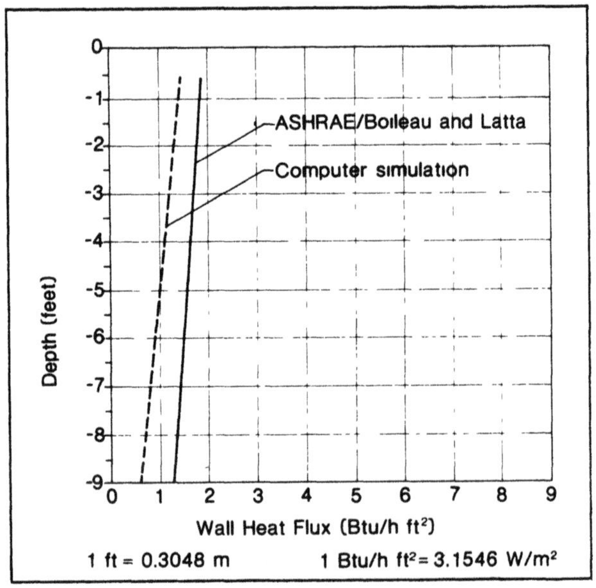

FIGURE 60
Comparison of ASHRAE and Computer-Predicted Heat Flux for Basement Wall with R-21 Insulation

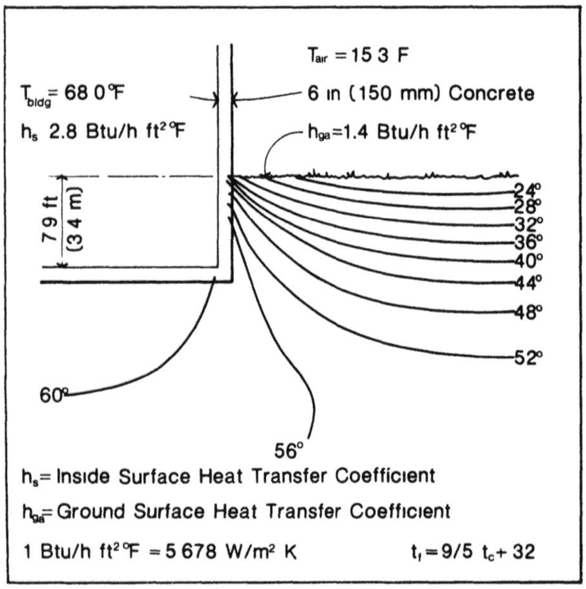

FIGURE 61
Computer-Predicted Temperature Profiles Around an Uninsulated Basement for Typical February in Minneapolis

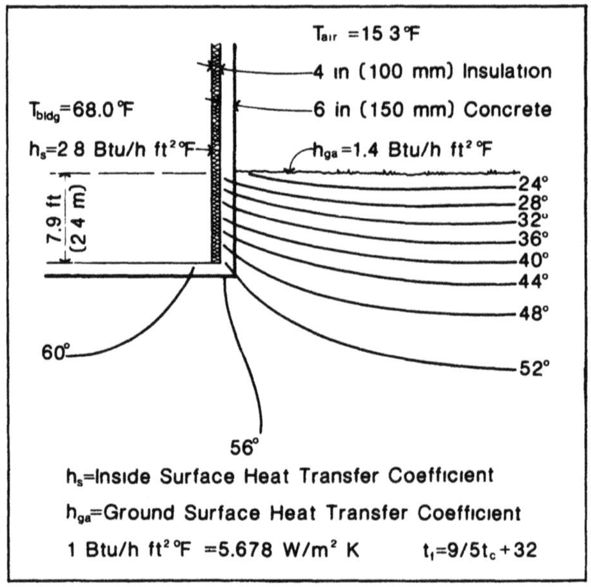

FIGURE 62
Computer-Predicted Temperature Profiles Around Basement with R-21 Insulation for Typical February in Minneapolis

dominant heat flux is in a vertical direction along the wall. Consequently the Boileau-Latta method overestimates the wall heat loss as depicted in Figure 62.

The inability of the ASHRAE/Boileau-Latta method to adequately model insulated basement walls, as well as the seasonal thermal responses of basement walls (since the ASHRAE technique is only capable of predicting average winter heat loss), clearly demonstrates the need for more sophisticated procedures. The Boileau-Latta method can be in error by as much as 30 percent. Computer simulations are required to accurately predict the basement wall heat fluxes for each day of the year. A transient two-dimensional finite-difference computer model is able to yield detailed results of thermal performance which are required for optimizing both insulation and soil mass. Note that existing simulation models, such as BLAST, NBSLD, and DOE2.1 currently do not account for these changes in soil temperature distribution and thus are unable to yield accurate results. Further, the ASHRAE/Boileau-Latta method is unable to estimate the beneficial wall heat losses during summer months.

b. Computer Model. The transient two-dimensional finite-difference computer program utilized in generating the Heat Flux Tables (Tables 21 through 42) presented in Appendix A is a direct extension of the program developed and documented by Speltz (see Reference 16, A Numerical Simulation of Transient Heat Flow, by Speltz, and Reference 17, The Thermal Benefits and Cost Effectiveness of Earth Berming, by Speltz).

The computer model solves the transient two-dimensional heat conduction equation in a Cartesian coordinate system using an explicit forward-differencing routine. To obtain the required accuracy in determining heat flux magnitudes and directions, node spacing is varied from 2 inches (50 mm) in regions of large temperature gradients to 24 inches (600 mm) in areas where small temperature gradients occur. The model considers a two-dimensional cross section of the building. By dividing the two-dimensional building down its vertical axis of symmetry, only half the building need be modeled and an adiabatic boundary condition is imposed at this axis as shown in Figure 63. The model is bounded by a second vertical adiabatic boundary sufficiently removed to minimize its impact on the heat loss of the structure. At the bottom of the calculation region, the temperature is set at a constant deep-ground temperature.

The boundary conditions at the interior (earth-contact) wall and floor surfaces are determined by calculating the convective and radiative heat flux between the specified air temperature of the building and the computed inside surface temperatures. A combined convective and linearized radiation heat transfer coefficient is used in this calculation (see Reference 18, Sensitivity of Room Thermal Response to Inside Radiation Exchange and Surface Conductance, by Buchberg). The boundary conditions at the ground surface provide for convective interchange with the air as well as radiative (solar and infrared) and evaporative exchange with the atmosphere. (For a detailed description of the heat balance at the ground surface, refer to Appendix B.) Changes in moisture content and heat transfer by convection and moisture migration in the soil are not explicitly calculated. Baladi (see Reference 19, Transient Heat and Mass Transfer in Soils, by Baladi) has shown that the pure heat conduction model is relatively accurate for dry and moist soils with calculation errors increasing as the moisture content and the soil porosity increases.

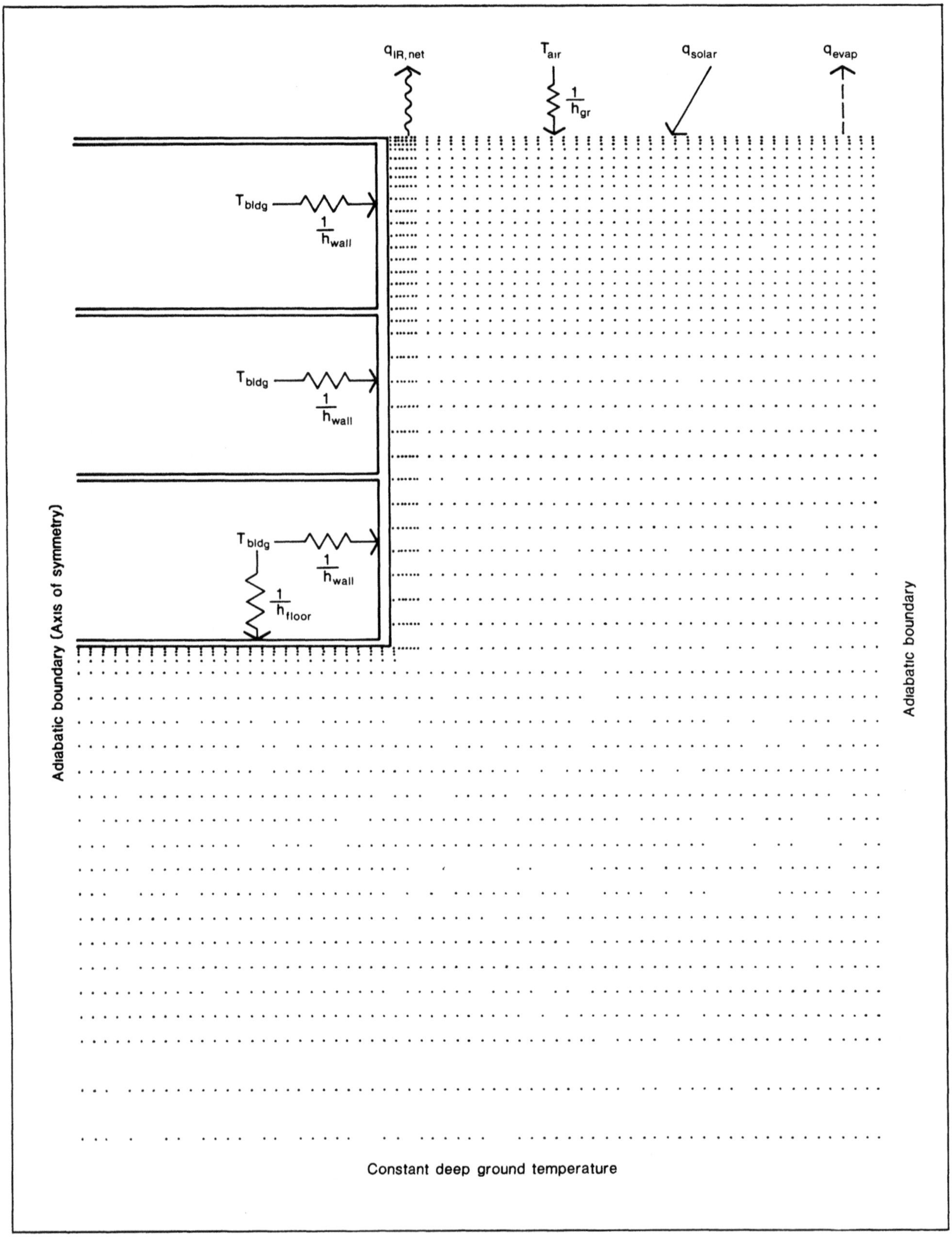

FIGURE 63
Distribution of Two-Dimensional Temperature Nodes and Boundary Conditions for Full-Berm, 3-story Building

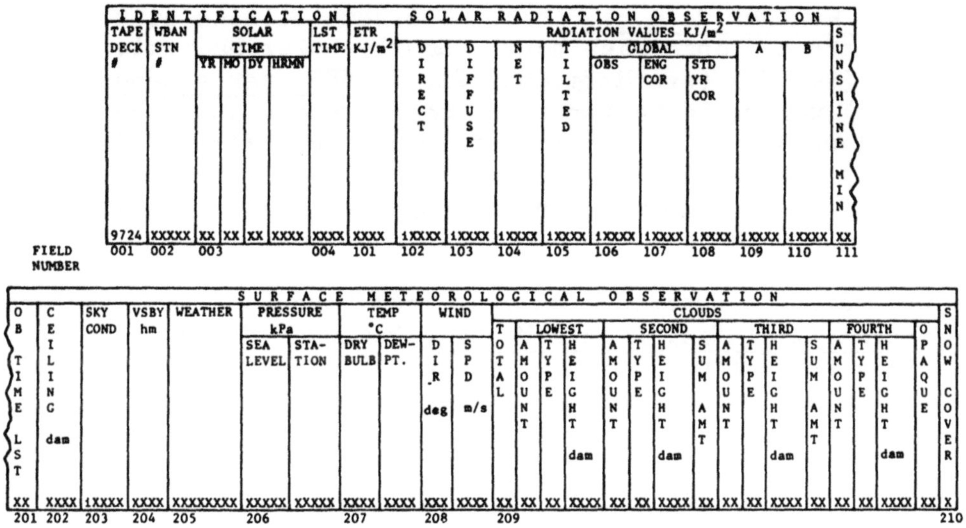

TAPE FIELD NUMBER	TAPE POSITIONS	ELEMENT
001	001 - 004	TAPE DECK NUMBER
002	005 - 009	WBAN STATION NUMBER
003	010 - 019	SOLAR TIME (YR, MO, DAY, HOUR, MINUTE)
004	020 - 023	LOCAL STANDARD TIME (HR AND MINUTE)
101	024 - 027	EXTRATERRESTRIAL RADIATION
102	028 - 032	DIRECT RADIATION
103	033 - 037	DIFFUSE RADIATION
104	038 - 042	NET RADIATION
105	043 - 047	GLOBAL RADIATION ON A TILTED SURFACE
106	048 - 052	GLOBAL RADIATION ON A HORIZONTAL SURFACE - OBSERVED DATA
107	053 - 057	GLOBAL RADIATION ON A HORIZONTAL SURFACE - ENGINEERING CORRECTED DATA
108	058 - 062	GLOBAL RADIATION ON A HORIZONTAL SURFACE - STANDARD YEAR CORRECTED DATA
109, 110	063 - 072	ADDITIONAL RADIATION MEASUREMENTS
111	073 - 074	MINUTES OF SUNSHINE
201	075 - 076	TIME OF COLLATERAL SURFACE OBSERVATION (LST)
202	077 - 080	CEILING HEIGHT (DEKAMETERS)
203	081 - 085	SKY CONDITION
204	086 - 089	VISIBILITY (HECTOMETERS)
205	090 - 097	WEATHER
206	098 - 107	PRESSURE (KILOPASCALS)
207	108 - 115	TEMPERATURE (DEGREES CELSIUS TO TENTHS)
208	116 - 122	WIND (SPEED IN METERS PER SECOND)
209	123 - 162	CLOUDS
210	163	SNOW COVER INDICATOR

FIGURE 64
SOLMET Tape Format

Note that the surface boundary condition model has been partially validated by comparing ground temperatures predicted by the model to experimental data recorded by the National Bureau of Standards for undisturbed soil plots in Washington, D.C. (see Reference 20, National Bureau of Standards Report, "Earth Temperature Beneath Five Different Surfaces").

The model requires input of hourly dry-bulb and wet-bulb temperatures, barometric pressure, wind velocity, cloud cover, direct normal, and diffuse solar radiation for a one-year period. The following documents these inputs:

c. Hourly Weather Data - SOLMET. The heat balance at the ground-air interface is performed once each hour. Hourly Solar Radiation-Meteorological (SOLMET) (see Reference 21, SOLMET User's Manual, National Climatic Center) weather data input is read directly into the computer program and allows an explicit calculation of the following heat flux components: (1) convection-heat transfer, (2) solar heat gain, (3) infrared radiation exchange with the atmosphere, and (4) latent heat of evaporation from ground cover.

The typical meteorological year data base (TMY) (see Reference 22, Typical Meteorological Year User's Manual, National Climatic Center) used in the simulations represent a 30-year average of solar insolation and meteorological data and is compiled directly from the SOLMET data base. (Typical months for the 30-year period were selected and then spliced together to make up one typical year--smoothing functions were used where each month was spliced together.) Figure 64 identifies the various insolation and weather parameters stored on the tape.

4. SUMMARIES AND HEAT FLUX TABLES. A principal concern of the Navy in utilizing earth-sheltered building design is return on investment. Consistent with this economic emphasis, this section provides tables for determination of below-grade heat flux as a function of climate, insulation thickness and depth, and degree of earth-sheltering. The primary objective of these tabulated results is to enable the system designer to predict with reasonable accuracy the monthly peak heat loads of earth-contact surfaces, thus allowing HVAC equipment to be sized. A secondary objective of the tabulation is to facilitate an overall understanding of the potential for energy conservation through earth-sheltered construction techniques--that is, to quantify the seasonal energy performance of below-grade walls and slabs for varying levels of earth-sheltering, insulation thickness and depth

These procedures should not be misconstrued as the definitive solution to predicting thermal performance of below-grade surfaces. If accurate thermal performance data is required for a specific building configuration and/or location, separate detailed hour-by-hour computer simulations must be carried out. Also, note that analysis methods are still under development. For example, the boundary condition at the ground surface is still being considered.

a. Description of Earth-Sheltered Building Configurations Modeled. Four building configurations were modeled: (1) slab-on-grade, (2) half-berm, (3) single-story full berm, and (4) three-story full berm. The configurations were selected so as to represent the full range of typically encountered earth-sheltered buildings. The four configurations were modeled under identical interior, weather, and soil conditions. Only the insulation thickness and depth is varied in the simulations (Refer to Figure 102 for an index which tabulates the simulations according to building location, installation thickness, and depth). In Section 6 a parametric analysis is carried out to determine the sensitivity of the building's thermal performance to soil type. Cross sections through the four building configurations are illustrated in Figures 65 through 68. These figures show the location of construction materials and dimensions.

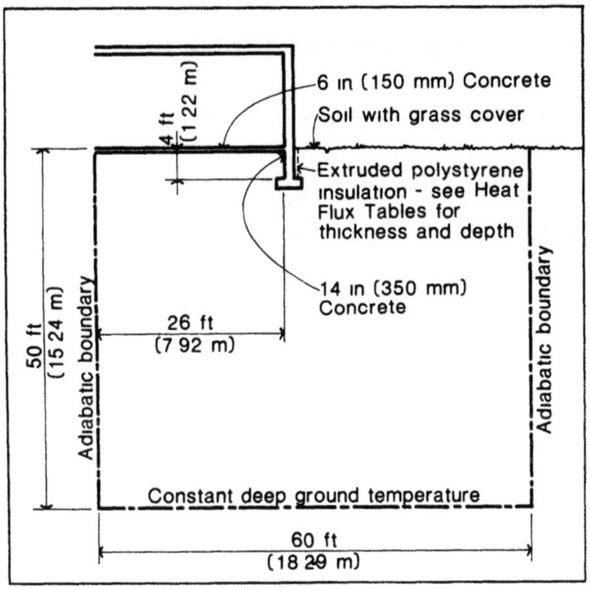

FIGURE 65
Schematic Cross-Section for Slab-on-Grade Building

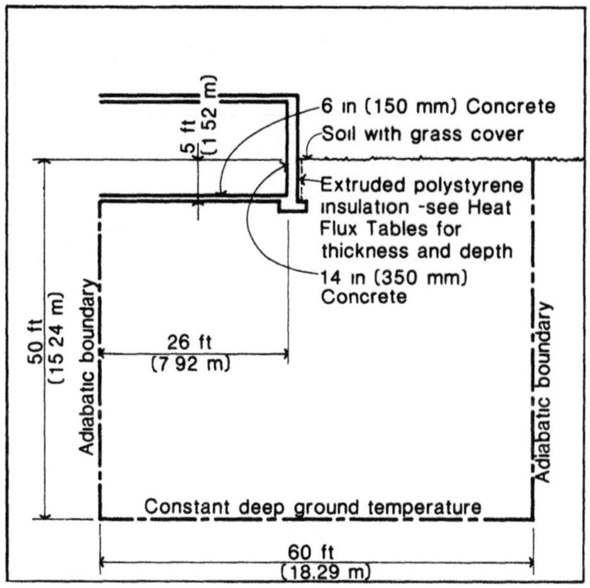

FIGURE 66
Schematic Cross-Section for Half-Berm Building

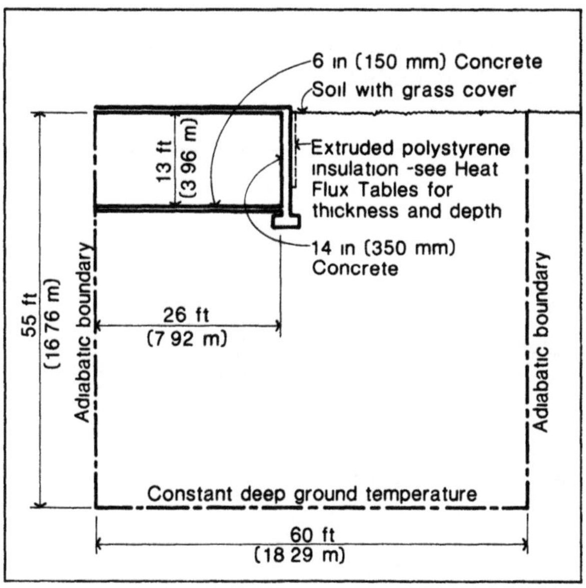

FIGURE 67
Schematic Cross-Section for Full-Berm Building

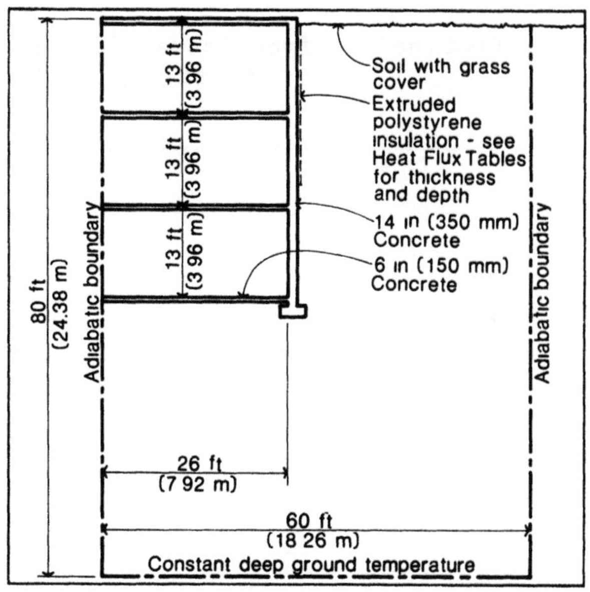

FIGURE 68
Schematic Cross-Section for Full-Berm, 3-Story Building

TABLE 5
Building and Soil Properties Used in Computer Simulation

MATERIAL		Density lb/ft^3	Density kg/m^3	Specific Heat Btu/(lb °F)	Specific Heat kJ/(kg °C)	Thermal Conductivity Btu/(ft h °F)	Thermal Conductivity W/(m °C)
Concrete		150.0	2403	0.230	0.963	0.900	1.558
Extruded Polystyrene		2.0	32	0.270	1.130	0.016	0.028
Soil:	Wet	120.0	1922	0.275	1.151	1.400	2.423
	Average	114.0	1826	0.275	1.151	1.100	1.904
	Dry	100.0	1602	0.275	1.151	0.400	0.692

TABLE 6
Combined Surface Convection and Linearized Radiation Heat Transfer Coefficients Used in Computer Model

Boundary at Building Interior	Direction of Heat Flow	Surface Conductance Btu/ft^2 h °F	Surface Conductance W/m^2 °C
Floor	Upward	0.78	4.4
	Downward	0.19	1.1
Wall		0.60	3.4

b. Input Parameters. Refer to Tables 5 through 8 for a summary of building and soil properties, heat transfer coefficients and building interior dry-bulb temperatures, and constant deep ground temperatures input into the computer model.

c. Heat Flux Tables. Computer simulation results of the dynamic thermal performance of the four building configurations analyzed are

TABLE 7
Specified Building Interior
Dry-Bulb Air Temperatures Used in Computer Model

Building Location	Dry-Bulb Air Temperatures (°F) J	F	M	A	M	J	J	A	S	O	N	D
Minneapolis	70.0	70.0	70.0	71.0	72.0	73.0	75.0	74.0	72.0	70.0	70.0	70.0
Boston	70.0	70.0	70.0	71.0	72.0	73.0	75.0	74.0	72.0	70.0	70.0	70.0
Seattle	70.0	70.0	70.0	71.0	73.0	74.0	75.0	75.0	73.0	71.0	70.0	70.0
Kansas City	70.0	70.0	70.0	71.0	73.0	74.0	75.0	75.0	73.0	71.0	70.0	70.0
Albuquerque	70.0	70.0	70.0	71.0	73.0	74.0	75.0	75.0	73.0	71.0	70.0	70.0
Phoenix	70.0	72.0	73.0	74.0	75.0	75.0	75.0	75.0	75.0	73.0	72.0	71.0
Miami	71.0	72.0	73.0	74.0	75.0	75.0	75.0	75.0	75.0	74.0	73.0	72.0

$1°F = 9/5°C + 32$

TABLE 8
Constant Deep Ground Temperatures
Used in Computer Model

Location	Deep Ground Temperature °F	°C
Minneapolis	45.1	7.28
Boston	52.3	11.28
Seattle	52.1	11.17
Kansas City	54.7	12.61
Albuquerque	57.8	14.33
Phoenix	71.3	21.83
Miami	76.5	24.72

presented in Tables 21 through 42. Henceforth, these results will be referred to as the "Heat Flux Tables." These are located in Appendix A. Seven U.S. cities have been analyzed:

- Albuquerque
- Kansas City
- Minneapolis
- Seattle
- Boston
- Miami
- Phoenix

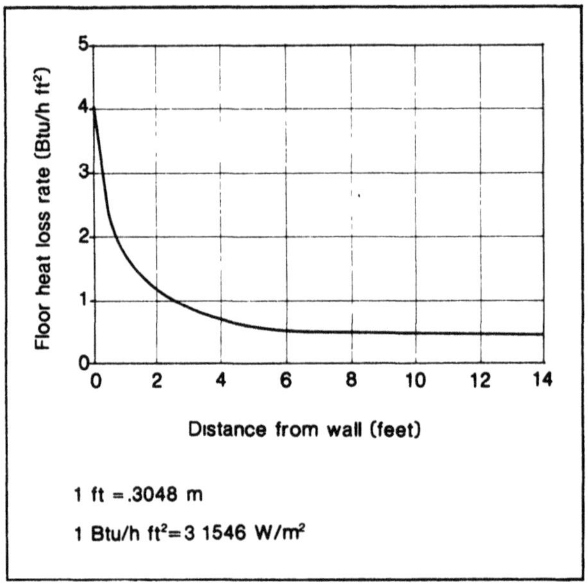

FIGURE 69
Floor Heat Loss for January, Minneapolis

The Heat Flux Tables are categorized by building location, insulation thickness, and depth. On the horizontal axis of the tables, the data is subdivided by surface type. The vertical axis is subdivided by building configuration. Average monthly heat fluxes, in addition to the monthly peak heat losses and gains, for each building surface are presented. The last two columns of each category--designated "HRS HT ON" and "HRS CL ON"--indicate the number of hours heat was lost or gained from the building surface.

The heat flux values presented for the walls represent the average flux for the entire surface (that is, average flux from floor to ceiling). The data for the slab is slightly more complex to interpret. The heat flux data under the heading "slab perimeter" represents the average heat loss or gain for the first 20 feet (6 m) of the slab. The "slab interior" represents the conditions at the slab's centerline (axis of symmetry).

To properly use the Heat Flux Tables, it is important for the user to understand the dynamic thermal performance of the slab-on-grade. For instance, consider a slab which is relatively large in width--approximately 50 feet (15 m) in width or greater. Because of the slab's large floor area, the heat loss, except near the perimeter, is not strongly coupled to the ground surface, but rather establishes a quasi-steady-state heat transfer rate with the deep-ground environment. Consequently, the thermal performance of large slabs can be described with reasonable accuracy by representing the slab at two regions: (1) the slab perimeter, outer 10 to 20 feet (3 to 6 m), and (2) the slab interior. This point is graphically depicted in Figure 69. As one moves inward from the slab perimeter, heat loss becomes essentially constant at the 10-foot (3-m) mark. On a residential scale, heat loss occurring at the corner of the building becomes significant when compared with the total loss of the earth contact slab and walls. As a consequence, the above strategies do not apply to small buildings.

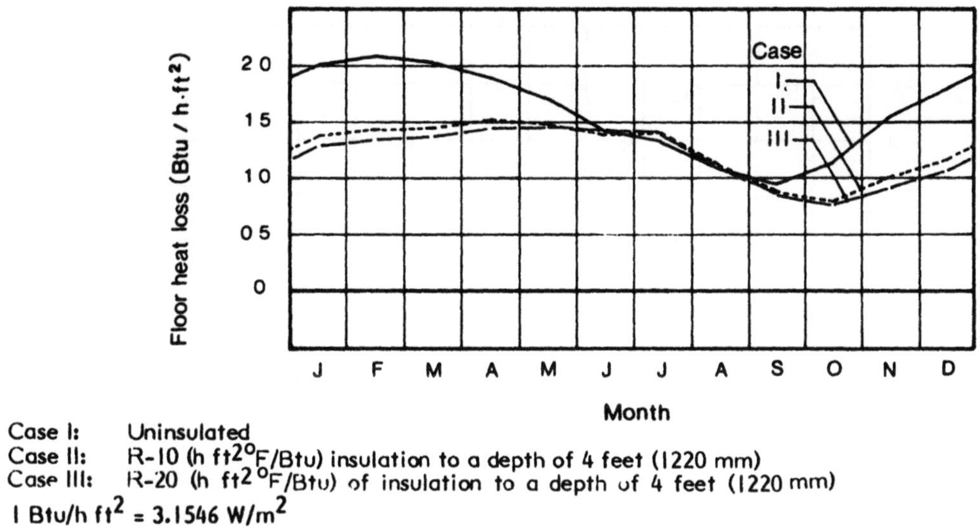

FIGURE 70
Annual Floor Heat Loss Minneapolis Slab-on-Grade

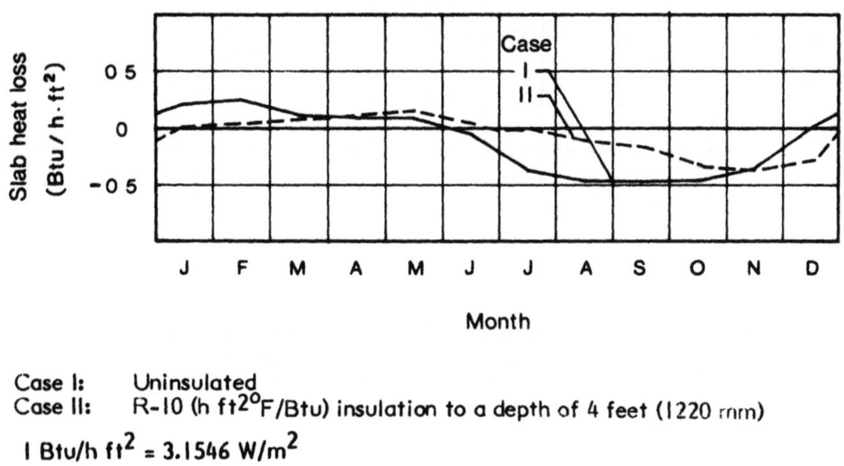

FIGURE 71
Annual Floor Heat Loss/Gain Miami, Slab-on-Grade Building

Figure 102 in Appendix A identifies seven major climatic regions of the U.S. Each region is represented by one of the seven cities analyzed. To obtain heat flux data for a particular location, the designer should: (1) find the region from Figure 102 which his building site falls within, (2) identify the representative city for that region (indicated in Figure 102), (3) refer to the appropriate heat flux tables in Appendix A for that city, and (4) use the Heat Flux Table with the nearest R-value for the insulation. It is important that the designer understand that this tabulated data only represents a set of specific climatic conditions and is

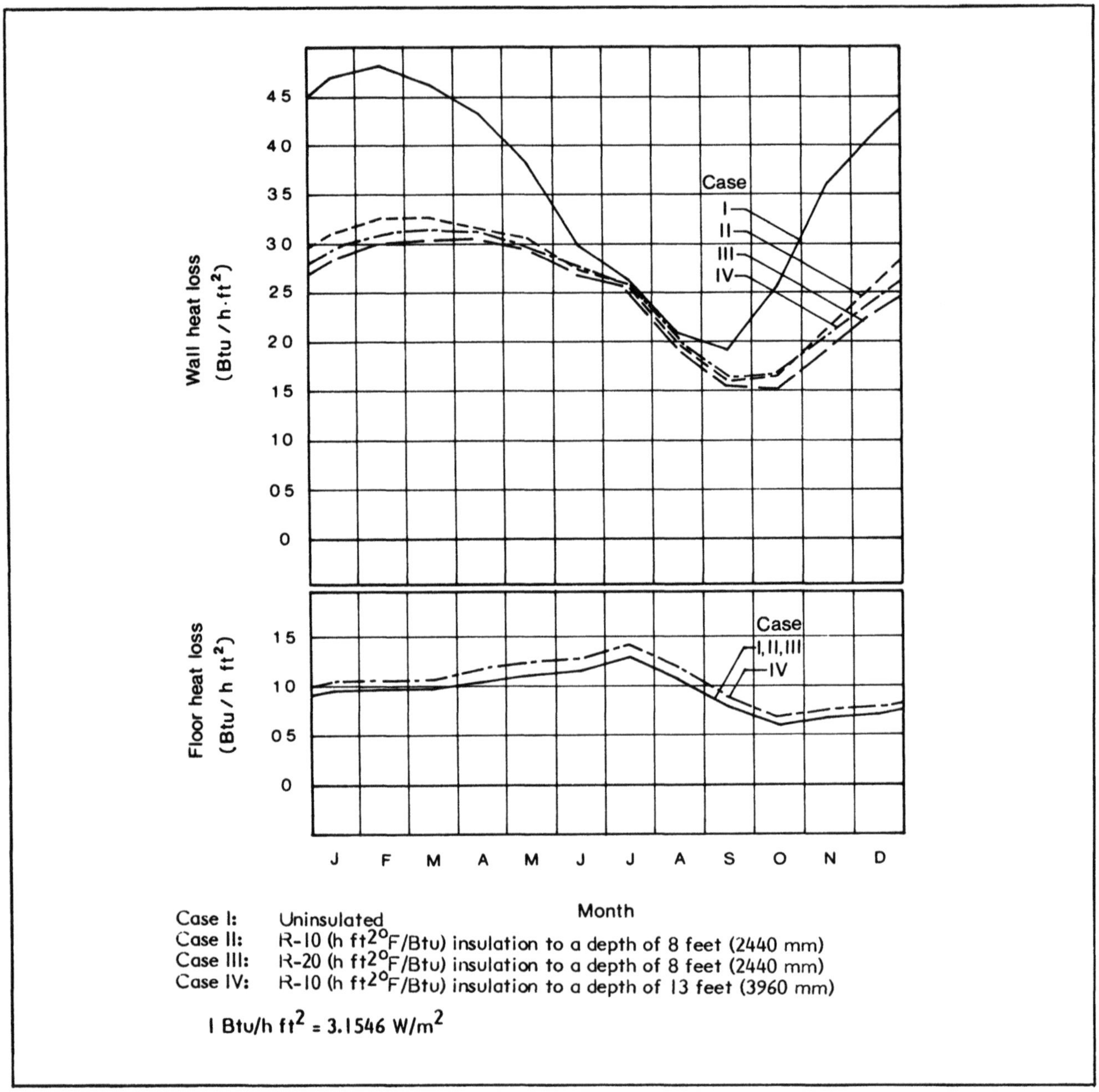

FIGURE 72
Annual Wall and Floor Heat Loss Minneapolis,
Full Berm, 1-Story Building

not intended to represent all microclimatic regions of the United States. However, the seven cities which were analyzed will provide a data base which is broad enough to: (1) optimize insulation thickness and depth, (2) provide heat flux information for sizing HVAC equipment, and (3) compute the integrated annual heat loss and gain of earth-contact surfaces.

d. Results and Discussion. Wall and floor thermal performance results for the slab-on-grade, full-berm one-story, and full-berm three-story buildings in Minneapolis and Miami are discussed.

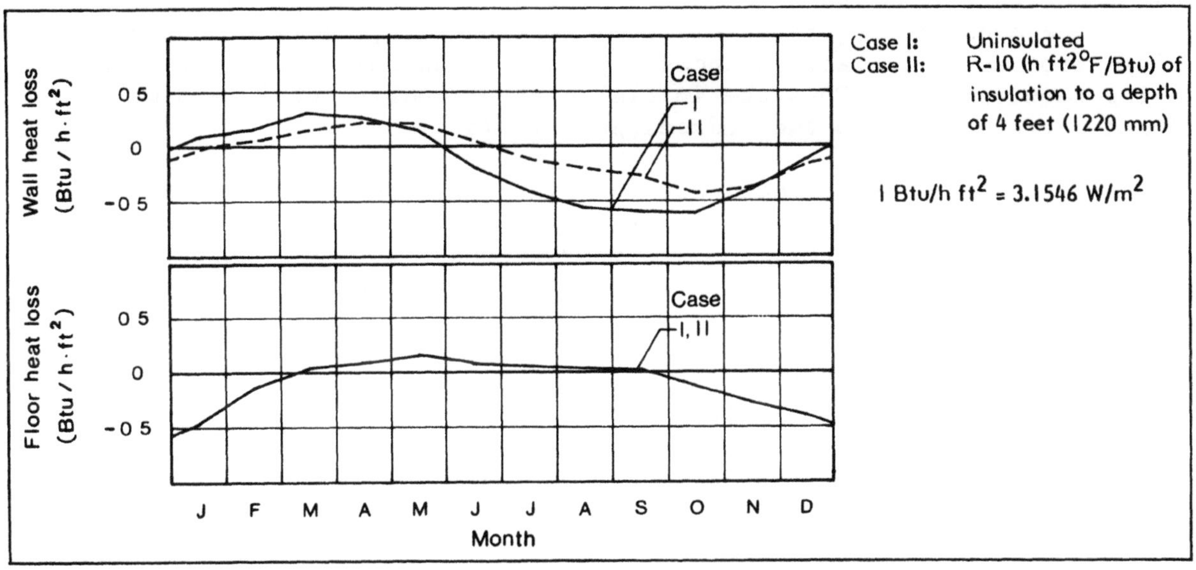

FIGURE 73
Annual Wall and Floor Heat Loss/Gain Miami,
Full Berm, 1-Story Building

Figure 70 shows that adding perimeter insulation to the slab-on-grade building in Minneapolis significantly reduces floor heat losses. In Miami, perimeter insulation reduces floor heat losses in the winter and gains in the summer (see Figure 71). However, these benefits are not nearly as significant as in the northern climates. A cost-benefit analysis would have to be carried out to determine if the reduction in building heating and cooling requirements offsets the capital investment in insulation.

Heat loss through the earth-covered walls for the full-berm one-story building in Minneapolis is substantially reduced by adding 2 inches (50 mm) of insulation to the exterior of the wall to a depth of 8 feet (2.4 m). See Figure 72. Additional levels of insulation produce only marginal improvements in thermal performance. Slab thermal performance is nearly identical for cases I, II and III. However, a significant reduction in slab thermal performance is observed in case IV. In this case, heat loss from the lower portion of the foundation wall is reduced by carrying insulation down to the footings, thus producing colder ground temperature. These colder ground temperatures increase heat loss from the slab. Insulation produces only marginal improvements in thermal performance in Miami (see Figure 73).

Several interesting points can be drawn from Figures 74 and 75 for the full-berm three-story building. First, ground coupling effects significantly reduce the amplitude of the wall's heat loss profile as one moves down from the upper to the lower levels. Second, the time-lag effects for the wall heat loss at levels two and three of the building are substantial. The lag can be months in duration. Peak heat losses do not occur until mid-July in Minneapolis. The thermal lag exhibited by cases I and II in Miami, demonstrate the beneficial effects of ground coupling. Slab heat loss offsets cooling requirements during the summer months while heat gains offset heating requirements during the winter months. However, these benefits may not be that significant when compared to the building's total space conditioning requirements. Nevertheless, cooling benefits derived from earth-contact surfaces may be very useful in reducing peak cooling requirements. Thus the initial cost of cooling equipment may be reduced corresponding to the reduced peak loads.

In general, the floor heat loss/gain has the maximum ground coupling of all the building configurations analyzed. The slab heat loss is very sensitive to the interior air temperature due to the relatively constant ground temperature compared with the variable outdoor air temperature.

5. SAMPLE LOAD CALCULATION PROCEDURES. This section presents methodologies for estimating the thermal performance of earth-sheltered buildings. Results of the detailed computer simulations, presented in the Heat Flux Tables provide a basis for estimating peak loads and annual energy performance of earth-contact surfaces.

a. Sizing of Heating and Cooling Systems.

(1) Space Heating Load. Heat loss calculations consist of estimating the maximum probable heat loss of each room or space to be heated while maintaining a selected indoor air temperature during periods of design outdoor weather conditions. Heat losses consist primarily of:

- o Transmission Losses. Heat transferred through exterior walls, windows, roof, floor, and earth-contact surfaces.
- o Infiltration Losses. Energy required to warm outside air which leaks through cracks around doors and windows.
- o Ventilation Requirements. Energy required to warm outside air to room temperature for ventilation purposes.

(a) Calculation of Transmission Heat Loss for Above-Grade Surfaces. See ASHRAE criteria (Reference 13) for tables and procedures for computing overall heat transmission coefficient for building materials for walls, floors, ceilings, glass, and doors. For all facilities, use outside design temperatures from 97-1/2 percent column in Army, Navy and Air Force manual, Engineering Weather Data, NAVFAC P-89 (see Criteria Sources). For inside design temperatures, refer to NAVFAC DM-3.3 (Chapter 3, Heating Systems).

(b) Calculation of Transmission Heat Loss through Earth-Contact Walls. To estimate the transient heat loss through basement and earth-contact walls, the Heat Flux Tables located in Appendix A must be used. Heat loss is estimated by means of:

$$q = (A)\ (LS_1) \qquad (1)$$

WHERE:

q = heat transfer through the earth-contact wall (Btu/h)
A = area of earth-contact wall (ft^2)
LS_1= peak heat loss (see Heat Flux Tables) (Btu/h ft^2)

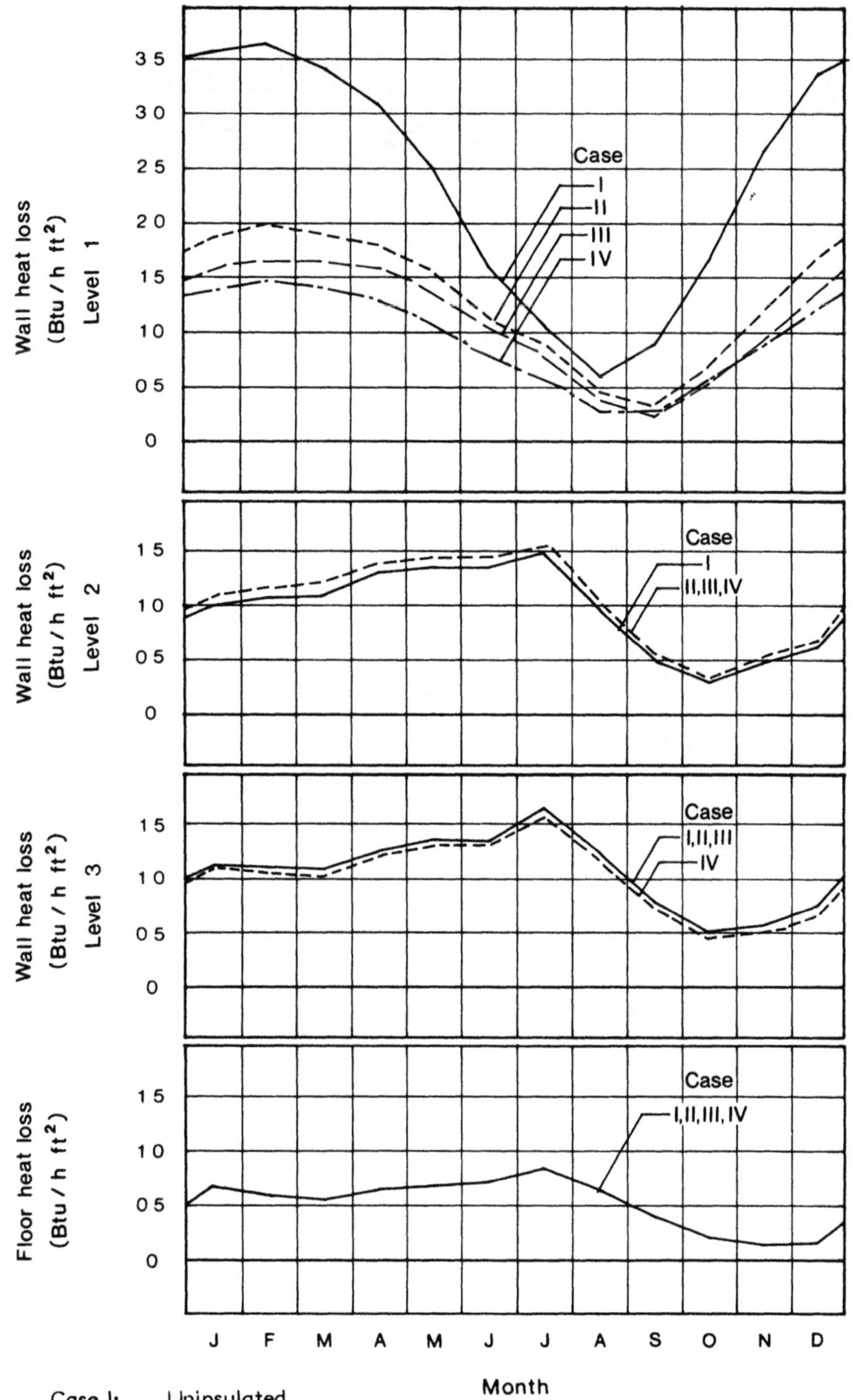

Case I: Uninsulated
Case II: R-10 (h ft^2°F/Btu) of insulation to a depth of 8 feet (2440 mm)
Case III: R-20 (h ft^2°F/Btu) of insulation to a depth of 8 feet (2440 mm)
Case IV: R-10 (h ft^2°F/Btu) of insulation to a depth of 16 feet (4880 mm)

1 Btu/h ft^2 = 3.1546 W/m^2

FIGURE 74
Annual Wall and Floor Heat Loss Minneapolis, Full Berm, 3-Story Building

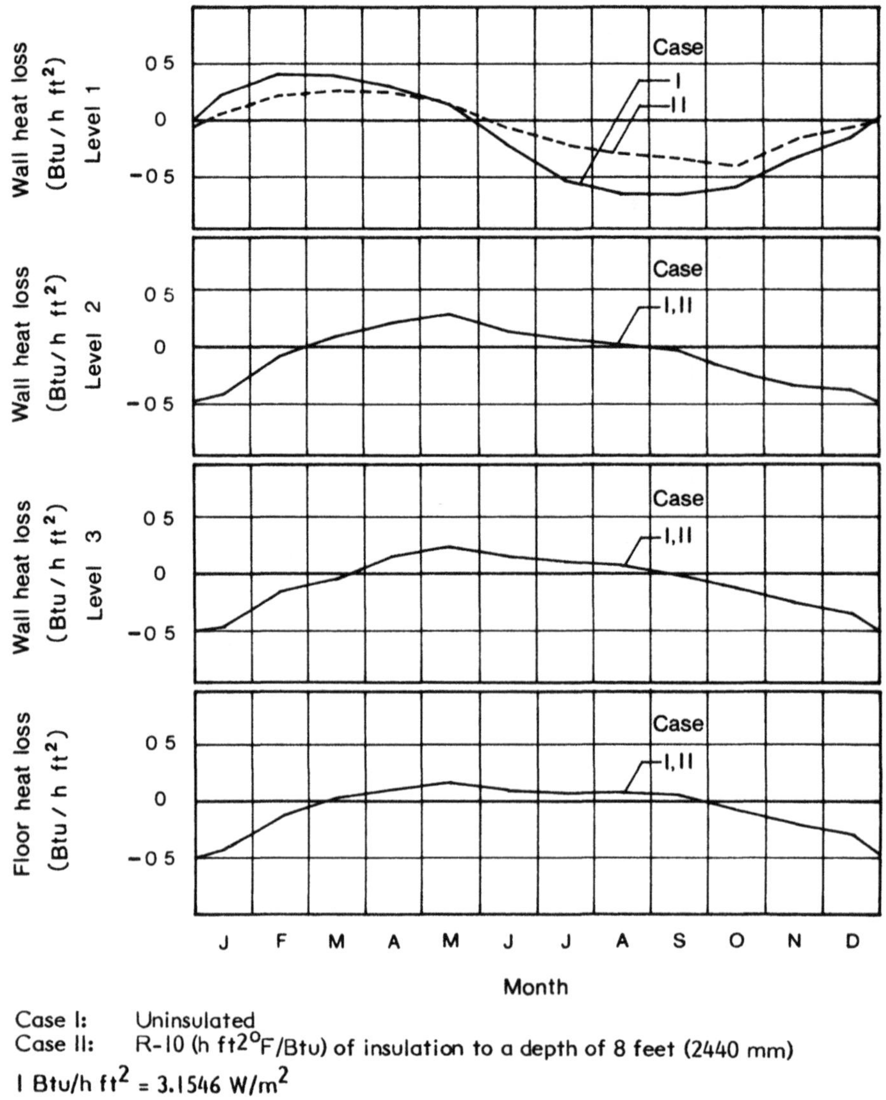

FIGURE 75
Annual Wall and Floor Heat Loss/Gain Miami, Full Berm, 3-Story Building

(c) Calculations of Transmission Heat Loss Through Earth-Contact Slabs. The transient heat loss through earth-contact slabs is calculated in a manner similar to that of earth contact walls. Slab heat loss is calculated by:

$$q = (A)\ (LS_2) \quad (2)$$

WHERE:

- q = heat loss through the floor (Btu/h ft^2)
- A = area of floor (ft^2)
- LS_2= peak heat-loss of the slab perimeter or interior (see Heat Flux Tables) in Appendix A (Btu/h ft^2)

The Heat Flux Tables only apply to buildings which are approximately 50 feet (15 m) in width or greater. If these methods are applied to conditions where the building is smaller in width, below-grade heat losses will be underestimated because of the neglected three-dimensional corner effects.

(d) Calculation of Sensible Heat Loss Due to Infiltration and Ventilation Air. Heat load due to infiltration and ventilation air is a function of quantity of air and design inside and outside temperature difference. Refer to NAVFAC DM-3.3 (Chapter 3, Heating Systems) and ASHRAE criteria (see Reference 13) for methods in estimating the quantity of infiltration and ventilation air.

The following example problem illustrates the use of the Heat Flux Tables for estimating the peak heating load of a partially earth-sheltered building. This analysis demonstrates the use of earth-sheltered design procedures in conjunction with above-grade procedures.

(2) Example Problem 1. Calculate the heat loss rate at design conditions of the two-story earth-sheltered office building (see Figures 76, 77, and 78) located in Minneapolis, Minnesota. From NAVFAC P-89, Chapter 1, design outdoor conditions are -16°F (-26.7°C) and 15 mile/hour (24 km/h) wind speed. Estimate infiltration losses by the air change method. Ignore internal heat gains. For earth-contact heat fluxes, use January design values from the Heat Flux Tables. Assume an indoor dry-bulb temperature of 70°F (21.1°C). The building is constructed as follows:

ABOVE-GRADE WALLS: 0.5 inches (13 mm) gypsum board, air space, 12-inch (300-mm) concrete block, 2-inch (50-mm) extruded polystyrene insulation, 1-inch (25 mm) air space, face brick. Overall heat transfer coefficiency = 0.0726 Btu/ft^2 h °F (0.4122 W/m^2 K).

ROOF: Acoustical tile, 24-inch (610-mm) air space, 20-gauge metal deck, 0.75-inch (19-mm) perlite insulation board, 2-inch (50-mm) isocyanurate foam insulation board, 4-ply built-up roof. Overall heat transfer coefficient = 0.0526 Btu/ft^2 h °F (0.2987 W/m^2 K).

WINDOWS: 1-inch (25-mm) insulating glass. Overall heat transfer coefficient = 0.58 Btu/ft^2 h °F (3.29 W/m^2 K).

BELOW-GRADE WALLS: 14-inch (350-mm) cast-in-place concrete with 2 inches (50 mm) of extruded polystyrene insulation board applied to the exterior wall to a depth of 8 feet (2438 mm).

EARTH-CONTACT FLOOR: 6-inch (150-mm) concrete slab.

SOIL: k = 1.10 Btu/h ft °F (1.904 W/m K), density = 114.0 lb/ft^3 (1,826 kg/m^3), C_p = 0.275 Btu/lb °F (1.1514 kJ/kg K).

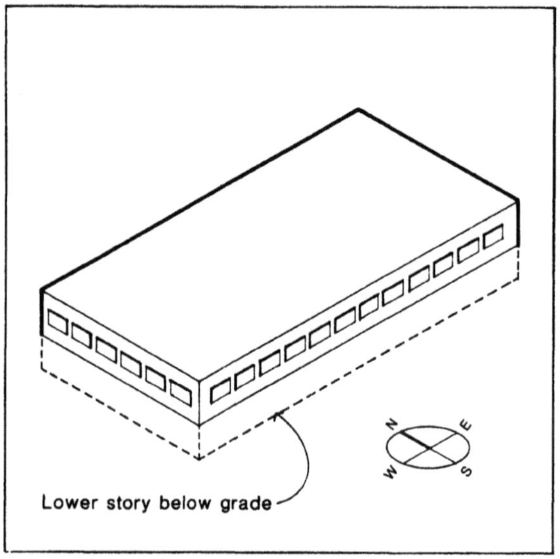

FIGURE 76
Isometric of Building Used for Example Problem 1

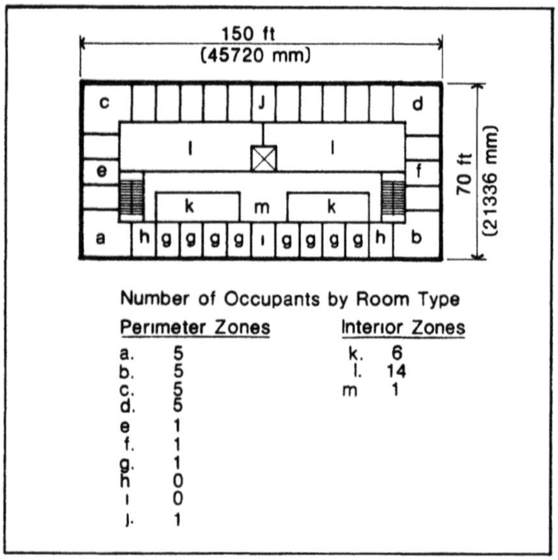

FIGURE 77
Diagrammatic Plan of Building Used for Example Problem 1--Exit Level

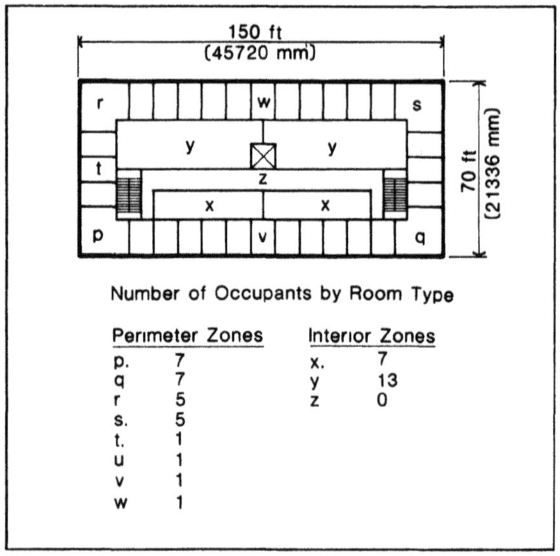

FIGURE 78
Diagrammatic Plan of Building Used for Example Problem 1--Below-Grade Level

SOLUTION: Calculations for this problem are presented in Table 9. The values in Column F of Table 9 were obtained by multiplying together the values in Columns B, C and D/E. See reference notes for Table 9 for further explanation of the data.

TABLE 9
Heat Loss Calculation for Example 1

A	B	C	D	E	F
Heat Loss Component	Net Exterior Area or Air Volume	U-Value Coefficient Btu/h ft2 °F	Temp. Diff. °F	Heat Loss Rate for Earth Contact Surface Btu/h ft2	Heat Loss Btu/h
Roof	10,500 ft2	0.0526	86		47,500
Above Grade Walls	4,400 ft2	0.0726	86		27,470
Below Grade Walls	5,280 ft2			3.23	17,050
Windows	880 ft2	0.580	86		43,900
Entry Door in South Wall	50 ft2	0.083	86		360
Earth Contact Floor Slab Perimeter	7,200 ft2			0.98	7,050
Earth Contact Floor Slab Interior	3,300 ft2			0.79	2,600
Infiltration	34,000 ft3/h	0.018	86		52,630
Ventilation Air	52,800 ft3/h	0.018	86		81,730
TOTAL LOAD					280,290

a. Calculated from gross wall area, less fenestration and doors.

b. Volume of infiltration, ft^3/h = (number of air changes/hour) x (floor area) x (ceiling height)

c. 0.018 = Btu/ft^3 °F

d. Outdoor air introduced for ventilation, ft^3/h = (number of occupants) x (5 ft^3/min./occupant) x (60 minutes/hour)

e. Value extracted from Table 30.

Conversion formulas:

i $ft^3 = 0.0283\ m^3$
1 $ft^2 = 0.0929\ m^2$
1 Btu/ft^3 °F = 67.0661 kJ/m^3K
1 Btu/h ft^2 °F = 5.6783 W/m^2K
1 Btu/h ft^2 = 3.1546 W/m^2
1°F = 9/5°C + 32
1 Btu/h = 0.2931 W

(3) Space Cooling Load. The space cooling load is the rate at which heat must be removed from a space to maintain the room air temperature at a constant value. The summation of all space instantaneous heat gains does not necessarily equal the cooling load for the space at that time. The thermal storage effect of the building mass tends to delay the transfer of heat to the room air. These thermal storage effects are important in determining cooling equipment capacity. For a comprehensive review of heat transfer principles related to sizing cooling equipment for buildings, refer to NAVFAC DM-3.3, Chapter 5, Air Conditioning and Dehumidifying System, and ASHRAE criteria (see Reference 13). The subsequent sections review cooling load calculation procedures as applied to earth-sheltered building. Calculation procedures for above-grade buildings are also included since many earth-sheltered buildings incorporate above-grade building spaces into the design.

(a) Calculation of Cooling Load Due to Transmission Heat Gains Through Above-Grade Walls and Roofs. Refer to ASHRAE criteria (Reference 13) for calculating transmission and solar loads. For "U" factor for type of construction involved, see ASHRAE criteria (Reference 13).

(b) Calculations of Space Cooling Load Due to Solar Heat Gain Through Windows. Refer to NAVFAC DM-3.3, Chapter 5, for design procedures for calculating solar heat gain through glass. Also, refer to ASHRAE criteria tables (Reference 13) for shade and storage factors.

(c) Calculation of Transmission Heat Loss/Gain Through Earth-Contact Walls. Use the Heat Flux Tables to estimate earth-contact wall heat fluxes during summer months. Heat loss or gain can be estimated by means of:

$$q = (A)\ (LS_3) \tag{3}$$

WHERE:

- q = heat transfer through the earth-contact wall (Btu/h)
- A = area of earth-contact wall (ft^2)
- LS_3 = peak heat loss/gain for summer design conditions from the Heat Flux Tables (Btu/h ft^2)

(d) Calculation of Transmission Heat Loss/Gain Through Earth-Contact Slabs. The slab heat loss, or gain, during summer design conditions is given as:

$$q = (A)\ (LS_4) \tag{4}$$

WHERE:

- q = heat loss/gain through the floor for summer design conditions (Btu/h ft^2)
- A = area of the floor (ft^2)
- LS_4 = peak heat loss/gain for summer design conditions. (Btu/h ft^2) See Heat Flux Tables in Appendix A.

(e) Calculations of Space Cooling Load Due to Lights and equipment. See ASHRAE criteria (Reference 13) for calculating heat load due to lights and equipment.

(f) Calculations of Space Cooling Load Due to People. For determination of space cooling load (sensible and latent) due to the people refer to ASHRAE criteria (Reference 13).

(g) Calculation of Space Cooling Load Due to Ventilation and Infiltration Air. For procedures in estimating the outside air load refer to NAVFAC DM-3.3, Chapter 5.

(4) Example Problem 2. Calculate the total cooling load, at design conditions, of the two-story, earth-sheltered office building (Figure 76, 77 and 78) located in Minneapolis, Minnesota. Refer to example problem 1 for a description of building construction. From NAVFAC P-89, Chapter 1, outdoor design conditions for Minneapolis are: dry-bulb temperature 92°F (33°C); daily range, 23°F (-5°C); wet-bulb temperature, 75°F (23.9°C); W_o = 0.0148 lb vapor/lb dry air. Indoor design conditions: Dry-bulb temperature, 78°F (25.6°C); wet-bulb temperature, 65°F (18.3°C); W_i = 0.0102 lb vapor/lb dry air. Occupancy: 176 office workers from 8 AM to 5 PM. Lights: 52,500 watts (2.5 watts/ft^2), non-vented fixtures, from 8 AM to 6 PM. Equipment: 10,500 watts (0.5 watts/ft^2). For earth-contact heat fluxes, use August design values from the Heat Flux Tables.

SOLUTION: Calculations for this problem are presented in Table 10. It may be necessary to make some trial cooling load calculations to determine the time of the maximum load. The cooling load results presented in Table 10 represent the total instantaneous space cooling load of the entire building. Cooling requirements for each zone must be calculated on an individual basis. See reference notes for Table 10 for further explanation of the data.

(5) Limitations of the Cooling Load Calculation Procedures Presented in this Section. One of the inherent limitations of the data presented in the Heat Flux Tables is that the computer simulations were performed with fixed interior air temperatures; that is, the interior air temperature was not allowed to float. Also, the simulations did not account for radiant heat transfer from people, lights, and equipment to the earth-contact surfaces. Consequently, the peak cooling loads, as estimated by the procedure presented in this manual, may be slightly overestimated since the cooling potential of earth-contact surfaces to remove the radiant portions of unwanted heat from internal sources is not modeled. Note that the cooling load factors (CLF's), which are utilized in this manual, do account for the thermal storage effects of the building mass. However, these factors are based on the thermal performance of above-grade structures.

Refer to Appendix C for Worksheets which will aid in the calculation of peak heating and cooling loads of earth-sheltered buildings.

If extremely accurate thermal performance data is required for a specific building and/or location, separate detailed hour-by-hour computer simulations must be carried out. In the near future, computer programs (for main frame computers) will be available for simulating the thermal performance of below-grade structures on an hour-by-hour basis. It should be noted that the use of these programs can involve a significant amount of training, preparation time for input data, and computing time.

TABLE 10
Cooling Load Calculations for Example 2

Heat Gain Component	Net Exterior Area and Volume	U-Value Coefficient Btu/h ft^2 °F	Internal Heat Gain Btu/h	Temp. Diff. °F	Cooling Load Temp. Diff. °F	Heat Flux Rate for Earth Contact Surface Btu/h ft^2	Shading Coefficient (SC)	Solar Heat Gain (SHG) Btu/h ft^2	Cooling Load Factor (CLF)	Cooling Load Btu/h
Roof	10,500 ft^2	0.0526			78.[f]					43,080
Below Grade Wall	5,280 ft^2					2.08[e]				(10,980)
Earth Contact Floor (Slab Perimeter)	7,200 ft^2					1.10[e]				(7,920)
Earth Contact Floor (Slab Interior)	3,300 ft^2					0.88[e]				(2,900)
Entry Door in South Wall	50 ft^2	0.083			48.9					200
Above Grade Walls[a]										
North	1,650 ft^2	0.0726			9.[g]					1,080
East	770 ft^2	0.0726			24.[g]					1,340
South	1,650 ft^2	0.0726			14.[g]					1,680
West	770 ft^2	0.0726			14.[g]					780
Windows - Conduction										
North	300	0.580			13.[h]					2,260
East	140	0.580			13.[h]					1,050
South	300	0.580			13.[h]					2,260
West	140	0.580			13.[h]					1,050
Solar Heat Gain										
North	300						0.60	33	0.80	4,750
East	140						0.60	211	0.19	3,370
South	300						0.60	180	0.49	15,800
West	140						0.60	211	0.71	12,580
Internal Sources										
Lights			179,130						0.90	161,215
Equipment			35,826						0.87	31,170
People			44,000						0.85	37,400
Ventilation and Infiltration										
Infiltration[b]	34,000 ft^3/h	0.018[c]		12						7,340
Ventilation[d]	52,800 ft^3/h	0.018[c]		12						11,400
							SUBTOTAL	Latent		318,005
Infiltration[b]	34,000 ft^3/h									13,060
Ventilation[d]	52,800 ft^3/h									20,280
People										44,000
							SUBTOTAL			77,340
							GRAND TOTAL LOAD			395,345

Footnotes:

a. Calculated from gross wall area, less fenestration and doors.
b. Volume of infiltration, ft^3/h = (number of air changes) x (floor area) x (ceiling height)
c. 0.018 = Btu/ft^3/°F
d. Outdoor air introduced for ventilation, ft^3/h = (number of occupants) x (5 ft^3/min./occupant) x (60 minutes/hour)
e. Value extracted fom Table 30.
f. Values extracted from Table 5A, Chapter 26, ASHRAE Handbook of Fundamentals.
g. Values extracted from Table 7A, Chapter 26, ASHRAE Handbook of Fundamentals.
h. Values extracted from Table 10, Chapter 26, ASHRAE Handbook of Fundamentals.

Conversion formulas:

1 ft^3 = 0.0283 m^3
1 ft^2 = 0.0929 m^2
1 Btu/ft^3 °F = 67.0661 kJ/m^3K
1 Btu/h ft^2 °F = 5.6783 W/m^2K
1 Btu/h ft^2 = 3.1546 W/m^2
1°F = 9/5°C + 32
1 Btu/h = 0.2931 W

b. Integrated Annual Energy Consumption. The calculation of annual energy requirements for large earth-sheltered buildings is somewhat more difficult to calculate than the peak load. A thorough analysis of the total energy balance must be made in order to accurately establish the energy requirements of the building. The degree-day concept is not accurate enough for determining annual energy requirements for large above-grade buildings since this method solely relies on the isolated item of heating. Cooling rather than heating is necessary in many winter situations within buildings of the modern multifunction type. This is a very typical situation in earth-sheltered buildings where envelope losses are minimized. It is becoming cost-effective to recover excess heat from the core of the building and utilize it in perimeter zones. These requirements further complicate calculation procedures.

Because of the complex nature of all factors involved, no simple method has yet been developed for estimating the energy requirements of large earth-sheltered buildings. Currently, the only accurate way of estimating these energy requirements is the use of computer simulation programs which perform hour-by-hour calculations. An alternative to computer simulation is the use of the "variable-bin" method. The "variable-bin" method consists of performing an instantaneous energy calculation at many different outdoor dry-bulb temperature conditions, and multiplying the result by the number of hours of occurrence of each calculation.

The remainder of this section presents methods for estimating space conditioning requirements of earth-sheltered buildings. The "variable-bin" concept is used in conjunction with the Heat Flux Tables. For a complete description of the variable method, refer to ASHRAE Systems Handbook (see Reference 23).

(1) Thermal Performance of Earth-Contact Surfaces. To estimate the annual heat loss/gain of earth-contact surfaces, refer to the Heat Flux Tables and then carry out the following procedure:

(a) Determine the location of the building and select the region which the building site falls within (see paragraph 4.c, and Figure 102).

(b) Select one of the four building configurations available which most closely approximates the building type under study (see paragraph 4a and Figures 65 through 68).

(c) Select the insulation thickness (see Figure 102).

(d) Enter the Heat Flux Tables at the appropriate building configuration and construction component. To obtain the heat loss for any monthly period multiply the area of the surface in question by the average heat loss (under the heading "AVE HEAT LOSS") by the number of hours heat is being lost (under the heading "HRS HT ON"). To obtain heat gains, use the heat flux values under the headings "AVE HT GAIN" and "HRS CL ON".

The above procedure can be repeated for other levels of earth-sheltering and insulation thickness to determine cost effectiveness of various building alternatives, that is, to perform an economic analysis based on energy performance.

(2) Total Space Conditioning Requirements. To accurately estimate the space-conditioning requirements of earth-sheltered buildings, the "variable-bin" method must be used in conjunction with the Heat Flux Tables. The "variable-bin" concept is based on the fact that energy used in a certain period of time equals the average load times the number of hours. The first principle to keep in mind when using this method is that the "bins" should be selected so that there is no qualitative change in the mode of operation within the bin. The second principle is that within each "bin" an average load is calculated with average climatic and building data. This average times the time length of the bin gives part of the consumption. This averaging gives results which are very similar to hour-by-hour calculations.

It is important to identify the bins which represent different usage functions. For example, the energy required to condition outside air would require a minimum of two bins: one bin for preheat (winter months) and a second for cooling (cooling months). Additional modes would be created by the various operating modes of the system or component. For instance, additional bins would be required if the fans are not operated during unoccupied hours or if the perimeter heating system is not used during summer months.

Note that the more bins created, the more time spent on calculations. For large buildings with variable load profiles (for example, a profile which has 100 percent internal heat gains during the day and 0 percent internal loads at night), two bins are required to describe these conditions. During the day, perimeter zones may not require any heating during winter months because of high interior heat gains, however, at night heating may be required to offset envelope heat losses. This same concept also applies to earth-sheltered buildings; envelope loads are reduced to such a point that internal heat gains may be more than adequate to offset these losses. Refer to the preceding section for calculation procedures on estimating the monthly thermal performance of earth-contact surfaces.

The bin method may be used with or without refinements such as coincident wet-bulb conditions, depending on the anticipated impact of the additional parameters. Weather data for use with the bin method is available in NAVFAC P-89.

The "variable-bin method" of energy estimating has the advantage of allowing building zones with complex operating schedules to be analyzed. This permits the user to accurately predict effects, such as reheat and "free cooling," that can only be guessed at with less sophisticated procedures.

Appendix C contains worksheets which are formatted to use the variable-bin method. To calculate the annual heating and cooling requirements of an earth-sheltered building, Worksheets A and B must be used. Depending on the number of different zone operating temperature schedules, Worksheet A (Climatological Data for Use in Calculating Energy Consumption) must be filled out a minimum of 12 times, once for each month. Worksheet B (Heating/Cooling Load Calcultions, Integrated Value) must be filled out for each zone of the building for each month and bin condition. The total annual heating and cooling load of the building is the summation of the totals on Worksheet B for each zone and bin condition.

Writing a computer program to perform these computations may be more expedient in the long run.

6. EARTH-COVERED ROOFS. The large thermal mass associated with earth-covered roofs dampens the continuous fluctuations in the external air temperature. Earth-covered roofs are more than capable of countering the radiant heat gains (solar) if plant cover is provided and well irrigated (See Reference 26). Vegetation-covered roofs can provide net cooling benefits during the cooling season. These benefits are significant, but typically will represent less than 1 percent of the total building cooling load (commercial- or institutional-scale buildings). Thus, an important distinction between earth-covered and conventional roofs is that an earth-covered roof may provide cooling benefits while a conventional roof would add to the total cooling load.

During the heating season, an earth-covered roof's annual energy performance will be similar to that of a well-insulated conventional roof. The difference in energy performance is directly dependent on the depth of earth cover and the amount of insulation applied to the conventional roof. Winter peak design loads of the two roof types will be similar unless large amounts of earth cover are applied to the roof (over 1 foot (300 mm)). It may be difficult to justify an earth-covered roof with small amounts of earth cover (6 inches (150 mm)) on a transient basis, since a roof of this type will nearly reach steady-state heat loss conditions over periods when cold fronts exist for days or weeks at a time. Earth-covered roofs with large amounts of earth cover should be analyzed by considering the thermal mass when determining the peak heating loads of the roof component.

If very accurate thermal performance data is required, hour-by-hour computer simulations must be carried out to predict the peak and annual energy performance of the roof component (heating and cooling).

7. SENSITIVITY OF BUILDING THERMAL PERFORMANCE TO SOIL TYPE. A parametric analysis of the thermal performance of earth-contact surfaces as a function of soil type is presented below. Tabulated results are presented for Minneapolis and Phoenix.

The thermal performance of earth-contact surfaces is very sensitive to soil thermal diffusivity. To partially quantify these effects, computer simulation results of the thermal performance of the slab-on-grade and full-berm one-story buildings, for various soil types, are presented in Tables 11 through 14. The two cases selected for analysis were modeled under identical interior and weather conditions. Only the soil type and level of insulation are varied. Simulation results (heat loss and gains) are presented on an integrated annual basis.

Several interesting points can be drawn from Tables 11 and 12 for the slab-on-grade building. First, note that the slab thermal performance, for the three soil types, varies substantially from the northern climates (Minneapolis) to southern climates (Phoenix). For example, annual heat

TABLE 11
Sensitivity of Slab Thermal Performance to Soil Type, Slab-on-Grade, Minneapolis

	Soil Type	Annual Heat Loss Btu/ft²	Normalized
Uninsulated			
	Average	13,187	100
	Dry	9,175	70
	Wet	14,982	114
Insulated			
	Average	10,572	100
	Dry	6,183	58
	Wet	11,873	112

Footnotes:
1. Insulation thickness = 4 in (102 mm), depth = 4 ft-0 in (2.44 m)
2. Refer to Table 7 for thermal properties of the three soil types.

1 Btu/ft^2 = 11.3559kJ/m^2

TABLE 12
Sensitivity of Slab Thermal Performance to Soil Type, Slab-on-Grade, Phoenix

	Soil Type	Annual Heat Loss		Annual Heat Gain	
		Btu/ft²	Normalized	Btu/ft²	Normalized
Uninsulated					
	Average	3,134	100	466	100
	Dry	2,868	92	105	22
	Wet	3,396	108	764	164
Insulated					
	Average	2,125	100	198	100
	Dry	1,638	77	108	55
	Wet	2,347	110	218	110

1. Refer to Table 5 for thermal properties of the three soil types.
2. Insulation thickness = 2 in (51 mm), depth = 4 ft-0 in (2.44 m).

1 Btu/ft^2 = 11.3559 kJ/m^2

loss in Minneapolis and Phoenix is reduced by approximately 40 percent and 20 percent, respectively, by going from the average to the dry soil. This discrepancy makes predictions for intermediate climates difficult. Second, the thermal performance of the slab during summer months is more sensitive to change in soil type than the thermal performance during winter months. This is largely attributable to the magnitudes of heat fluxes involved. In Phoenix, slab thermal performance during the summer months (heat gains) will vary up to 80 percent depending on soil type. Examination of Tables 13 and

TABLE 13
Sensitivity of Slab and Wall Thermal Performance to Soil Type, Full Berm, 1-Story Minneapolis

	Soil Type	Annual Heat Loss: Wall Btu/ft^2	Wall Normalized	Slab Btu/ft^2	Slab Normalized
Uninsulated					
	Average	30,808	100	8,301	100
	Dry	23,004	75	4,533	55
	Wet	32,737	106	9,563	115
Insulated					
	Average	21,517	100	8,550	100
	Dry	17,227	80	4,612	54
	Wet	22,720	106	9,886	116

Footnotes:
1. Insulation thickness = 4 in (102 mm), depth = 8 ft-0 in (4.88 m).
2. Refer to Table 5 for thermal properties of the three soil types.

1 Btu/ft^2 = 11.3559 kJ/m^2

16 reveals that slab thermal performance is influenced to a greater degree by changes in soil type than by the wall construction of the full-berm one-story building. This is largely due to the stronger coupling of the floor to the ground environment and the magnitudes of the heat fluxes involved.

In short, the simulation results presented in Tables 11 through 14 can be applied with reasonable accuracy to correct the heat flux data for Minneapolis and Phoenix Heat Flux Tables located in Appendix A for very wet or very dry soil conditions. Caution should be exercised when applying the corrections for other locations. For example, correction factors could be obtained for other locations by interpolation based on heating and cooling degree days. However, this methodology assumes that the correction is linear, when in reality it would be non-linear.

If extremely accurate thermal performance data is required for soil conditions other than those simulated, separate simulations must be carried out.

TABLE 14
Sensitivity of Slab and Wall Thermal Performance to Soil Type
Full Berm, 1-Story, Phoenix

	SOIL TYPE[1]	ANNUAL HEAT LOSS			
		WALL		SLAB	
		Btu/ft[2]	Normalized	Btu/ft[2]	Normalized
Uninsulated	Average	7090	100	541	100
	Dry	4432	63	911	59
	Wet	8041	113	1763	115
Insulated	Average	4393	100	1570	100
	Dry	3318	76	921	59
	Wet	4880	111	1803	115

	SOIL TYPE[1]	ANNUAL HEAT GAIN			
		WALL		SLAB	
		Btu/ft[2]	Normalized	Btu/ft[2]	Normalized
Uninsulated	Average	1329	100	477	100
	Dry	175	13	336	70
	Wet	1737	131	501	105
Insulated	Average	374	100	463	100
	Dry	127	34	340	73
	Wet	535	143	494	107

1 Refer to Table 5 for thermal properties of the three soil types.
2 Insulation thickness 2" (51 mm), depth 8'-0" (2.44 m).

1 Btu/ft^2 = 11.3559 kJ/m^2

Section 19. DESIGN EXAMPLES

1. GENERAL. The design example presented in this Section is restricted to a single program and site, a Naval and Marine Corps Reserve Center in Cincinnati, Ohio. Two basic configurations, a half-bermed and full-bermed structure, are developed schematically. The life-cycle analysis addresses the major cost variables and utilizes the Heat Flux Tables in Appendix A to determine the energy performance of each configuration. One of the two configurations is carried through to final schematic design.

2. PROGRAM REQUIREMENTS. The program divides into two sections: the administrative functions and the training functions. In terms of their impact on space conditioning, the administrative portion is used daily and has a normal thermostat setting--78°F (25.6°C) for cooling and 68°F (20°C) for heating. Except for some storage functions with heating to 55°F (12.8°C), the training functions have the same temperature requirements as the administrative functions. However, the schedule of occupancy for the training functions is intermittent, normally with weekend occupancy only. A thermostat setback to 55°F (12.8°C) is required for these functions.

Table 15 lists the required area for each training function according to location relative to an exterior wall. Classrooms, for instance, must have natural light. (They must also have windows for fire rescue and ventilation if the building does not have an automatic sprinkler system.) On the other hand, the assembly room, Navy storage, flammable storage, garage, and mechanical room, must be located on an exterior wall for vehicular access. Table 16 provides similar data for the administrative functions. A vehicle maintenance facility that is separate from the main building is also required. The functions for this part of the program are given in Table 17. A summary which includes a projected estimate of total gross area is given in Table 18.

3. TOPOGRAPHIC FEATURES. The site is on the same zoning lots as the existing reserve center. The existing main building and ancillary buildings must remain functional until the completion of the new reserve center. The site, in general, slopes from east to west (see Figure 79). The western edge has a slope of over 40 percent.

It would be impractical to locate the building near the lease line because surface drainage would be towards the building with little room for a drainage swale to divert the water around the building. Due to the volume of water coming off the slope, catch basins would probably be required to prevent erosion even if a drainage swale could be developed. This location would also tend to increase the amount of underground water against the backwall of the building.

Significant excavation of the sloped area will probably increase construction costs in terms of both excavation costs and construction access. Measures would have to be taken to avoid erosion of the slope during construction. The stability of the slope must also be taken into consideration. A visual examination of the actual site reveals evidence of slope instability in the form of old landslide scars, slumps, and bowed trees.

TABLE 15
Space Allocation by Mechanical, Natural Light, and Access
(Training Functions)

NET AREA IN SQUARE FEET

Functions with Intermittent Use or Low Ambient Temperature

	Windows or Exterior Doors	No Windows or Exterior Doors
Assembly	2,250	
Classroom	800	
Classroom	500	
Classroom	620	
Classroom	825	
Project Room		60
Dentist		130
Medical Administration		520
Doctor		90
Storage (Medical)		120
Examination Room		80
Toilet Room (Medical)		40
Multimedia, Navy	400	
Multimedia storage		60
Training aids, Navy		700
Training aids, Marine		400
Marine room		120
Storage, Navy	950	
Storage, Marine		475
Armory, Marine		400
Lockers, Marine		1,660
Lockers, Navy		760
Drying Room		70
Shower Room		120
Men's Toilet Room		280
Women's Toilet Room		120
Women's Shower		40
Women's Drying Room		25
Women's Locker Room		110
Janitor, Navy		80
Janitor, Marine		80
Flammable Storage	70	
Garage	500	
Mechanical Equipment	800	
Walls and Partitions (5%)	386	327
Subtotal (each column)	8,101	6,867
Subtotal (both columns)		14,968
Circulation (15 percent)		2,245
	TOTAL	17,213

1 $ft^2 = 0.0929\ m^2$

TABLE 16

Space Allocation by Mechanical, Natural Light, and Access (Administrative Functions)

NET AREA IN SQUARE FEET

Functions with daily use and normal thermostat setting.

	Windows or Exterior Doors	No Windows or Exterior Doors
Recruiting, Navy	300	
Recruiting, Marine	260	
Conference Room	1,520	
Crew's Lounge	450	
Navy Active Administration	370	
Navy Unit Administration	1,290	
C.O. Office, Navy	240	
Ex.O. Office, Navy	130	
Reserve Office, Navy	135	
Reserve Office, Navy	135	
Office, Navy	200	
Office, Navy	210	
Office, Navy	100	
Marine Active Administration	690	
Marine Unit Administration	1,860	
Reserve Office	150	
Reserve Office	150	
I and I Office	105	
I and I Office	105	
Men's Toilet Room		90
Women's Toilet Room		40
Janitor		30
Walls and Partitions (5 percent)	370	88
Subtotal (each column)	7,770	168
Subtotal (both columns)		7,938
Circulation (15 percent)		1,190
TOTAL		9,128

1 $ft^2 = 0.0929\ m^2$

TABLE 17
Program Area Tabulation for Vehicular Maintenance Facility

Vehicle Maintenance Facility	Gross Area in Square Feet
2 Bays Vehicle Maintenance	960
Maintenance Office	90
Tool Room	90
Toilet and Shower	50
Lead/Acid Battery Room	75
Flammable Storage	65
Silver/Zinc Battery Room	60
Electrical Maintenance	240
Jeep Radio Room	170
Radio Issue	45
Communication Storage	575
General Storage	390
Mechanical	50
TOTAL NET	2,860

$1\ ft^2 = 0.0929\ m^2$

4. CONTEXTUAL CONSIDERATIONS. In general, all of the steeply sloped areas have trees with some undergrowth. The site is much higher than the surrounding terrain to the west and offers an exceptional view in this direction. Land use to the east as well as to the north (see Figure 79) is residential. Because the site is within view of the residential areas, it will be visually appropriate to minimize the industrial aspects of the facility (the vehicle maintenance and storage yards, in particular). The roof of any single-story structure within the site will be visible from the approaches to the east. If the building elevation is set low enough to allow a full berm, the roof may even be visible from the entrance road. Consequently, the attractiveness of the roof will be a major consideration.

5. SITE ACCESS. Whether or not the final building is substantially earth-sheltered, the slope of this site allows handicapped access to a two-level building without the use of elevators. Assuming that the training functions are located on a lower level and the administration functions are located on an upper level, pedestrian and vehicular access to the lower level will be from the northwest corner of the site while administration (and the main entrance) can be accessed from the east.

6. BUILDING ORIENTATION. There are basically three building alignments that are suggested by the existing site features. These are: alignment with the top of the slope, alignment with any existing construction that may be retained, and, alignment with the approach road to the north.

TABLE 18
Summary of Program by Zones

AREAS IN SQUARE FEET

Functions which require daily heating and air-conditioning at normal ambient temperatures (administrative functions)

Exterior wall location required		7,770
Interior location suitable		168
Circulation (estimated)		1,190
	Subtotal gross area	9,128

Functions which require only intermittent heating and air-conditioning to normal temperatures and functions which require only low-ambient temperatures (training functions)

Exterior wall location required		8,101
Interior location suitable		6,867
Circulation (estimated)		2,245
	Subtotal gross area	17,213
	Total Gross Area (main building)	26,341
	Total Gross Area (vehicle maintenance)	2,860
	Total Gross Area for Project	29,201

1 $ft^2 = 0.0929\ m^2$

a. Alignment with the Slope. Alignment with the top of the slope will probably provide the most natural integration of the building and the terrain whether or not the building is earth-sheltered. Alignment with the slope also affords the best angle for view to the west. Diagramatically, Figure 80 shows a two-story rectangular building superimposed on this portion of the site. Figure 81 diagrams the same building as a fully-bermed structure. Note that grading and excavation are minimal in the half-berm version. Figure 81 suggests, however, that extensive retaining walls for the full berm building may be required in the final schematic design.

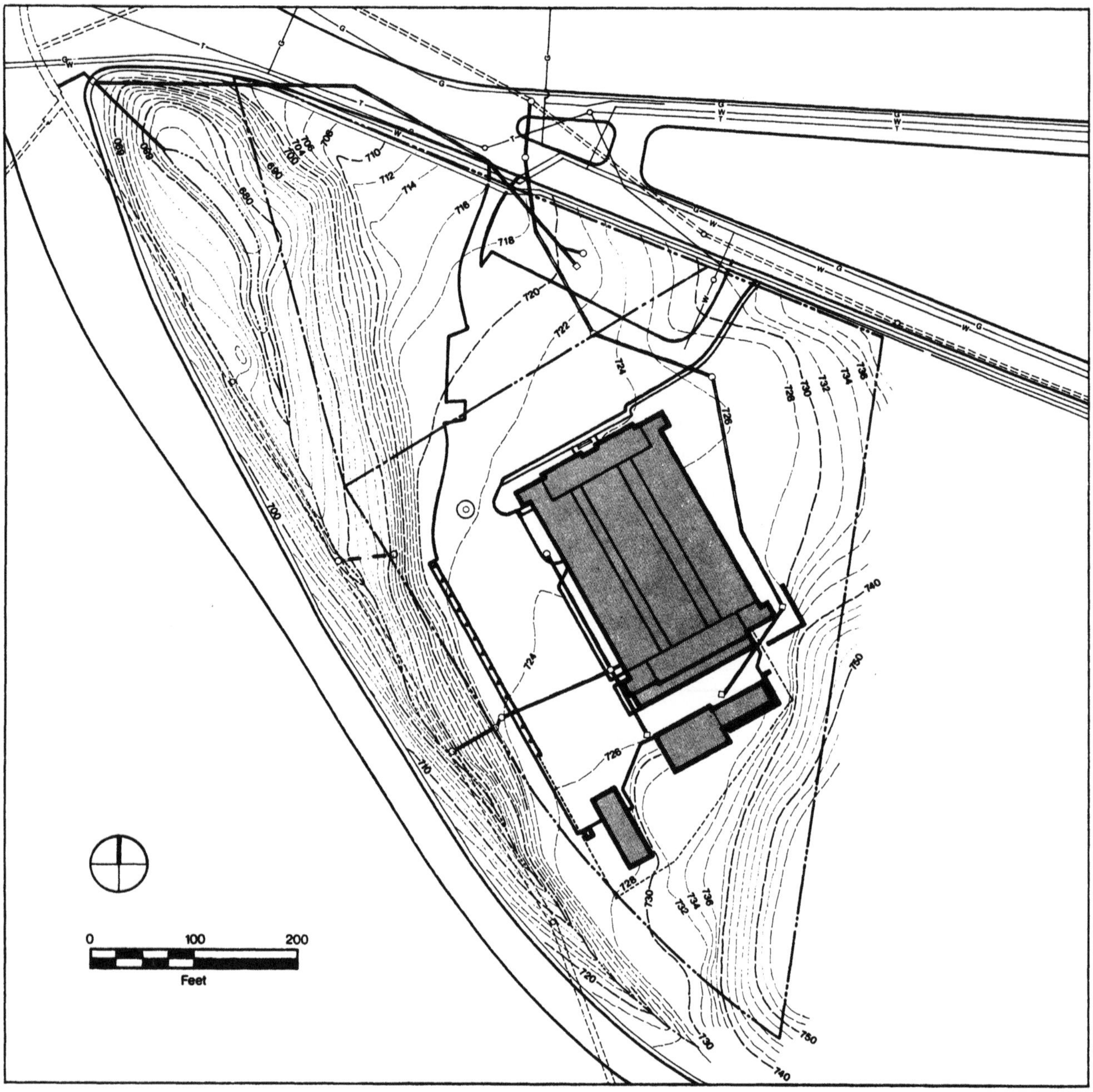

FIGURE 79
Existing Site for Design Example

b. Alignment with the Road. Alignment with the road will require greater amounts of earthwork than would be required for alignment with the slope (refer to Figures 82 and 83). The building would cut further into the slope. In addition to excavating the building area, additional land area would have to excavated to allow access and provide natural light to the lower floor. This orientation, however, has greater potential for passive solar adjustments than the previous alignment. Also, if the building is earth-covered (and therefore potentially ambiguous architecturally) alignment with the road may strengthen the architectural definition of the building.

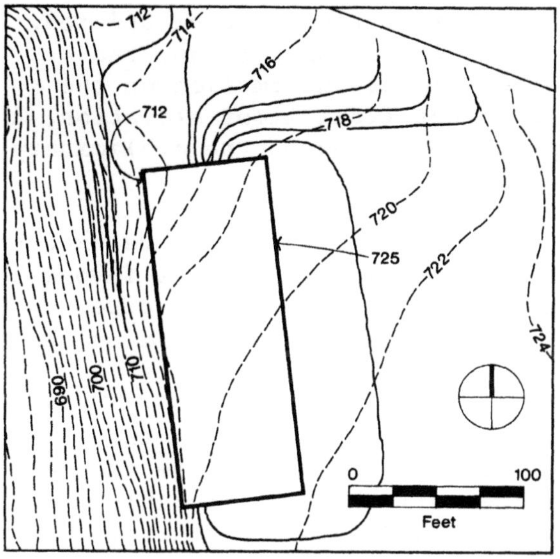

FIGURE 80
Half-Berm Building
Aligned with Slope

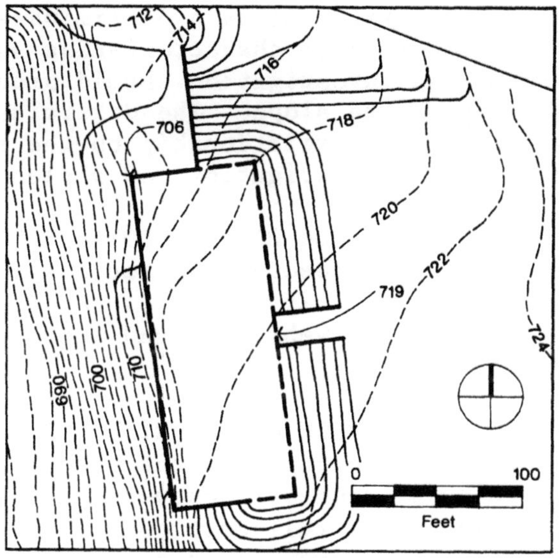

FIGURE 81
Full-Berm Building
Aligned with Slope

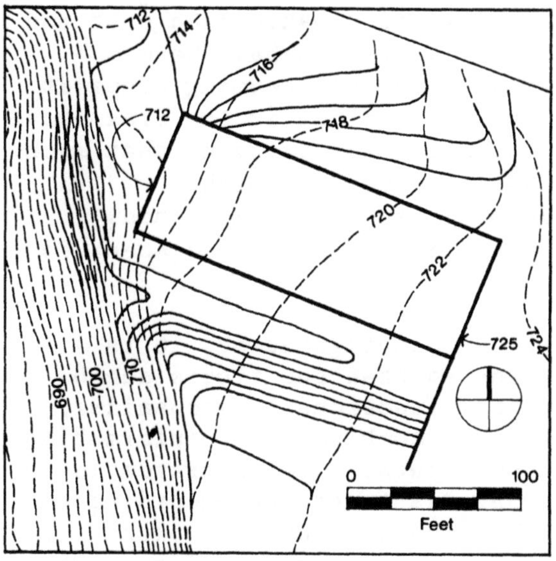

FIGURE 82
Half-Berm Building
Aligned with Road

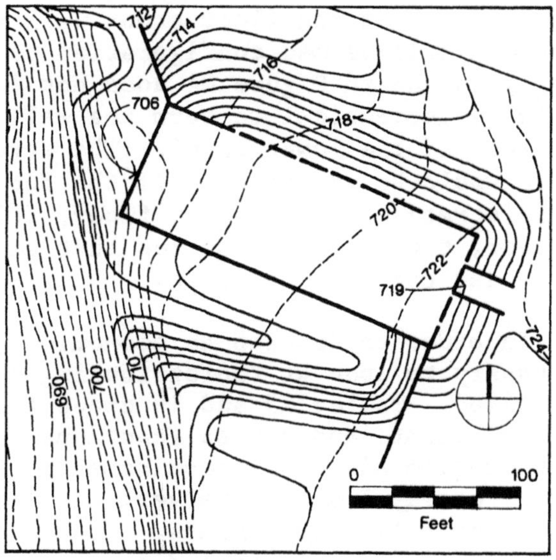

FIGURE 83
Full-Berm Building
Aligned with Road

c. Alignment with Retained Buildings and Site Features. The location of the existing vehicle maintenance building and the access requirements suggest that as much of the existing vehicle maintenance facility be reused and modified as possible. Other features that might be saved include the flag pole, the stone retaining walls behind the vehicle maintenance buildings, and the guardrail to the southwest of the reserve center. Alignment with retained site features is not diagrammed. Such an orientation would share in the advantages and disadvantages of the other two alignments.

d. Preliminary Geotechnical Report. It is unlikely that the building will have the same plan as assumed in the foregoing analysis. However, this level of analysis allows a preliminary geotechnical report to be ordered. In addition to soil characteristics and ground water information, the geotechnical investigation should provide additional information on the stability of the slope. The determination of the final floor elevation, and therefore of the relationship to the slope, will depend on the preliminary geotechnical investigation and a plan diagram.

7. ORIENTATION FOR PASSIVE SOLAR BENEFIT. Exterior walls that are exposed for light, fire rescue, and ventilation access should also have a favorable solar orientation if possible. Access functions which do not require windows, such as vehicular and mechanical access can be positioned on the northwest side.

8. PROTECTION FROM WINTER WIND. Prevailing winter wind is from the northwest. Unfortunately the exterior wall at this point must be exposed for vehicular access. However, the retaining wall which is probably necessary in the full-berm configuration may provide some sheltering. Evergreen trees are also a possible buffer.

9. EQUALIZATION OF FLOOR AREAS IN THE FULL-BERM CONFIGURATION. The lower- and upper-level floors have significantly different areas (see Figure 84). Equalization of the floor areas will reduce initial construction cost and enable a full-bermed configuration to be developed. The half-berm configuration in Figure 84 does not lend itself to full berming since it would be impractical to berm the wall of the upper floor in the area above the roof of the lower level. The structural costs for carrying that much earth would be excessive. If the functions are reversed, with administration on the lower floor, full-berming can be accomplished. However, the program requires that administration be located adjacent to the lobby and the main entrance.

In order to develop an economical full-berm configuration, it is necessary to look at redistributing functions so that the floors can be equalized. Tables 15 and 16 group the program functions by mechanical, natural light, and access functions. Figure 85 combines this information with program adjacency requirements. The horizontal line separates low and normal energy loads. The vertical line separates functions according to parimeter wall requirements. Figure 86 takes these groups and places them on different floors, according to the adjacency determined in Figure 85. Resultant floor areas and estimated lineal feet of access required for each floor level are then tabulated. Schematic section number 1 in Figure 86 is acceptable as a half berm configuration. Schematic Section 2 does not work on the site because administration is not on the entrance floor. Schematic Section 3 will probably not allow optimum full berming because of the large amount of natural light access required at the upper floor. Schematic Section 4 has a similar problem. Schematic Section 5 is geometrically acceptable for full-berming. The window and access requirements for both floors are more evenly matched and the floors are nearly the same size. However, the adjacency of the administration to lockers is undesirable without visual, acoustical, and psychological separation.

10. ALLOCATION OF FUNCTIONS BY MECHANICAL REQUIREMENTS. Section 9, Space Planning, and Programming, provides guidance for locating the mechanical

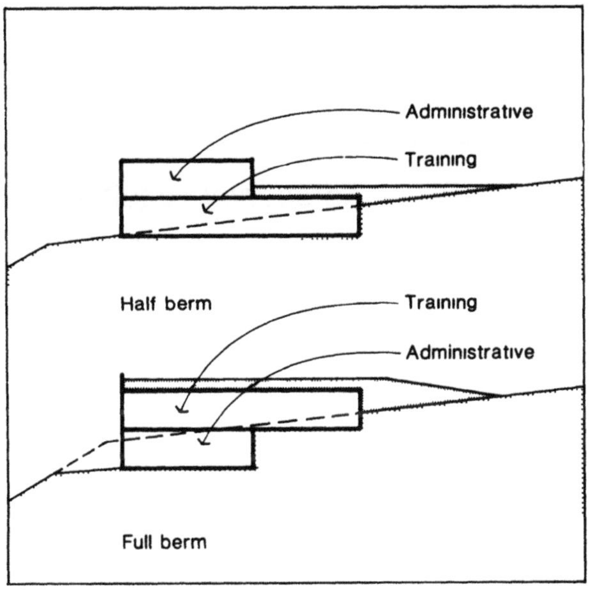

FIGURE 84
Schematic Section of Training and Administrative Floors

zones for optimum thermal benefit. In the case at hand, it must be determined whether a storage function with a low thermostat setting or a function such as offices, with a normal thermostat setting, should be located on an earth-sheltered wall for maximum thermal benefit. On one hand, it would appear that the storage function would buffer the office function against the colder temperatures near the ground surface. On the other hand, the office can obtain greater cooling benefit by being located on an earth-sheltered wall (in this particular climate). To get a rough idea of which arrangement is better from an energy standpoint, it is necessary to look at the relative energy loads of each of the two arrangements.

This is quickly done by utilizing the Heat Flux Tables in Appendix A. Figure 102 in Appendix A indicates that Cincinnati is on the border between zones 2 and 4. The corresponding cities for these zones are Kansas City and Boston. Turning to Table 38 in Appendix A for Kansas City with R-20.8 insulation, the appropriate heat flux values are selected for the various floor and wall components for January and July for "full-berm, one-story." (Average, rather than dry or wet, soil conditions are assumed for this purpose.)

Although one space is heated to 55°F (12.8°C) and the other to 68°F (20°C), the reduction in heat flux will not be proportional to the difference in temperatures due to the thermal mass of the earth. An approximate estimate of the heat fluxes for a space with a 55°F (12.8°C) thermostat setting can be obtained by slightly reducing the listed values, say by 5 percent, but not by 20 percent, which would be proportional and, therefore, characteristic of a more steady-state heat transfer (above-grade) rather than an earth-sheltered situation. Accuracy at this point is not of great concern since the magnitude of the below-grade heat loss is very low. Because the peak heat losses shown in Table 38 are near the average instantaneous (hourly) values, the instantaneous values will be sufficient

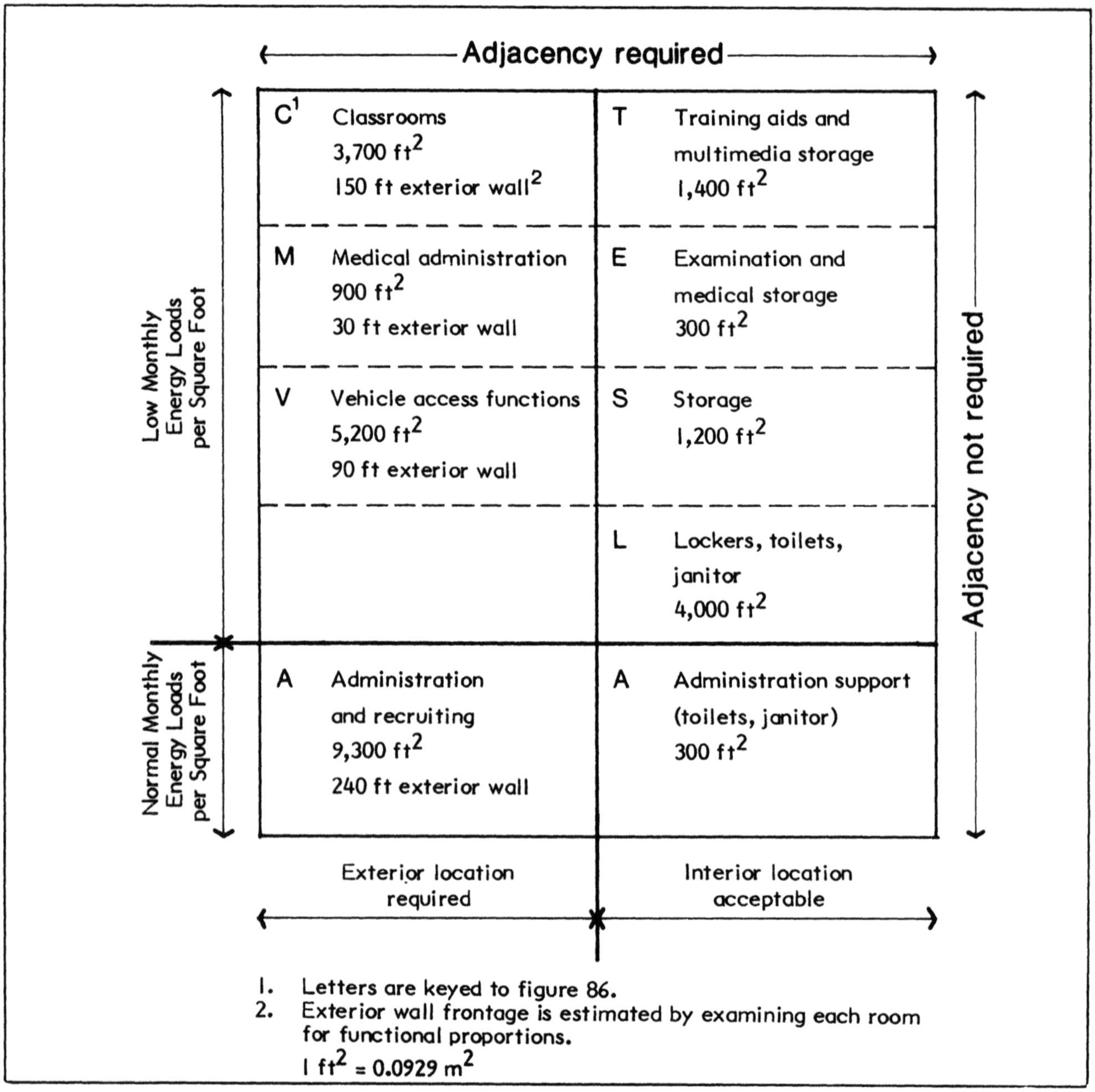

FIGURE 85
Space Allocation by Program Requirements

for this comparison. Also, ventilation air need not be considered for this purpose (ventilation loads will tend to be independent of the location of the zone).

The total hourly loads for the two selected months are obtained as shown in Figures 87 and 88. The results (which are approximate) are summarized in

Schematic Section (letters keyed to figure 85)	Area (ft^2)	Exterior wall required (ft)
1 	9,600	240
	16,700	270
(acceptable for half-berm configuration)		
2 	16,700	270
	9,600	240
(programmatically does not work on site)		
3 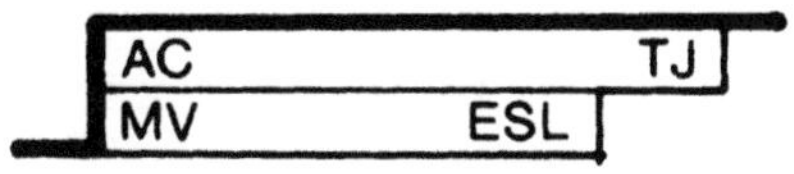	14,700	390
	11,600	120
(required exterior wall length too long at upper floor)		
4 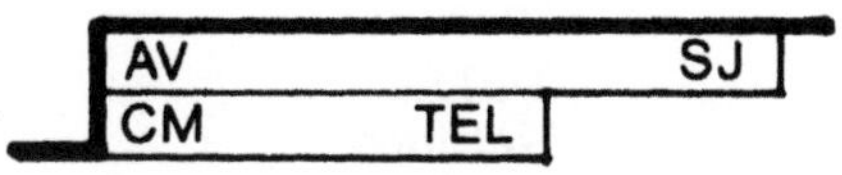	16,000	330
	10,300	180
(required exterior wall length too long at upper floor)		
5 	13,600	240
	12,700	270
(geometrically acceptable for full berm but upper floor adjacency is questionable)		

$1\ m^2 = 0.7639\ ft^2$

$1\ mm = 0.0328\ ft$

FIGURE 86
Space Allocation to Equalize Floor Areas

Figure 89. Placing the office function, instead of storage, on an earth-sheltered wall, yields a small energy savings for January and July. Examination of the other months will yield a similar conclusion. If Table 36 in Appendix A is used (Boston), the advantage of the perimeter location for the office will be more pronounced. This can be seen by visual inspection of the values in Table 36 and by comparing these to the values in Table 38. If less insulation is used, the same conclusion holds. This also is evident by visual inspection of the values in Tables 29 and 31.

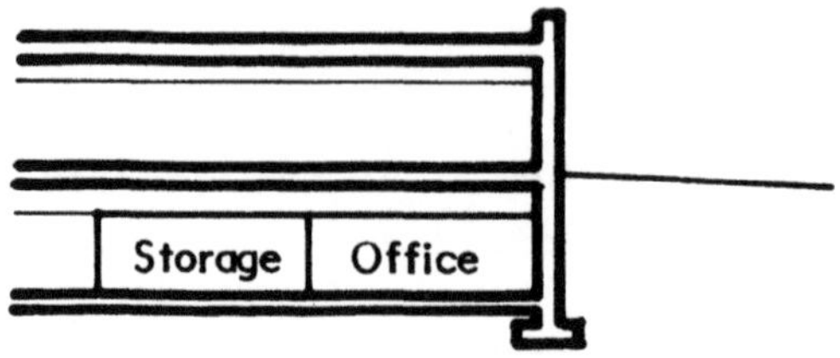

Hourly heat flux for OFFICE SPACE = Average heat flux from Table 36, Btu/h ft^2 X Surface area, ft^2

Hourly heat flux for STORAGE SPACE = Average heat flux from Table 36, Btu/h ft^2 X Correction Factor X Surface Area', ft^2

JANUARY

OFFICE			Btu/h
wall level-1	(2.10) (135)	=	+284
perimeter slab	(.55) (200)	=	+110
internal gain (lights and equipment)	(-10.24) (200)	=	-2,048
internal gain (people)	(-250) (2)	=	-500
	TOTAL	=	-2,150
	Net load	=	0
	(Use economy cycle)		
STORAGE			
interior slab	(.42)(.95)(200)	=	+78
interior gain (lights)	(-1.7) (200)	=	-340
	TOTAL	=	-262
	Net load	=	0
	(Use economy cycle)		

JULY

OFFICE			Btu/h
wall level-1	(1.74) (135)	=	+235
perimeter slab	(.93) (200)	=	+186
internal gain (lights and equipment)	(-10.24) (200)	=	-2,048
internal gain (people)	(-250) (2)	=	-500
	TOTAL	=	-2,127
	Net cooling load	=	-2,127
STORAGE			
interior slab	(.75) (.95) (200)	=	+143
internal gain (lights)	(1.7) (200)	=	-340
	TOTAL	=	-97
	Net cooling load	=	0
	(allow higher room temp.)		

1 ft^2 = 0.0929 m^2
1 Btu/h ft^2 = 3.1546 W/m^2
1 Btu/h = 0.2931 W

FIGURE 87
Instantaneous Below-Grade Heat Flux with Office on Earth-Contact Wall

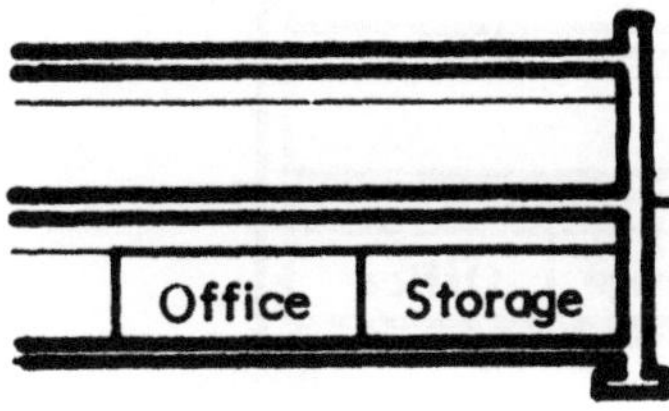

Hourly heat flux for OFFICE SPACE = Average heat flux from Table 36, Btu/h ft^2 X .Surface area, ft^2

Hourly heat flux for STORAGE SPACE = Average heat flux from Table 36, Btu/h ft^2 X Correction Factor X Surface Area,* ft^2

JANUARY

			Btu/h
OFFICE			
interior slab	(.41) (200)	=	+82
internal gain (lights and equipment)	(-10.24) (200)	=	-2,048
internal gain (people)	(-250) (2)	=	-500
	TOTAL	=	-2,446
	Net load	=	0
	(Use economy cycle)		
STORAGE			
perimeter slab	(.55)(.95)(200)	=	+105
wall level-1	(2.10)(.95)(135)	=	+269
interior gain (lights)	(-1.7) (200)	=	-340
	TOTAL	=	+34
	Net heating load	=	34

JULY

			Btu/h
OFFICE			
interior slab	(.75)(200)	=	+150
internal gain (lights and equipment)	(-10.24) (200)	=	-2,048
internal gain (people)	(-250) (2)	=	-500
	TOTAL	=	-2,398
	Net cooling load	=	-2,398
STORAGE			
perimeter slab	(.93)(.95)(200)	=	+177
wall level-1	(1.74)(.95)(135)	=	+223
internal gain (lights)	(1.7) (200)	=	-340
	TOTAL	=	+60
	Net heating load	=	60

1 ft^2 = 0.0929 m^2
1 Btu/h ft^2 = 3.1546 W/m^2
1 Btu/h = 0.2931 W

FIGURE 88
Instantaneous Below-Grade Heat Flux with Storage on Earth-Contact Wall

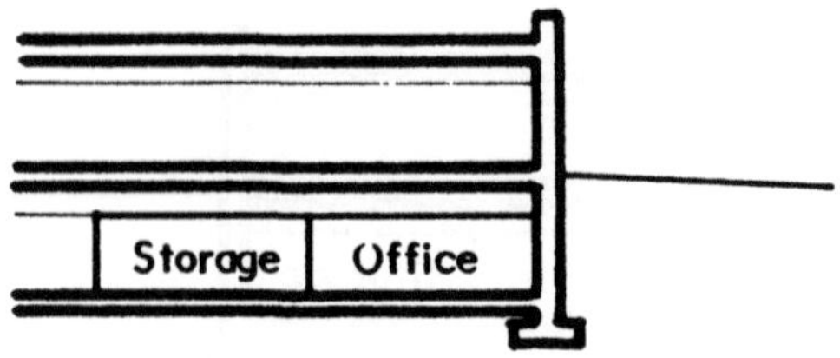

Cooling	Heating
2127 Btu/h (office)	0

Relative energy costs for January and July (excluding ventilation) = $9.27

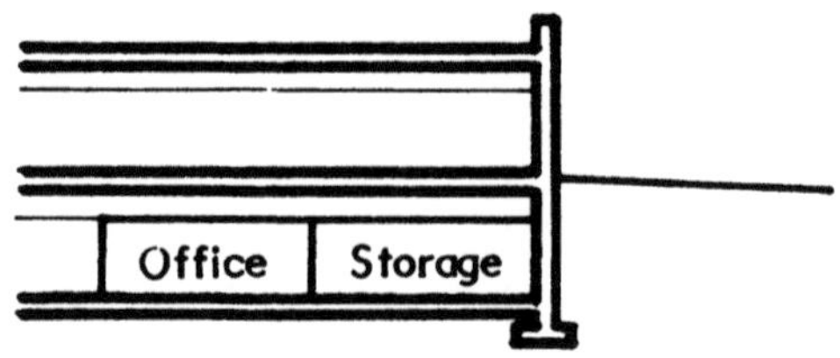

Cooling	Heating
2398 Btu/h (office)	94 Btu/h (storage)

Relative energy costs for January and July (excluding ventilation) = $10.69

1 Btu/h = 0.2931 W

FIGURE 89
Comparison of Energy Costs for
Different Functions Located on Earth-Contact Wall

In addition to the calculated benefit, placing the office function on the earth-sheltered wall will improve the human comfort level since the surface temperature of the earth-sheltered wall is likely to be warmer in winter and cooler in summer than an above-grade wall. The same is true in a southern climate except that that the mechanical cooling benefit to be derived from the earth-sheltered wall is minimal.

Based on the foregoing analysis, functions with cooling loads should be located against the earth-sheltered wall--if there are no other factors (such as programmatic requirements) that would suggest a different location.

11. OPTIMIZATION OF SURFACE-TO-VOLUME RATIO AS A FUNCTION OF MECHANICAL ZONES. Where office functions are involved and there is no thermostat setback, the preceding analysis would indicate that the interior zones should be minimized and that the perimeter earth-sheltered zones, maximized. This would increase the surface-to-volume ratio. However, for the functions with no or very little internal gains for five days of the week (such as the training functions), heating loads will probably be incurred, making it more beneficial to reduce the surface to volume ratio. On the other hand, on the weekends, when these functions are occupied, the greater surface-to-volume ratio would be more beneficial. To quantify this trade-off, an analysis similar to the preceding one can be performed for similar zones, one on an earth-sheltered wall and the other in an interior zone. Such an analysis will show that it takes very little internal gain to offset the heat loss in winter for the perimeter location. In this case, approximately 0.05 to 0.1 watt/ft^2 (0.5 to 1.1 watt/m^2) (of floor area) is sufficient to do this.

However, the surface-to-volume ratio of the below-grade spaces which have a significant cooling requirement will be primarily a function of programmatic requirements and construction costs (see paragraph 12, below). But, if the storage spaces, which have no cooling, end up in the perimeter earth-sheltered zones for programmatic reasons, the surface-to-volume area of these zones can be reduced to lower the minimal heating requirement of these zones.

12. SURFACE-TO-VOLUME RATIO AS A FUNCTION OF CONSTRUCTION COST. Costs per lineal foot of earth-sheltered wall may be at least $150 per lineal foot of a 14-foot high wall including costs for concrete, footings, drain tile, waterproofing, insulation, excavation outside the wall, and backfill against the wall. The preceding analysis for placing storage or offices in a perimeter zone suggests that the energy savings to be derived from putting functions with cooling requirements in an earth-sheltered perimeter zone, rather than an interior zone will be less than $1 per lineal foot ($1/300 mm) of wall per year. Also, the length of mechanical ducts may have to be increased with a less compact plan. On the other hand, departing from a square or compact plan will tend to diminish the number of interior columns, and depending on the firefighting access provided, may eliminate the requirement for a sprinkler system.

Consequently, for a building of this scale, energy considerations do not affect the life-cycle benefit of increasing the surface-to-volume ratio for the earth-sheltered portions of the building. The main determinant of aspect ratio, in terms of life-cycle costs, will be the trade-off of reduced wall costs and the additional costs of an automatic fire protection system--which would be required for the building if some areas were more than 75 feet (22 860 mm) from an exterior access panel or window (see Section 12, Fire Protection and Life Safety, Paragraph 4).

13. SURFACE-TO-VOLUME RATIO LIMITATIONS. A low aspect ratio will be limited by the program requirements for exterior access. Another factor is

life-safety access. It is possible to design the building without an automatic sprinkler system provided that the requirement, of Section 12, paragraph 4 are met.

14. DETERMINATION OF DEPTH OF EARTH-SHELTERING AS A FUNCTION OF THERMAL LAG. Other factors notwithstanding, it is desirable to place the lower floor elevation so that the average wall and floor slab area is sufficiently removed from the ground surface for thermal buffering yet is close enough to the surface to take advantage of seasonal thermal lag. This cannot be accurately determined with the available data; the following analysis can be done, however.

a. Zones With Heating Only. The further from the ground surface, the lower the heat loss will be for the zones with heating only.

b. Zones With Cooling and Heating Without Thermostat Setback. A comparison of a half-berm and full-berm structure is possible based on the data in the Heat Flux Tables. Consider the sample office space in Figure 87. If similar calculations are performed using Table 38 for a half-berm configuration--5 feet (1530 mm) of depth as opposed to 13 feet (3960 mm) of depth--the following result is obtained for July: the average heat loss (due to thermal lag) for the slab and wall (level-1) is 248 Btu/hour. This is equivalent to 0.36 watts/ft^2 of floor area. In January, the average heat loss for the office space would be 315 Btu/hr or 0.45 watts/ft^2 of floor area. By comparison, the values calculated in Figure 87 are equivalent to 0.62 watts/ft^2 and 0.58 watts/ft^2 for July and January respectively.

Thus it appears likely that internal gains (of 3 watts/ft^2 of floor area during occupied hours) will more than offset the below-grade heat loss regardless of the depth. If the "wall level-1" heat loss values in Table 38 are examined for all of the months for the 13 foot depth ("full-berm one-story" values in the table), it will be seen that the greatest heat loss is in April. This would be in May if Table 36 (Boston) were used. The highest heat loss for the perimeter slab would be in July.

c. Zones With Both Cooling and Heating and Temperature Setback. Zones which require cooling but also have a thermostat setback will behave similarly to both types of zones examined above in terms of below-grade heat loss. However, the temperature setback will increase the heat-sink capacity of the earth mass and will result in a greater beneficial heat loss during the cooling mode.

15. DETERMINATION OF DEPTH OF EARTH-SHELTERING AS A FUNCTION OF CONSTRUCTION COSTS. Construction costs will not be quantified at this point. However, the following observations can be made:

First, the preliminary geotechnical report (which was obtained immediately after the building site was chosen) indicates that a thick layer of fill covers the site. This fill would have to be removed in the area of the building. This would place the bearing elevation of the footings for the half-berm structure on weathered glacial till which should support the anticipated loads with a long-term settlement that would be within tolerable limits (approximately 1 inch (50 mm)). In order to provide a maximum practical thickness for this stiffer glacial till, the lowest floor elevation should be approximately 713 feet (217 m) above sea level. Below this layer, at the bearing elevation of a full-berm structure, the footings would rest on less compact strata. Building settlement would probably exceed 1 inch (50 mm), especially with earth loads on the roof. The report recommends that thin-wall steel-shell cast-in-place concrete piles or augered intrusion grout piles be used below the footings. Drilled piers would encounter ground water at the depths required for penetration and, therefore, are not recommended.

16. UTILITIES. The selected building location is near the existing utilities to the north (see Figure 79). The invert elevations of the storm and sanitary sewers are low enough to avoid sumps and pumping for floor elevations as low as 700 feet (213 m) above sea level.

17. STRUCTURAL SYSTEMS. Appropriate structural systems are evaluated for economy and suitability to earth cover. Considering the lateral bracing of the foundation wall, the extensive amount of concrete for below-grade walls, and local construction factors, a concrete structural system is chosen. A pan-joist system is most economical for the larger spaces. Rather than mixing systems, the pan-joists will be used in the other areas. Reuse of forms in these areas should make pan-joists more economical than a mix of pan-joist and waffle slab construction.

Preliminary structural design must be done on representative conditions to allow costing for life-cycle analysis. Structural slabs, foundation walls, columns, footings, and piles must be designed for the bay sizes and loads indicated by the preliminary plans.

18. DIAGRAMMATIC PLAN FOR 2-STORY FULL-BERM BUILDING.

a. Full-Berm Configuration. The preliminary plan for the full-berm configuration is based on schematic section number 5 shown in Figure 86. The classroom, medical and administrative functions are located in a two-story wing to the southwest. The orientation of this wing takes advantage of the view to the west while reducing afternoon summer solar gain by facing southwest rather than west. The lockers, storage functions, garage, and vehicle assembly (functions which do not require daylighting or views) are located in a two-story block to the northeast. The orientation of this block allows vehicle access to the garage, flammable storage, and the assembly function. The two wings are separated by a linear two-story skylighted atrium.

An interior ramp in the southeast end of the atrium allows the main floor elevation to be depressed below grade without any exterior ramps or depressed entrances. The ramp enclosure also eliminates snow and ice on the ramp. Because the roof is earth-covered, the linear nature of the atrium and entrance helps organize the building visually. The relationship to the other site features is reinforced by aligning the continuous linear form with other linear site elements. Thus a primary design axis is set up. The entrance road, flagpole, and vehicular turnaround define a secondary access (see Figure 90).

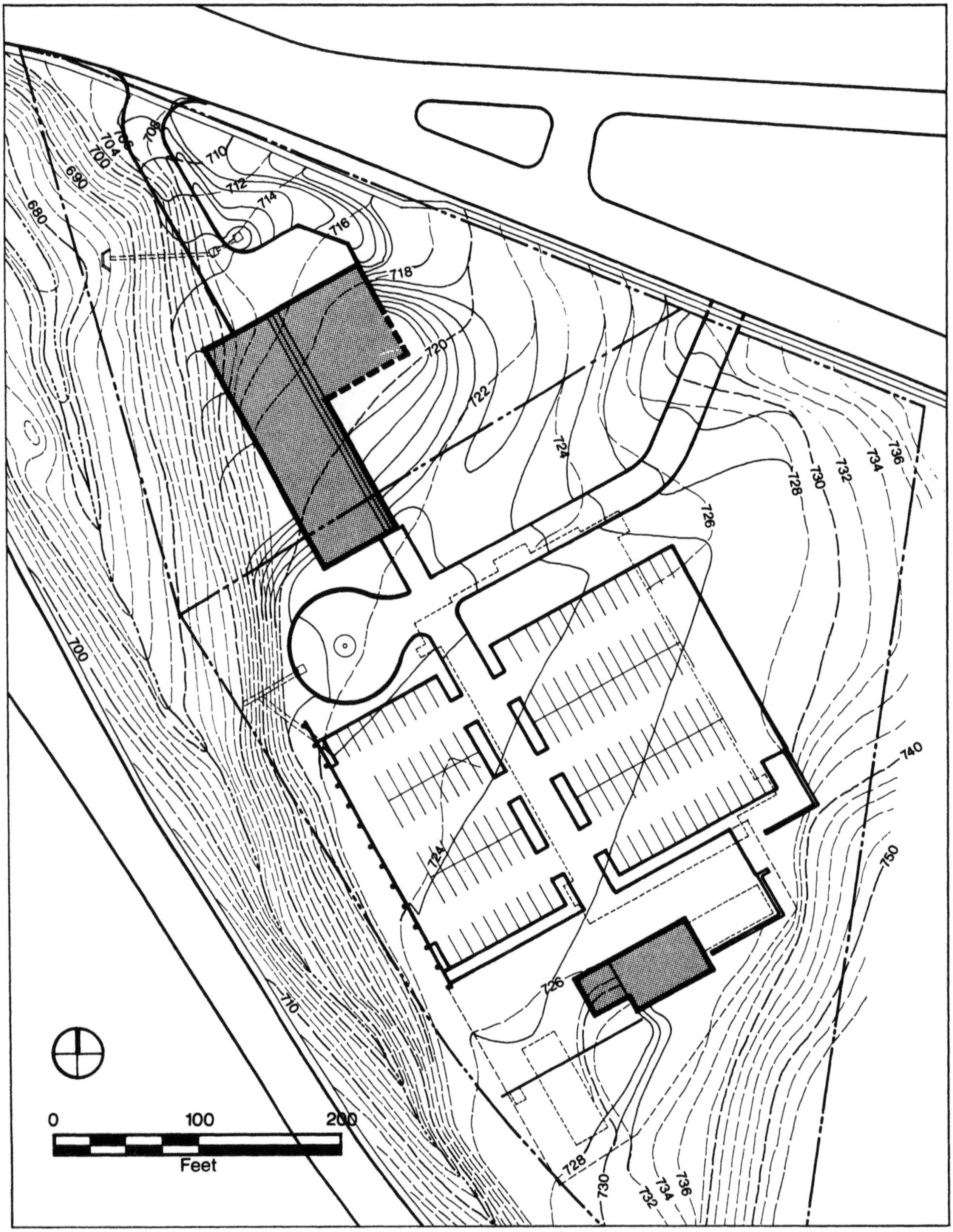

FIGURE 90
Preliminary Site Plan for Full-Berm Building

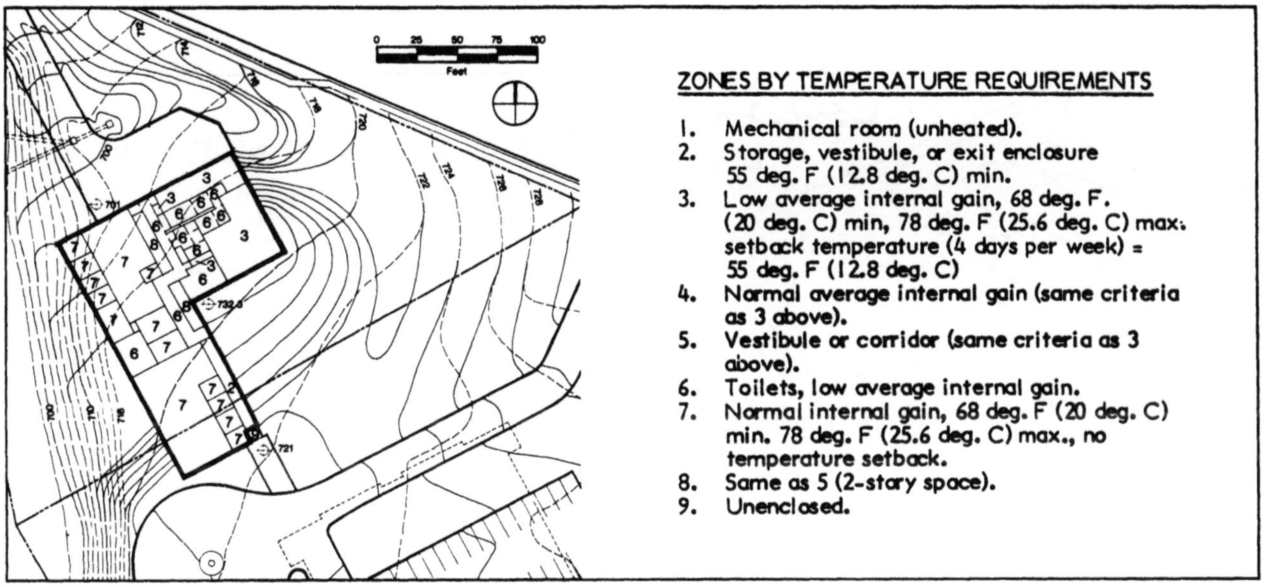

FIGURE 91
Preliminary Schematic Plan of Full-Berm Building--Upper Level

All fire exiting is either downward or horizontal. The horizontal depth of the building is such that fire access can be provided in lieu of an automatic sprinkler system.

Access for maintenance and fire-fighting is provided by maintaining a relatively flat area on the southwest side of the building. This also allows a setback from the top of the slope. This reduces the load superimposed on the top of the slope. The setback from the slope may have to be increased further, however, depending on the recommendations of the final geotechnical report (to be ordered after life-cycle costing). The roads are designed with an inverted crown. All surface drainage employs catch basins to avoid erosion of the slopes. The catch basin at the service drive on the northwest carries water to an energy dissipater to be located at the toe of the slope. Cut and fill has not been balanced at this point. Excess material may be used to form berms to help screen the vehicle maintenance facility or parking, however. Or, the berms may be used to help reinforce the formal cross-axis arrangement of the site. Planting within the parking area can be used to reinforce the primary axis.

The diagrammatic floor plans, Figures 91 and 92, will be used to determine the relative life-cycle cost of this configuration.

b. Half-Berm Configuration. The preliminary site plan for the half-berm building (refer to Figure 93) is similar to the site plan for the full-berm building in terms of access and surface drainage. However, the organization of the site and the building are substantially different. In terms of site development, it is not essential that parking and vehicle maintenance reinforce the building entrance in the way it does for the full-berm building. Rather, in this case, enough of the building is exposed above grade that the building can adequately define the entrance without assistance from the site. Nevertheless, the site plan organization for the

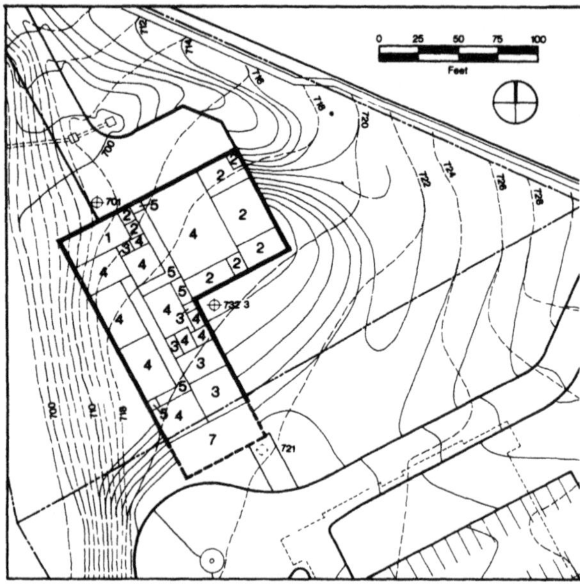

ZONES BY TEMPERATURE REQUIREMENTS

1. Mechanical room (unheated).
2. Storage, vestibule, or exit enclosure 55 deg. F (12.8 deg. C) min.
3. Low average internal gain, 68 deg. F. (20 deg. C) min, 78 deg. F (25.6 deg. C) max. setback temperature (4 days per week) = 55 deg. F (12.8 deg. C)
4. Normal average internal gain (same criteria as 3 above).
5. Vestibule or corridor (same criteria as 3 above).
6. Toilets, low average internal gain.
7. Normal internal gain, 68 deg. F (20 deg. C) min. 78 deg. F (25.6 deg. C) max., no temperature setback.
8. Same as 5 (2-story space).
9. Unenclosed.

FIGURE 92
Preliminary Schematic Plan of Full-Berm Building--Lower Level

full-berm building could be beneficially used for the half-berm building. This could be further considered after the life-cycle cost comparison is complete.

The basis for the plan of the half-berm building is schematic section number 1 shown in Figure 86. Of the five sections in Figure 86, section 1 has the best allocation of functions in terms of adjacency requirements. Because of this and because equalization of floor area is not as critical for the life-cycle economy of the half-berm configuration, the lower floor is considerably larger than the upper floor. The lower floor is below-grade on two sides and exposed on the southwest and northwest sides for view, natural light, and access. The upper floor, housing the administrative functions, is completely above-grade. The entrance to the upper floor is on grade. The approach to the entrance is across an open courtyard that is also the roof of the lower floor. The wall around the courtyard, in addition to defining the entrance forecourt or courtyard, prevents vehicles from driving onto the earth-sheltered roof. This will result in lower structural costs, although the savings will be partially offset by the cost of the courtyard wall. The upper floor is positioned so that the view to the west is not across a roof. The administrative spaces are divided into Navy and Marine functions and placed in separate wings with the lobby functions located in the middle (refer to Figure 94). A lightweight arcade runs along the courtyard side of each wing. The arcade on the north wing provides passive shading; the arcade along the south wing provides a covered walk to the entrance. The arcades are directly above the corridors of the lower floor (refer to Figure 95). One-half of the width of each arcade consists of a continuous skylight which daylights the corridor below.

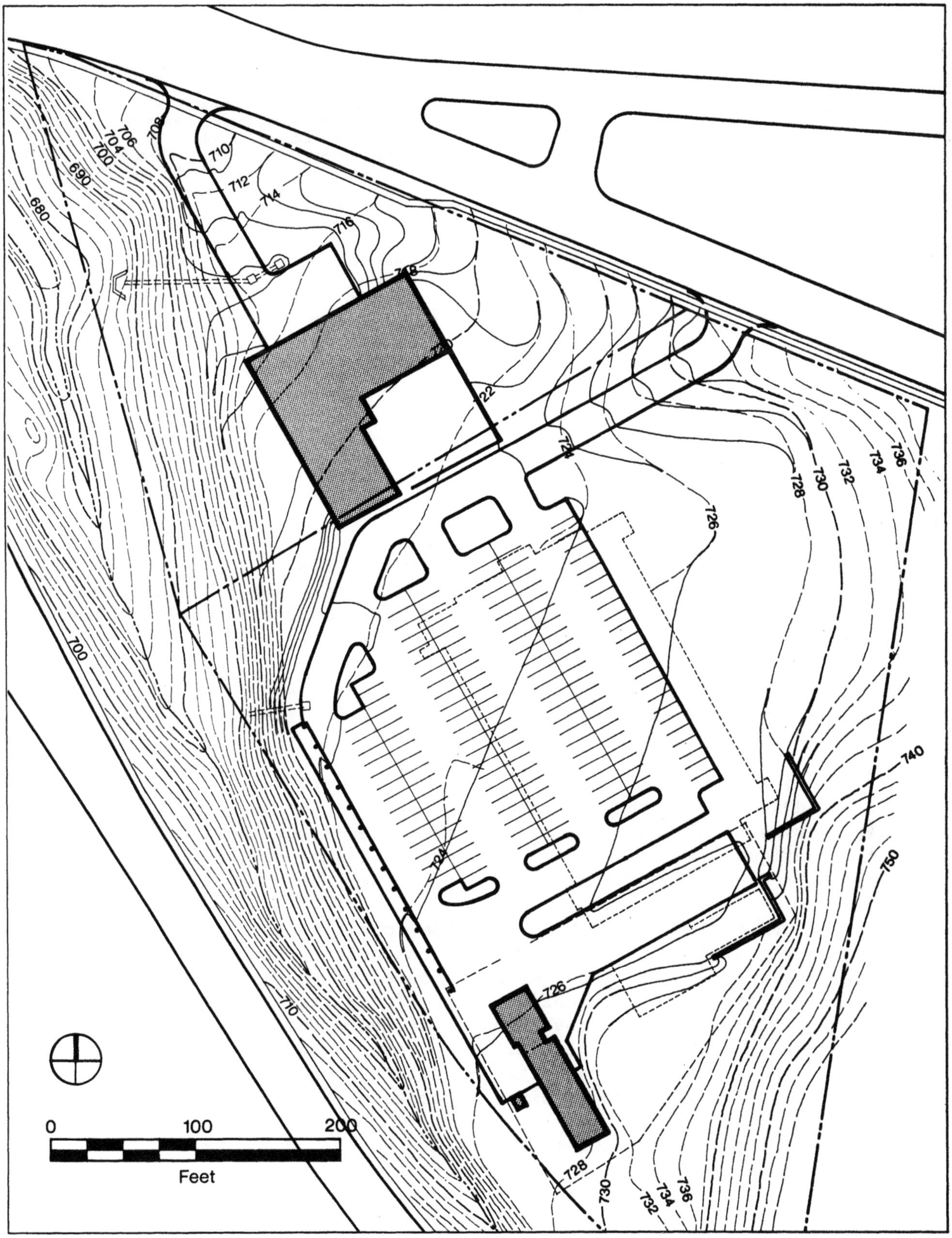

FIGURE 93
Preliminary Site Plan for Half-Berm Building

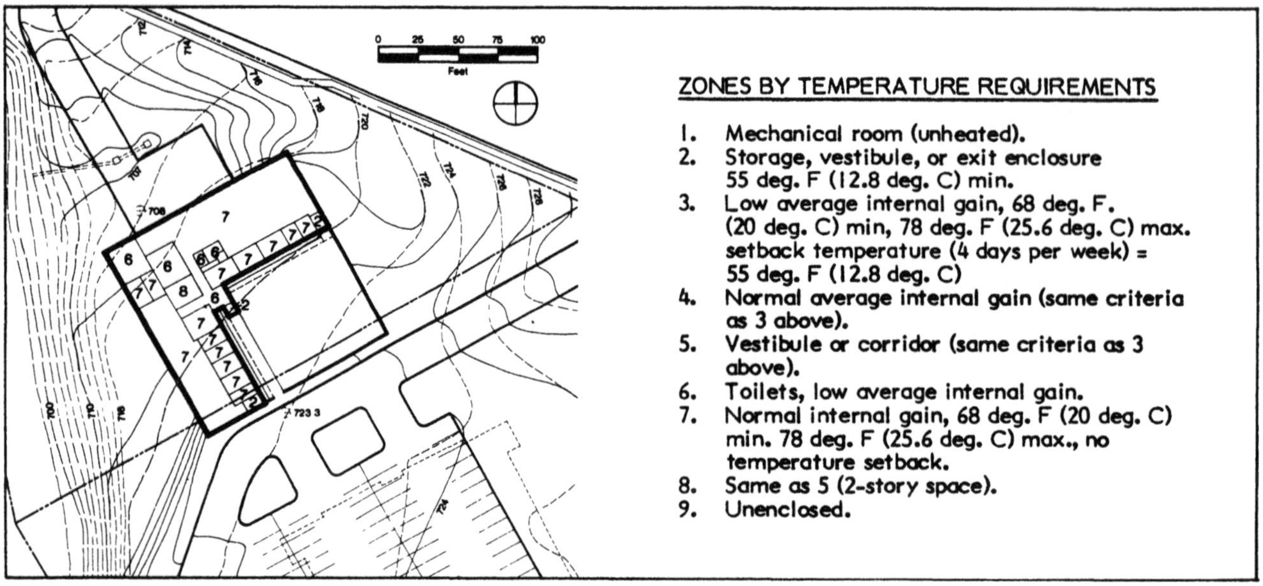

FIGURE 94
Preliminary Schematic Plan of Half-Berm Building--Upper Level

The square plan of the lower floor level (Figure 95) is primarily a result of the organizational concept of the building. Aspect ratio, as previously determined, is not an energy consideration for the below-grade spaces. However, almost one-half of the exterior wall is exposed above grade for light and access. The classroom spaces which are normally placed with the long room dimension parallel to the windows are turned perpendicular to the windows to reduce heat loss, primarily during the unoccupied hours.

Exiting from the lower floor is upward except for supplementary exits located at the assembly room and the garage. The exit at the assembly room also serves as a handicap access to the lower floor level. Unlike the full-berm configuration, the half-berm configuration must be sprinklered because of the depth from the exterior exposed wall and direction of exiting.

19. AREA TAKEOFF AND UNIT PRICING. Only the conditions and quantities which vary between the alternatives are costed. The cost of the interior partitions and stairs, for instance, are assumed to be invariant. Two alternatives to the half-berm configuration are initially examined: a full-berm building with earth cover and a full-berm building without earth cover. This allows the total cost of earth cover--which affects structure, mechanical systems, energy consumption, maintenance and repairs, as well as the roof system itself--to be examined separately.

20. ENERGY PREDICTIONS AND MECHANICAL SYSTEMS COSTS. Wall systems and insulation thicknesses are first defined. The insulation thickness chosen for above-grade walls and roof areas are based on minimum NAVFAC requirements for the region. Insulation for below grade walls is determined from Table 3 in Section 14, Insulation. Other insulation thicknesses may be analyzed later following the initial life-cycle analysis.

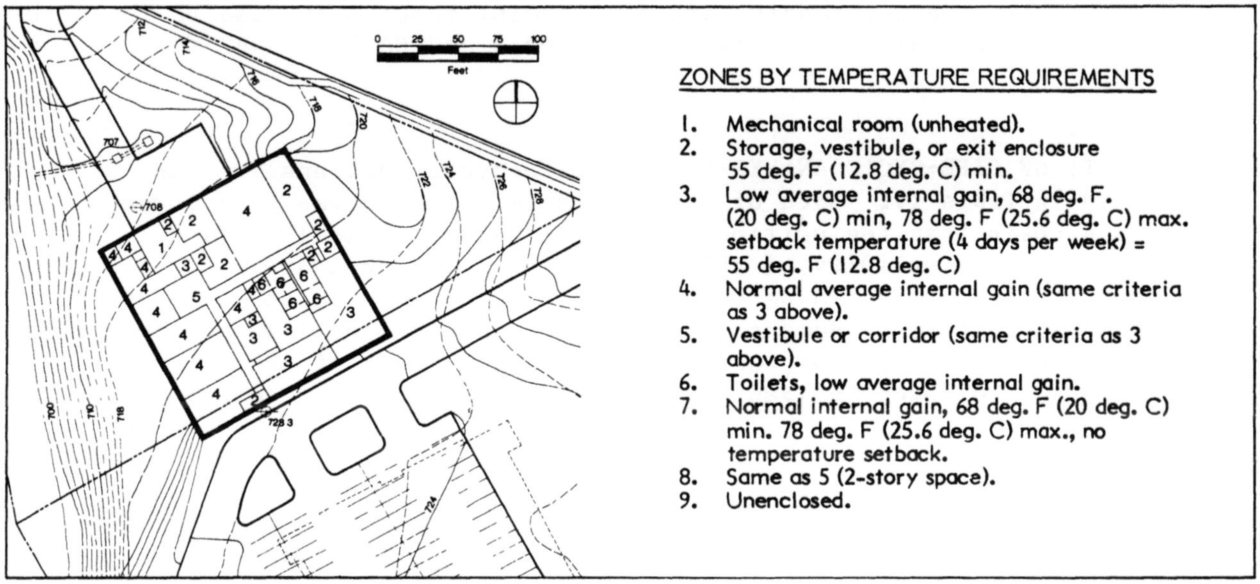

FIGURE 95
Preliminary Schematic Plan of Half-Berm Building--Lower Level

The temperature requirements and internal loads for each room or space are identified in plan (see Figures 91, 92, 94 and 95).

For this particular building, the energy calculations are the most laborious aspect of the life-cycle cost analysis. Energy requirements for the representative zones are determined using the variable bin method described in Section 18, Energy Calculations. For each zone (above-grade and below-grade), this must be done for each month and each bin condition. To facilitate the calculations, the worksheet method described in Section 18 can be computerized. The worksheet method for one zone with temperature set backs and a varying occupancy schedule may take as much as eight manhours of work if done by hand. Writing the computer program may be more expedient in the long run.

The zone energy requirements (load calculations) should be saved as an intermediate result. These results can then be modified to predict new zone energy requirements when certain building parameters are altered--for instance an earth-covered roof that is replaced with a conventional roof.

After obtaining the total annual energy requirements, the estimated peak loads must be calculated based on the procedure described in Section 18. From the peak loads, the equipment, and system costs can be obtained.

21. LIFE-CYCLE COST FACTORS. Life-cycle cost factors are obtained directly from NAVFAC for each project. These include discount factors, spending escalation rates and the life-cycle cost period. It is extremely important that these factors be carefully considered to reflect the best economic predictions currently available from NAVFAC. See Section 5, Life-Cycle Costing, for a sensitivity analysis. See also, NAVFAC P-442 for the prescribed procedure for life-cycle analysis.

Certain maintenance, repair, and replacement costs must also be included in the life-cycle analysis. Refer to Section 5 and NAVFAC P-442 for criteria.

22. RESULTS OF LIFE-CYCLE ANALYSIS. The life-cycle cost analysis is summarized in Table 19. Both of the full-berm alternatives have lower life-cycle costs than the half-berm alternative as initially defined. In order to check the validity of this comparison, certain adjustments are now made to the half-berm building to arrive at new life-cycle costs for the full-berm alternatives. Because these are changes to the life-cycle basis they, in effect, increase or decrease the relative cost of the alternatives. To lower the basis, the following changes are made:

a. Elimination of Sprinkler System. To eliminate the sprinkler system, the classification of the building as an "underground building" must be dropped. To do this, the configuration of the building must be changed to reduce the building depth from exterior fire access (according to the criteria in Section 12, paragraph 4). For now, it is assumed that the earth-covered portion of the building can be relocated on the west side of the building. The unit prices and unit energy requirements previously obtained are applied to the new configuration. The configuration is not worked out in plan but is assumed to be workable for now. The net result (after considering additional energy and maintenance costs) increases the relative life-cycle cost of each of the full-berm alternatives by $1,000 as shown in Table 20. This is labeled a "loss" for the full-berm configurations.

b. Elimination of Skylight and Arcade. With the redistribution of the half-berm building, the skylights which once served to light deep interior space are no longer as important. The skylight can now be omitted. However, the intangible loss to the habitability of the building should be recognized. The skylight in the full berm configuration cannot be so easily eliminated because natural light would then be absent in much of the building.

With the earth-covered lower floors relocated, the vehicular dropoff can be located close to the main entrance. This reduces the importance of the arcade. Consequently, the arcade, as well as the skylight, can be omitted. The net result is a $35,000 savings for the half-berm building--or in terms of the full-berm alternatives, a $35,000 loss.

c. Elimination of Earth-Cover on the Half-Berm Building. Finally, by examining the energy and cost figures for the earth-sheltered roof for the full-berm alternatives, a rough estimate may be obtained for eliminating the earth-cover in the half-berm configuration. The life-cycle discounted energy savings for earth-cover are about $500 more than the life-cycle discounted costs for maintaining the grass on the roof. However, the initial architectural and structural savings obtained amount to $37,000. Thus the relative cost of the full-berm alternatives is increased by $37,000

TABLE 19 (Continued on next page)
Summary of Life-Cycle Cost Analysis

ASSUMED ECONOMIC FACTORS FOR THIS PROJECT (Obtained from NAVFAC)

Life-Cycle Period	= 25 years
General Inflation Rate	= .09
Discount Factor	= .07 + general inflation rate

	Full Berm Without Earth on Roof	Full Berm With Earth on Roof
I. INITIAL CONSTRUCTION COSTS RELATIVE TO HALF BERM BUILDING		
Exterior Above-Grade Wall	$ -6,300	$ -6,300
Exterior Below-Grade Wall, including Waterproofing and Insulation	+38,800	+38,800
Windows	-29,100	-29,100
Skylight and Skylight Curb	+53,700	+53,700
Metal Roof, including Insulation	+29,800	+29,000
Built-Up Roof, including Roof Insulation and Roof Structure	+42,800	-143,800
Roof Flashing, Parapets	-2,300	-4,700
Earth Covered, including Soil Gravel Layer, Insulation, Waterproofing Roof Structure and Irrigation System	-109,400	+122,900
Footings at Foundation Wall	-11,000	-11,000
Slab on Grade	-6,100	-6,100
Structural Floor	+19,900	+19,900
Sprinkler System	-38,500	-38,500
Canopy and Sun Screen	-11,500	-11,500
Retaining Wall	+6,900	+6,900
Columns and Footings, including Piles (when applicable)	-1,600	+40,600
Bulk Excavation	+11,700	+11,700
Berm (Fill)	+18,000	+18,000
Foundation Backfill	+7,600	+7,600
Mechanical	-46,000	-60,000
Total Construction Saving/Loss	-$33,600 (savings)	+$38,100 (loss)

TABLE 19 (Continued)
Summary of Life-Cycle Cost Analysis

Gas Spending Escalation Rate:	= .08 + general inflation rate
Electricity Spending Escalation Rate:	= .07 + general inflation rate
Maintenance Spending Escalation Rate:	= .00 + general inflation rate
Midpoint of funding to start of savings accrual	= .5 years

		OPERATIONAL COSTS	
		Full Berm Without Earth on Roof	Full Berm With Earth on Roof
2.	ENERGY COSTS RELATIVE TO HALF BERM BUILDING		
	Natural Gas	-$ 820/year	-$ 851/year
	Electricity	-1,300/year	-1,904/year
	Total 25-Year Cumulative Discounted Energy Savings/Loss	-$54,500 (savings)	-$69,900 (savings)
3.	MAINTENANCE RELATIVE TO HALF BERM BUILDING		
	Grass Fertilization	-$200/year	+$190/year
	Grass Mowing	-270/year	+250/year
	Watering	-200/year	+190/year
	Window Cleaning	+70/year	+70/year
	Total 25-Year Cumulative Discounted Maintenance Savings/Loss	-$7,300 (savings)	+$8,500 (loss)
4.	MAJOR REPAIRS RELATIVE TO HALF BERM BUILDING		
	Total 25-Year Cumulative Discounted Repairs for Built-Up Roof Replacement, Painting, Equipment Replacement	-$1,000 (savings)	-$5,000 (savings)
5.	TOTAL LIFE-CYCLE SAVINGS/LOSS	-$96,400 (savings)	-$28,300 (savings)

TABLE 20
Adjustments to Life-Cycle Cost Analysis

COST ADJUSTMENTS TO FULL BERM BUILDINGS RELATIVE TO HALF BERM BUILDING

	Full Berm Without Earth Cover on Roof	Full Berm With Earth Cover on Roof
1. Change configuration of half-berm building to eliminate sprinkler system.		
Additional Backfilling	-$4,000	-$4,000
Additional Windows	-6,000	-6,000
Additional Wall Area	-20,000	-20,000
Eliminate Sprinkler System	+38,000	+38,000
Energy, Maintenance Repairs	-7,000	-7,000
SUBTOTAL	+1,000 (loss)	+1,000 (loss)
2. Omit skylight and arcade on half-berm building.		
Initial Construction	+34,000 (loss)	+34,000 (loss)
Energy, Maintenance, Repairs	+1,000 (loss)	+1,000 (loss)
SUBTOTAL	+35,000 (loss)	+35,000 (loss)
3. Omit earth-cover on lower roof of half-berm, and use built-up roof.		
Initial Construction	+37,000 (loss)	+37,000 (loss)
Energy, Maintenance, Repairs[1]	0	0
SUBTOTAL	+37,000 (loss)	+37,000 (loss)
TOTAL FOR ALL ADJUSTMENTS	+$73,000 (loss)	+$73,000 (loss)
TOTAL FROM PREVIOUS LIFE-CYCLE SAVINGS/LOSS	-$96,400 (savings)	-$28,300 (savings)
FINAL LIFE-CYCLE SAVINGS/LOSS Using Adjusted Half-Berm Configuration as the Basis	-23,400 (savings)	+44,700 (loss)

[1] Cumulative discounted energy savings due to earth cover on the roof are approximately equal to additional discounted costs for maintenance of grass on the roof.

d. Final Adjusted Life-Cycle Costs for Full-Berm Alternatives. After adding the above cost adjustments to the relative life-cycle cost of the full-berm buildings, the full-berm configurations become approximately $73,000 more expensive than previously calculated. The full-berm building without earth-cover on the roof still shows a present worth savings, however. These savings are approximately $23,000. Neither this full-berm configuration nor the stripped-down half-berm configuration have earth-cover on the roof. The primary cost differences are the berm height, the piles and foundation, and the aspect ratio as defined by wall-to-floor area. The life-cycle analysis shows that the berm more than pays for itself. Per square foot of building area, the piles, footings, and foundation system for the full-berm are more expensive than the spread footings and foundation system for the half-berm. However, with a lower aspect ratio and fewer columns, the full berm building proves to be more economical. This outcome, in favor of the full-berm building is attributible to the strategy of equalizing the floor areas as well as berming the exterior walls.

23. REFINEMENTS TO INSULATION THICKNESS AND DEPTH. Additional analyses are required to predict building energy performance as a result of modifying the thickness and depth of thermal insulation. The original energy calculations, utilizing the variable bin method, must be modified to reflect changes in insulation thickness and depth. This effort will require less computation time than the original effort because only the zones affected by the changes must be considered. The building's energy components (roof, walls, windows, floor, air infiltration, ventilation, internal heat gains) cannot be analyzed independently of each other--a complete heat balance must be recomputed for each bin. If this calculation procedure is computerized, time required for carrying out the analysis will be reduced dramatically.

24. EVALUATION OF INTANGIBLE DIFFERENCES. At this point, the full-berm configuration without earth-cover could be recommended on the basis of the life-cycle cost analysis. However, the following intangibles (some of which can be translated into costs) should be observed:

a. Security. Even with special provisions such as parapets or landscaping, the northern portion of the roof will be accessible from the berm. The berm height could be reduced and the parapet height increased to help solve this problem. Based on the unit costs already obtained, it can be shown that this will increase the relative life-cycle cost of the building. As a compromise, a hedge along the top of the berm may be used.

b. Natural Light. Although the functions requiring daylighting are fairly well accommodated in all the schemes, the full-berm scheme has greater variety and interest because of the skylight. The skylight and atrium also tend to create a better organizational clarity for the full-berm building.

c. Exterior Noise Reduction. Exterior noise reduction should be better for the full berm building.

d. Radiation Protection. As a fallout shelter, the radiation protection factor will also be higher for the lower level areas adjacent to the earth-berm.

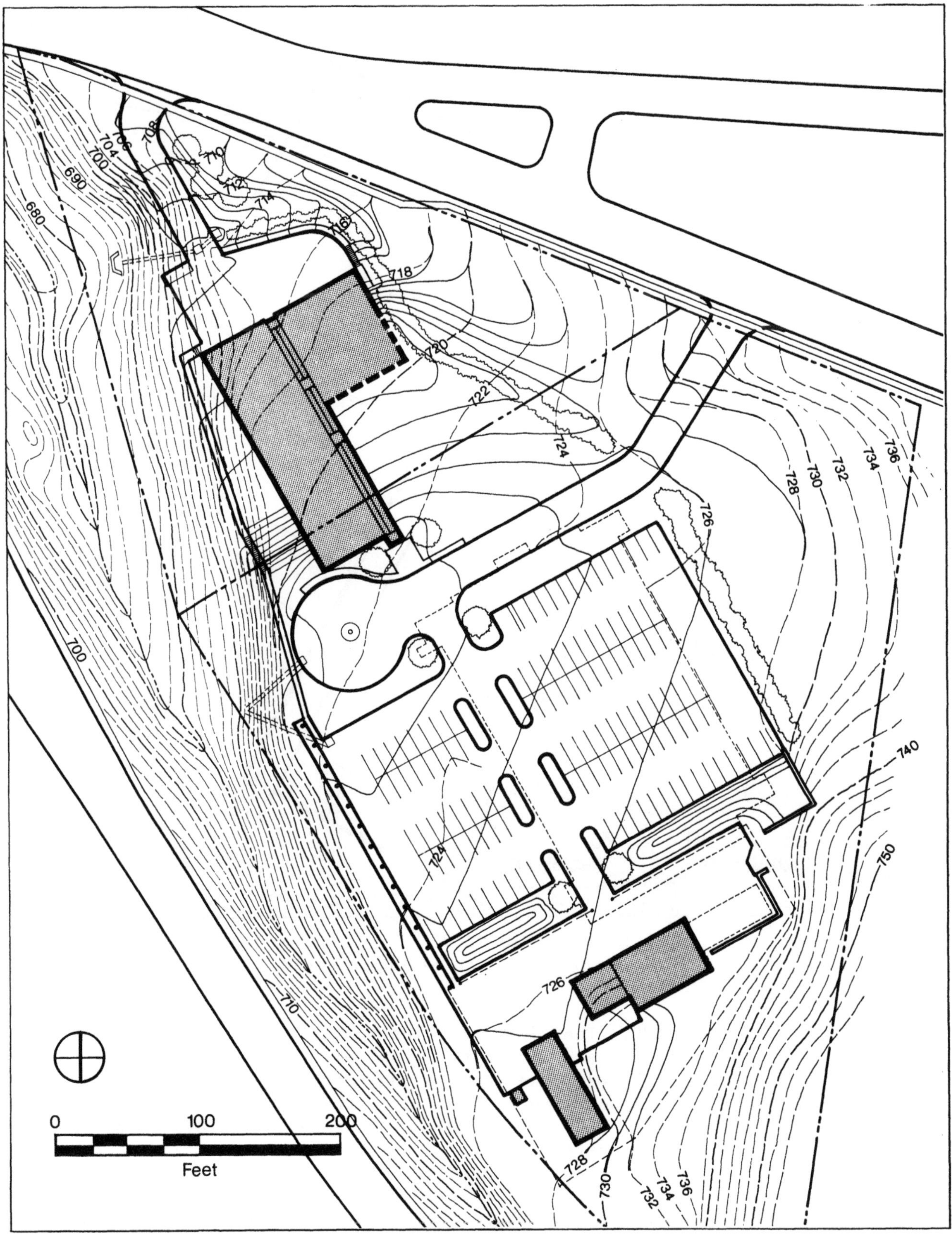

FIGURE 96
Final Schematic Site Plan

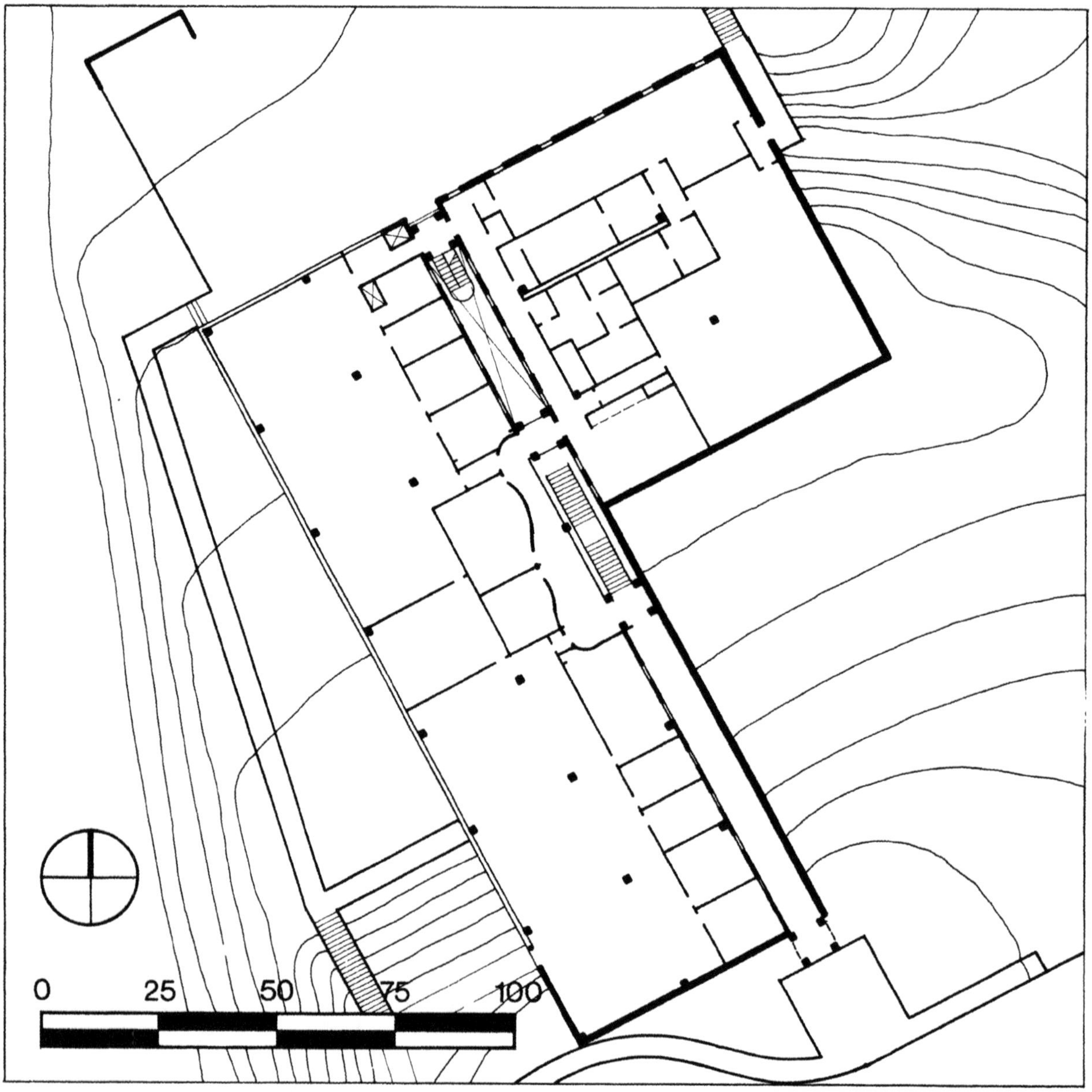

FIGURE 97
Final Schematic Plan of Upper Level

25. FINAL SCHEMATIC DESIGN. The final solution to the visibility of a non-earth-sheltered and the security/safety problem of berming up to the roof edge is solved as follows. The roof to the north of the skylight is earth-covered to improve appearance and to reduce the potential for vandalism or damage to roof (see Figure 96). A guardrail is provided at the roof edge where the earth-covered roof is accessible. Unauthorized access to the roof of the south wing is prevented by the linear skylight. The north side (accessible side of this skylight) is insulated metal roofing to improve thermal efficiency and to reduce the potential for vandalism. (The roof of the south wing is not earth covered because of the negative life-cycle benefit, as determined in Paragraph 22, above.)

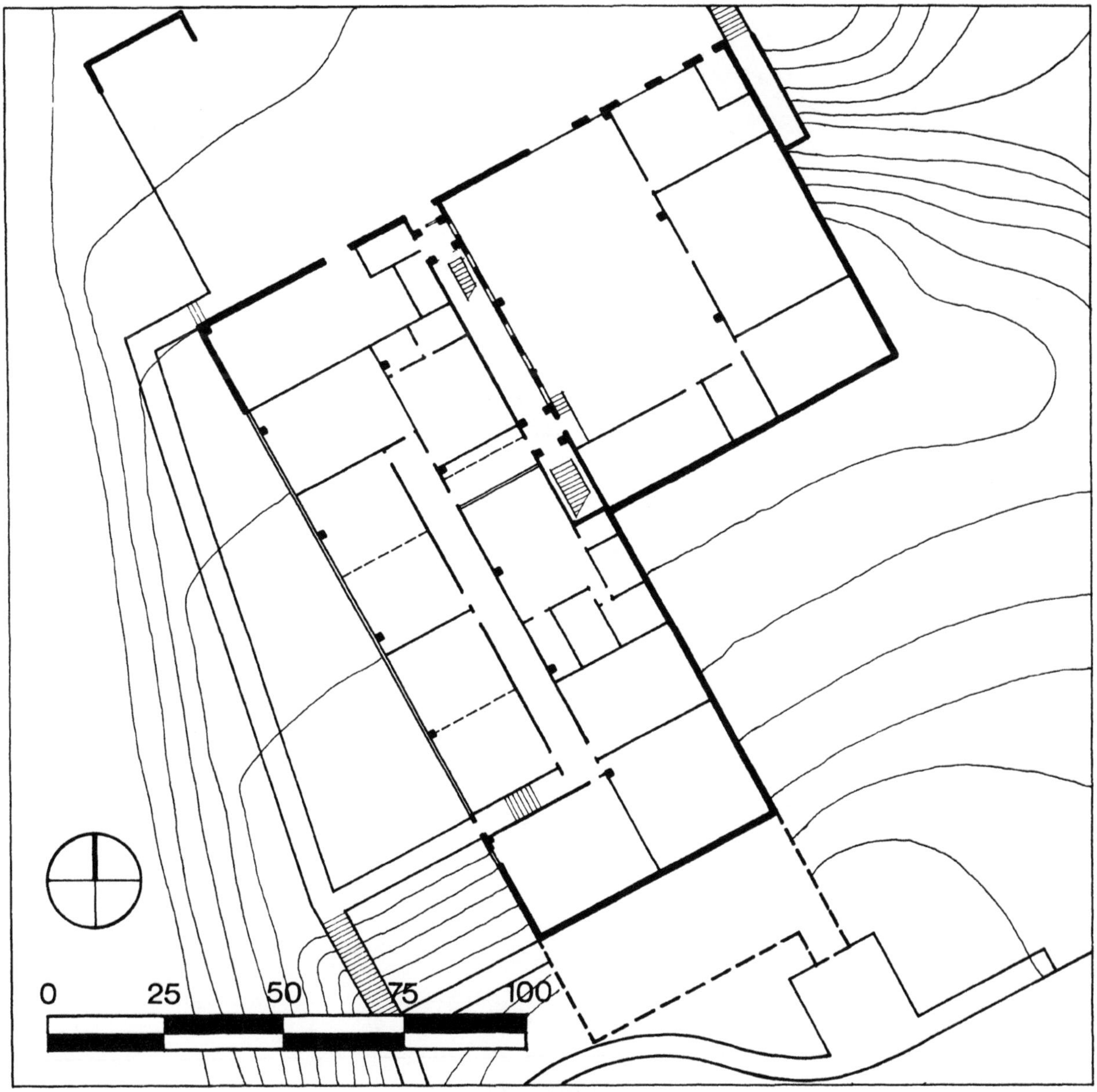

FIGURE 98
Final Schematic Plan of Lower Level

An evergreen species is selected for a windbreak along the north side of the building. The planting is continued along the north side of the parking area to partially screen the parking from the residential area across the street and to help define exterior space adjacent to the building (see Section 6, Site Context, Paragraph 2F). The break in the evergreen planting marks the turn in the approach road as well as the change in gradient.

Excess excavated material is distributed on the site in areas not requiring compaction. Most of this is used to extend the earth berm at the building and to create small berms in front of the vehicle maintenance facility.

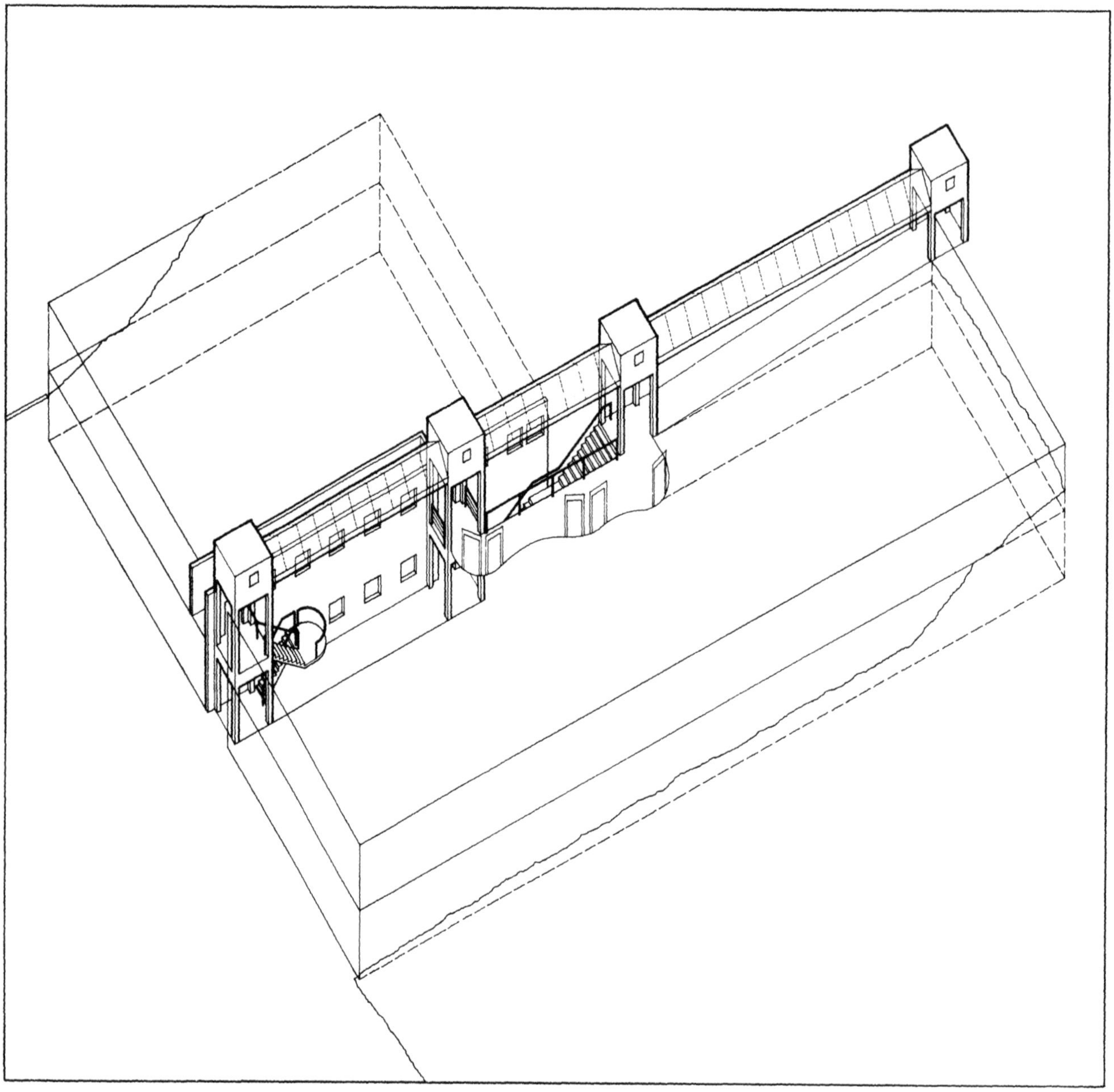

FIGURE 99
Axonometric of Circulation Space

The final schematic plans for the upper and lower floors are shown in Figures 97 and 98. Figure 99 depicts the main circulation and the atrium. The interruptions in the skylight are required for excess heat exhaust for the atrium as well as other mechanical functions such as cooling units and exhaust for toilets and lockers. The skylight (insulated glass with heat-strengthened laminated safety glass for the inside lite) has no thermal shuttering. However, motorized blinds can be used to reduce heat gain. The walls facing the atrium are masonry to increase thermal mass for direct solar gain benefit (see Section 11, Natural Lighting, Skylights and Passive Solar).

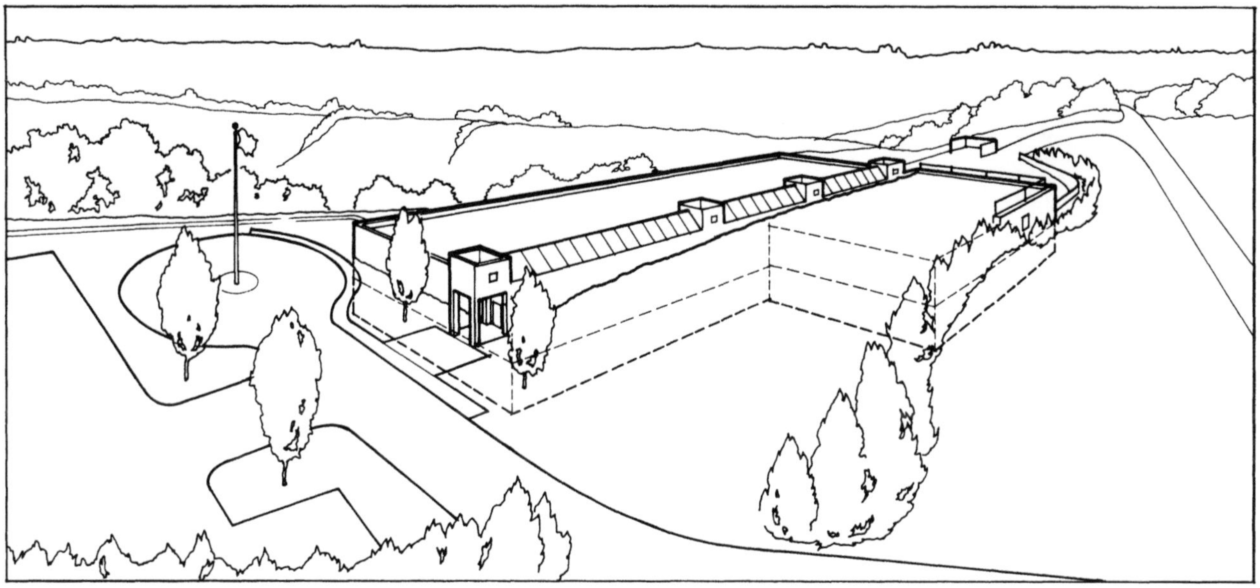

FIGURE 100
Perspective Looking West

FIGURE 101
Perspective Looking East

The aerial perspective looking west (Figure 100) shows the building from the main entrance. The aerial perspective looking east (Figure 101) shows the solution to the fenestration. The lower-level windows are screened from late afternoon summer sun by the trees along the top of slope. Both upper-level and lower-level windows utilize a reflective coating to reduce solar gain.

REFERENCES

1. Labs, K., "Regional Analysis of Ground and Above-Ground Climate," Underground Space, Vol. 6., No. 6 and Vol. 7, No. 1, Pergamon Press, New York, 1972.

2. Olgyay, V., Design with Climate: Bioclimatic Approach to Architectural Regionalism, Princeton, New Jersey, Princeton University Press, 1963.

3. Givoni, B., Man, Climate and Architecture, 2nd Ed., London, Applied Science Publisher, Ltd., 1976.

4. Robinette, G. O., Plants, People and Environmental Quality, G.P.O. Stock No. 2405-0479, 1972.

5. ASHRAE, Inc., The American Society of Heating, Refrigerating, and Air Conditioning Engineers, New York, NY, Standard 62-1981, Ventilation for Acceptable Indoor Air Quality, 1981.

6. Haliburton, T.A., Use of Engineering Fabric in Transportation Related Applications, Office of Development, Federal Highway Administration, Washington, D.C.

7. Balcomb, J.D., and Environmental Action, Inc. Passive Solar Design Handbook, Vols. 1 and 2, DOE Publishing, Inc., 1980.

8. Meixel, G., "Effect of Insulation Placement on the Heat Transfer Through Underground Walls" ASCE Technical Conference, Energy Conservation in Building Design, Construction and Management, 1982.

9. Houghten, F.C., Taimuty, D.I. Gutberlet, C. and C.J. Brown, "Heat Loss Through Basement Walls and Floors," Transactions American Society of Heating and Ventilating Engineers, No. 1213, pp. 369-384, 1942.

10. Dechow, F.J. and Epstein, K.A., Laboratory and Field Investigations of Moisture Absorption and its Effects on Thermal Performance for Various Insulation, ASTM Symposium on Advances in Heat Transmission Measurements, Philadelphia, PA, 1977.

11. Tobiasson, W. and Ricard, J., Moisture Gain and Its Thermal Consequence for Common Roof Insulation, U.S. Army Cold Regions Research and Engineering Laboratory, Hanover, NY, Proceedings, 5th Conference on Roofing Technology, April 19-20, 1979.

12. Department of Energy Report, The Measurement of Apparent Thermal Conductivity of Soils in the Field and Laboratory Using Thermal Conductivity Probes, Contract DE-AC03-805F11508, 1983.

13. ASHRAE, American Society of Heating, Refrigerating, and Air-Conditioning Engineers, Handbook of Fundamentals, New York, NY, 1981.

14. Boileau, G.G., and Latta, J.K., Calculation of Basement Heat Losses, National Research Council of Canada, Division of Building Research, Ottawa, Canada, Report No. NRC 10477 Dec. 1968.

15. Meixel, G.D. Shipp, P.H., Bligh, T.P., "The Impact of Insulation Placement on the Seasonal Heat Loss Through Basement and Earth-Sheltered Walls," Underground Space, Vol. 5, pp. 41-47, Pergamon Press, 1980.

16. Speltz, J., "A Numerical Simulation of Transient Heat Flow in Earth-Sheltered Buildings for Seven Selected U.S. Cities," M.S. Thesis, Trinity University, 1980.

17. Speltz, J. and Haves, P., "The Thermal Benefits and Cost Effectivenss of Earth Berming," Proceedings, 5th National Passive Solar Conference, October 1980, AS/ISES, Amherst, Massachusetts, pp. 337-341.

18. Buchberg, H., "Sensitivity of Room Thermal Response to Inside Radiation Exchange and Surface Conductance," Building Science Vol. 6, pp. 133-149, Pergamon Press, 1971.

19. Baladi, J.Y., Schoenhals, R.J., and Ayers, D.L., "Transient Heat and Mass Transfer in Soils," AIAA-ASME Thermophysics and Heat Transfer Conference, Palo Alto, California, May 24-26, 1978.

20. Kusuda, T., "Earth Temperature Beneath Five Different Surfaces" Report 10-373, National Bureau of Standards, Washington, D.C., OCD Work Order DAH 20-68-W-0106, 1971.

21. National Climatic Center, "Hourly Solar Radiation-Surface Meterological Observations," SOLMET User's Manual, Vol. 1, TD-9724, Asheville, NC, 1980.

22. National Climatic Center, Typical Meteorological Year User's Manual, TD-9734, Asheville, NC, May 1981.

23. ASHRAE, American Society of Heating, Refrigerating and Air-Conditioning Engineers, 345 East 47th Street, New York, NY 10017, Systems Handbook, 1980.

24. Clark, G. and Allen, C.P., "The Estimation of Atmospheric Radiation for Clear and Cloudy Skies," Proceedings of the 2nd National Passive Solar Conference, University of Pennsylvania, March 16-18, 1978.

25. Kreith, F. and Sellers, W.D., General Principles of Natural Evaporation, Heat and Mass Transfer in the Biosphere, Part I Transfer Processes in the Plant Environment, Scripta Book Co., 1975.

26. Speltz, J. and Meixel, G., "A Computer Simulation of the Thermal Performance of Earth Covered Roofs," The Potential of Earth-Sheltered and Underground Space, Proceedings, Underground Space Conference and Exposition, June 8-10, 1981, Kansas City, Pergamon Press, New York, pp. 91-108, 1981.

BIBLIOGRAPHY

The entries in this bibliography are divided into groups by topic. These categories are listed below.

	PAGE
General Reference	2
General Works	3
Construction and Life-Cycle Costs	3
Economic/Insurance	3
Building Code/Zoning	4
Legal Aspects	4
Planning	4
Site Planning/Landscape Design	4
Climate	5
Passive Solar Design	5
HVAC Systems	6
Condensation/Humidity	6
Air Quality	6
Thermal Environment of Underground Shelters	6
Heat Transfer	7
Thermal Properties	8
Heat Storage	8
Blast and Earthquake Resistance	8
Acoustics	9
Lighting	9
Human/Physical/Psychological Design	10
Mined Construction	10
Structural Analysis	10
Construction Methods	11
Construction Materials	11
Drainage	11
Waterproofing (General)	11
Upside-Down Roof Systems	12
Bituminous Waterproofing	12
Bentonite Waterproofing	12
Sheet Membrane Waterproofing	12
Liquid-Applied Membrane Waterproofing	13
Moisture Problems in Insulation	13

BIBLIOGRAPHY

GENERAL REFERENCE

Alternatives in Energy Conservation: The Use of Earth Covered Buildings, Proceedings and Notes of a Conference Held in Forth Worth, Texas, July 9-12, 1975, NSF/RA-760006.

Boyer, L.L., ed.; Proceedings, Earth Sheltered Building Design Building Innovations, Architectural Extension, Oklahoma State University, Oklahoma City, Oklahoma, April 18-19, 1980.

Boyer, L.L., ed.; Proceedings, Earth Sheltered Performance and Evaluation Conference, Architectural Extension, Oklahoma State University, Tulsa, Oklahoma, October, 1981.

Earth Sheltered Housing Design, Underground Space Center, University of Minnesota, Minneapolis, 1978.

Holthusen, T.L., ed.; The Potential of Earth Sheltered and Underground Space Proceedings, Underground Space Conference and Exposition, June 8-10, 1981, Kansas City, Pergamon Press, New York, 1981.

Jones, L.K.; Underground Construction (bibliography), Council of Planning Librarians, Chicago, Illinois, 1978.

Legal, Economic and Energy Considerations in the Use of Underground Space, National Academy of Sciences/National Committee on Tunneling Technology, Washington, D.C., 1974.

Moreland, F., ed.; Earth Covered Buildings and Settlements, Proceedings, Alternatives in Habitat: the Use of Earth Covered Buildings and Settlements, GPO No. CONF-7805-138-P2. (Also available from NTIS.)

Moreland, F.; Higgs; Shih, eds.; Earth Covered Buildings: Technical Notes, Proceedings, Alternatives in Habitat: the Use of Earth Covered Buildings and Settlements, GPO No. CONF-7805-138-P1. (Also available from NTIS.)

Need for National Policy for the Use of Underground Space: Engineering Foundation Conference, Berwick Academy, South Berwick, Maine, June 25-29, 1973.

Scalise, J.W., ed.; Earth Integrated Architecture, College of Architecture, Arizona State University, Tempe, 1975.

Stauffer, T., Sr.; Underground Utilization: A Reference Manual of Selected Works, Department of Geosciences, University of Missouri, Kansas City, Missouri, 1978.

Sterling, R.L.; Earth Sheltered Community Design, Van Nostrand Reinhold, New York, 1981.

Sterling, R.L.; Earth Sheltered Residential Design, Van Nostrand Reinhold, New York, 1982.

Underground Space, Journal of the American Underground-Space Association, University of Minneapolis, Minnesota, Minnesota.

Wells, M.; Underground Designs, P.O. Box 1149, Brewster, Massachusetts, 02361, 1977.

GENERAL WORKS

Bligh, T.P.; "Energy Conservation by Building Underground," Underground Space Vol. 1, No. 1, Pergamon Press, New York, 1976, pp. 19-33.

Fairhurst, C.; "Going Under to Stay on Top," Underground Space Vol. 1, No. 2, Pergamon Press, New York, 1976, pp. 71-86.

Harrison, Lt., L.; "Is it Time to Go Underground?" The Navy Civil Engineer XIV:3, 1973.

Jones, L.G.; "Nontraditional Military Uses of Underground Space," Underground Space Vol. 2, No. 3, Pergamon Press, New York, 1978, pp. 153-156.

Labs, K.; "The Architectural Underground, Part I," Underground Space Vol. 1, No. 1, Pergamon Press, New York, 1976, pp. 1-8.

Labs, K.; "The Architectural Underground, Part II," Underground Space Vol. 1, No. 2, Pergamon Press, New York 1976, pp. 133-156.

Van Der Meer, W.J.; "Subterranean Housing in Arid Zones," Planning and Designing of Innovative Houses in Arid Zones: Past and Present, Gideon Golany, ed., Halsted Press Division, John Wiley & Sons, New York, 1980.

CONSTRUCTION AND LIFE-CYCLE COSTS

"Cost and Code Study of Underground Buildings: A Report to the Minnesota Energy Agency," Underground Space Vol. 4, No. 3, Pergamon Press, New York, 1979, pp. 119-136.

ECONOMIC/INSURANCE

De Saventhem, E.M.; "Insuring Risks Underground--Some General Considerations," Underground Space Vol. 2, No. 1, Pergamon Press, New York, 1977, pp. 19-25.

Sterling, R.L.; Earth Sheltered Housing: Code, Zoning, and Financing Issues, Van Nostrand Reinhold, New York, 1982.

BUILDING CODE/ZONING

Sterling, R.L.; Earth Sheltered Housing: Code, Zoning, and Financing Issues, Van Nostrand Reinhold, New York, 1982.

LEGAL ASPECTS

Tarlock, D.A.; "Legal Aspects of the Use of the Underground," Legal, Economic and Energy Considerations in the Use of Underground Space, National Academy of Sciences, Washington, D.C., 1974.

Thomas, W.A.; "Ownership of Subterranean Space," Underground Space Vol. 3, No. 4, Pergamon Press, New York, 1979, pp. 155-164.

PLANNING

Hagman, D.G.; Planning the Underground Uses: Legal, Economics and Energy Considerations in the Use of Underground Space, National Academy of Sciences, Washington, D.C., 1974.

Jansson, B.; "City Planning and the Urban Underground," Underground Space Vol. 3, No. 3, Pergamon Press, New York, 1978, pp. 99-116.

Jansson, B. and Winqvist, T.; Planning of Subsurface Use, Pergamon Press, New York, 1977.

Labs, K.; Planning for Underground Housing, NSF SED 10100 Staffer/2129-2228.

LaNier, R.; "Assessing Environmental Impact of Earth Covered Building," Underground Space Vol. 1, No. 4, Pergamon Press, New York, 1977, pp. 309-315.

Legget, R.F.; Cities and Geology, McGraw-Hill, New York, 1973.

Legget, R.F.; "What Every City Should Do About Geology--and Its Underground," Underground Space Vol. 4, No. 5, Pergamon Press, New York, 1980, pp. 275-278.

SITE PLANNING/LANDSCAPE DESIGN

Friedberg, P.M.; "Roofscape," Architectural and Engineering News, September, 1969, pp. 24-37.

Morgan, W.; "Buildings as Landscape: Five Current Projects by William Morgan," Architectural Record, September, 1972, pp. 129-136.

Robinette, G.O.; Plants, People, and Environmental Quality, GPO, Stock No. 2405-0479.

Wirth, T.E.; "Landscape Architecture Above Buildings," Underground Space Vol. 1, No. 4, Pergamon Press, New York, 1977, pp. 333-346.

CLIMATE

Geiger, R.; The Climate Near the Ground, Harvard University Press, 1965.

PASSIVE SOLAR DESIGN

Bowen, A.; Passive Cooling, Proceedings, International Passive and Hybrid Cooling Conference, Miami Beach, AS/ISES, 1981.

Boyer, L.; "Passive Energy Design and Habitability Aspects of Earth Sheltered Housing in Oklahoma," Underground Space Vol. 4, No. 6, Pergamon Press, New York, 1980, pp. 333-340.

Franta, G., ed.; Proceedings, Fourth Passive Solar Conference, American Section of the International Solar Energy Society, Inc., 1979.

Hayes, J. and Snyder, eds.; Proceedings, Fifth National Passive Solar Conference, American Section of the International Solar Energy Society, Inc., Newark, Delaware, 1980.

Mazria, E.; The Passive Solar Energy Book, Rodale Press, Emmaus, Pennsylvania, 1979.

Miller, H.; Passive Cooling Applications Handbook, American Section of the International Solar Energy Society, Inc., Newark, Delaware, 1980.

Passive Solar Design: An Extensive Bibliography, AIA Research Corp., 1978. (Available from National Technical Information Center, Rept. No. HCD/CS-4113-3.)

Passive Solar Design Handbook Vol. 1 Passive Solar Design Concepts, Total Action, Inc., for DOE, 1980, DOE/CS-0127/1, US-59.

Passive Solar: State of the Art, Proceedings, Second National Passive Solar Conference, Philadelphia, Pennsylvania, March 16-18, 1978, American Section of the International Solar Energy Society.

Passive Systems, Proceedings of the 1978 Annual Meeting of the American Section of the International Solar Energy Society, Inc., Denver, Colorado, August 28-31, 1978, Vol. 2.2, Sec. 1.

Third Passive Solar Conference Proceedings, San Jose, California, January 11-13, 1979, American Section of the International Solar Energy Society, Inc.

HVAC SYSTEMS

Hay, H.R.; International Aspects of Air Conditioning with Movable Insulation," Solar Energy, December, 1969, pp. 427-28.

CONDENSATION/HUMIDITY

Croone, D.J., and Sherratt, A.F.C.; Condensation in Building, Applied Science Publishers Ltd., London, 1972.

Latta, J.K., and Beach, R.K.; "Vapor Diffusion and Condensation," Canadian Building Digest, Report No. CBD-57, National Research Council of Canada, Division of Building Research (243950), Ottawa, Canada.

Seiffert, K.; Damp Diffusion in Buildings, Applied Science Publishers Ltd., London, 1979.

AIR QUALITY

Budnite, R.J.; Human Diseases from Radon Exposure: The Impact of Energy Conservation in Buildings, Energy and Environment Division, Lawrence, Berkeley Laboratory, University of California at Berkeley, 1978.

Recirculation of Exhaust Air, U.S. Dept. of Health, Education, and Welfare, Division of Physical Sciences and Engineering, 1976.

Sollid, E.; A Study on Dehumidification of Shelter Space, Office of Civil Defense, Battlecreek, Michigan, 1961.

THERMAL ENVIRONMENT OF UNDERGROUND SHELTERS

Achenbach, P.R.; "Mathematical Analysis of Thermal Environment in Underground Shelters;" ASHRAE Symposium Bulletin (Survival Shelters), 1962.

Drucker, E.E., and Haines, J.T.; "A Study of Thermal Environment in Underground Survival Shelter Using an Electronic Analog Computer," ASHRAE Transactions Vol. 70, No. 1857, 1965.

Ducan, G.J.; Natural Ventilation of Underground Shelters, OCDOS62114, MRD Division, General American Transportation Corp., Niles, Ohio, 1964.

Heating and Air Conditioning of Underground Installations, Dept. of the Army Technical Manual, TMS-855-4, August 1965.

Kiser, E.R.; Selective Bibliography on Environmental Control Inhabitability of Survival Shelters, College of Engineering, New York University, New York, 1962.

Kusuda, T., and Achenbach, P.R.; "Numerical Analysis of the Thermal Environment of Occupied Underground Spaces with Finite Cover Using a Digital Computer," ASHRAE Transactions Vol. 69, No. 1853, 1963.

Stephenson, J.M.; Cooling Analysis for Protection Structures Located Above and Below Ground, TRR493, U.S. Naval Civil Engineering Lab, Port Hueneme, California, 1968.

Summary of Natural Ventilation of Underground Shelters, OCD062114, MRD Division, General American Transportation Corp., Niles, Ohio, 1964.

HEAT TRANSFER

Akridge, J.M.; "A Decremented Average Ground Temperature Method for Estimating the Thermal Performance of Underground Houses," Passive Cooling, Proceedings, International Passive and Hybrid Cooling Conference, Miami Beach, 1981, American Section of the International Solar Energy Society, Inc., Newark, Delaware, pp. 141-145.

Bircher, T.L.; "The Thermal Performance of Earth Covered Buildings in Hot, Arid Regions," Proceedings, Fifth National Passive Solar Conference, American Section of the International Energy Society, Inc., Newark, Delaware, 1980, pp. 332-336.

Eckert, E., Bligh, T., and Pfender, E.; "Energy-Exchange Between Earth Sheltered Structures and the Surrounding Ground," Energy Vol. 4, No. 2, 1979, pp. 171-181.

Houghten, F.C.; "Heat Loss Through Basement Walls and Floors," Transactions American Society of Heating and Ventilating Engineers, No. 1213, 1942, pp. 369-384.

Latta, J.K., and Boileau, G.G.; "Heat Losses from House Basement," Canadian Buildings, No. 10, National Research Council of Canada, No. NH31, Ottawa, Canada, 1969.

Meixel, G.D.; "Energy Use of Non-Residential Earth-Sheltered Buildings in Five Different Climates," The Potential of Earth Sheltered and Underground Space, Proceedings, Underground Space Conference and Exposition, June 8-10, 1981, Kansas City, Pergamon Press, New York, 1981, pp. 227-257.

Shipp, P.H.; "Analysis and Measurement of the Thermal Behavior of the Walls and Surrounding Soil for a Large Underground Building," Underground Space Vol. 5, No. 2, Pergamon Press, New York, 1980.

Speltz, J., and Haves, P.; The Thermal Benefits and Cost Effectiveness of Earth Berming," Proceedings, Fifth National Passive Solar Conference, American Section of the International Energy Society, Inc., Newark, Delaware, 1980, pp. 332-336.

Szydlowski, R.G., and Kuehn, T.H.; "Analysis of Transient Heat Loss in Earth-Sheltered Structures," Underground Space Vol. 5, No. 4, Pergamon Press, New York, 1980, pp. 237-246.

THERMAL PROPERTIES

Fitton, E.M., and Brooks, C.F.; "Soil Temperature in the United States," Monthly Weather Review Vol. 59, January, 1931.

Fluker, G.J., "Soil Temperature," Soil Science Journal, 1958.

Gilpin, R.R.; "Study of Some Factors That Influence Ground Temperature," Canadian Congress of Applied Mechanics, No. 4 Proceedings. Available from Ecole Polytech, Montreal, Quebec, 1973.

Jumikis, A.R.; Thermal Soil Mechanics, Rutgers University Press, New Brunswick, New Jersey, 1966.

Kersten, M.S.; "Thermal Properties of Soils," Minnesota Engineering Experiment Station Bulletin, LII:28,21, Minneapolis, Minnesota, 1949.

Kusuda, T., and Achenbach, P.R.; "Earth Temperature and Thermal Diffusivity of Selected Stations in the United States," Transactions of the American Society of Heating, Refrigerating, and Air-Conditioning Engineers, Vol. 71, 1965, pp. 61-75.

Peters, A.; Soil Thermal Properties: An Annotated Bibliography, The Franklin Institute, Philadelphia, Pennsylvania, 1962.

HEAT STORAGE

Catani, M.J., and Goodwin, S.E.; "Heavy Building Envelopes and Dynamic Thermal Response," ACI Journal, February 1976, pp. 83-86.

Peavy, B.A.; Dynamic Thermal Performance of an Experimental Masonry Building, National Bureau of Standards Building Science Series 45, COM-73-50725, NTIS, July, 1973.

BLAST AND EARTHQUAKE RESISTANCE

Earth Covered Buildings: An Exploratory Analysis for Hazard and Energy Performance, Moreland Associates, Report to Federal Emergency Management Agency, MAI-81, November, 1981.

"Earthquake Laws Force Schools Underground," American Schools and Universities Vol. 45, No. 11, pp. 23-25, July, 1973.

Duke, C.M., and Leeds, D.J.; "Effects of Earthquakes on Tunnels," Protective Construction in a Nuclear Age, pp. 303-328.

"Final Report for Protective Blast Shelter System Analysis," Bechtel Corp. for Civil Defense, Detroit, Michigan, 1968.

Kiesling, E.W.; "Economical Wind Protection--Underground," Underground Space Vol. 4, No. 5, Pergamon Press, New York, 1980, pp. 279-282.

LeDoux, J.C.; Shielding Factors for Underground Shelters of Various Geometric Shapes, TROSO, U.S. Naval Civil Engineering Laboratory, Port Hueneme, California, April 1961.

Rongued, P.I.; "Underground Structures Subject to Air Overpressure," American Society Civil Engineering Proceedings, EM 1 No. 2751, 1961.

Sisson, G.N.; "Underground for Nuclear Protection," Underground Space Vol. 4, No. 6, Pergamon Press, New York, 1980, pp. 341-348.

Soil-Structure Interaction, Proceedings of the Symposium on, Engineering Research Laboratory, University of Arizona, Tucson, 1964.

Wright, R.N.; Building Practices for Disaster Mitigation, National Bureau of Standards Report BSS 46, COM-73-50188, NTIS.

ACOUSTICS

Beranek, L.L.; Noise and Vibration Control, McGraw-Hill, New York, 1971.

Donato, R.J., "Insulating Homes Against Aircraft Noise," Acoustical Society America Vol. 53, No. 4, April, 1973, pp. 1025-1027.

Goome, D.J., Noise, Buildings and People, Pergamon Press, Oxford, England, 1977.

Low, H.K.; "Prediction and Measurements of Airborne Sound Transmission Through Concrete Structures," Architectural Science Review, Vol. 17, No. 1, March, 1974, pp. 5-9.

Miller, F.N.; "Isolation of Railroad, Subway Noise and Vibration," Progressive Architecture, Vol. 46, April, 1965, pp. 203-208.

LIGHTING

Bennett, E.; "Solar Optics: Light as Energy: Energy as Light," Underground Space Vol. 4, No. 6, Pergamon Press, New York, 1980, pp. 349-354.

Skylight Energy Performance, Center for Industrial and Institutional Development, University of New Hampshire, Durham, New Hampshire, 1975.

Systems Analysis for Skylight Energy Performance, Center for Industrial and Institutional Development, University of New Hampshire, Durham, New Hampshire, 1975.

HUMAN/PHYSICAL/PSYCHOLOGICAL DESIGN

Collins, B.L., ed.; Windows and People: A Literature Survey, Psychological Reaction to Environments With and Without Windows, National Bureau of Standards, Gaithersburg, Maryland.

Longmore, J. and Ne'eman, E.; "The Availability of Sunshine and Human Requirements for Sunlight in Buildings," Journal of Architectural Research Vol. 3, No. 2, May, 1974, pp. 24-29.

Lutz, F.W.; "Studies of Children in an Underground School," Underground Space Vol. 1, No. 2, Pergamon Press, New York, 1976, pp. 131-134.

Lutz, F.W.; Abo Revisited, Defense Civil Preparedness Agency, Contract No. DAHCZO-72-C-0115, June, 1972.

Lutz, F.W. and S.B.; Interim Report of the Abo Project, Office of Health, Education and Welfare, Contract No. OE-3-99-003, January, 1964.

Paulus, P.B.; "On the Psychology of Earth Covered Buildings," Underground Space Vol. 1, No. 2, Pergamon Press, New York, 1976, pp. 127-130.

Ruys, T.; "Windowless Offices," M.A. Thesis, University of Washington, Seattle, 1970.

Sommer, R.; Tight Spaces: Hard Architecture and How to Humanize It, Prentice-Hall, Inc., Englewood Cliffs, New Jersey, 1974.

MINED CONSTRUCTION

Nilsson, S.; "Feasibility of Underground Storage Plants in Metropolitan Areas: A Study of Major Economic Factors," Underground Space Vol. 4, No. 3, Pergamon Press, New York, 1979, pp. 153-158.

Occupance and Use of Underground Mined Out Space in Urban Areas: An Annotated Bibliography, Council of Planning Librarians, Exchange Bibliography No. 602, 1974.

Stauffer, T.P., Sr.; "Kansas City's Use of Limestone Mines for Business, Industry and Storage," Rockstore '77, Pergamon Press, Oxford, England, 1977.

Stauffer, T.P., Sr.; "Occupancy and Use of Underground Mined Space in Urban Areas," Critical Review in Environmental Control Vol. 7, No. 3, pp. 227-308.

STRUCTURAL ANALYSIS

Balsara, J.P.; Pressure Distribution on a Buried Flat Plate Subjected to Static and Airborne Overpressure, U.S. Army Engineering Waterways Experiment Station, Corps of Engineering, Vicksburg, Michigan, 1968.

Flathan, W.J.; "Research in Soil Structure Interaction," American Society of Civil Engineers Proceedings, 91 (St. 4 No. 4427), 1965, pp. 35-45.

Komendant, A.E.; "Earth Covered Structures," Underground Space Vol. 3, No. 6, Pergamon Press, New York, 1979, pp. 279-284.

Shih, J.C.; "Simplified Design Procedure of Underground Shells," Underground Space Vol. 4, No. 1, Pergamon Press, New York, 1979, pp. 45-50.

CONSTRUCTION METHODS

Optimum Subsurface and Underground Shell Structures for Better Housing in Hot-Arid Lands, U.S. Army Research Office, Research in Some Aspects of Fluid and Solid Mechanics, Interim Technical Report No. 1, Duke Station, Durham, North Carolina, 1970.

Standard Specifications (With Design Guide and Instructions to the Designer) for the Construction of Magazine, Steel Oval-Arch-Earth-Covered, U.S. Army Corps of Engineers, Omaha, Nebraska, May 1978.

Sterling, R.L.; "Structural Systems for Earth-Sheltered Housing," Underground Space, Vol. 3, No. 2, Pergamon Press, New York, 1978, pp. 75-82.

CONSTRUCTION MATERIALS

McDonald, J.E., and Liu, T.C.; Concrete for Earth-Covered Structures, Army Engineer Waterways Experiment Station, Report No. WES-MP-C-78-15, Vicksburg, Michigan, 1978.

DRAINAGE

Edwardsen, K.I.; New Method of Drainage of Basement Walls, National Research Council of Canada, Ottawa, Canada, 1972.

Elmrotli, A., and Hoglund, I.; New Basement Wall Design for Below-Grade Living Space, Canada Institute for Scientific and Technical Information, Ottawa, Canada, No. 271.

"Soil Considerations in Subgrade Waterproofing," Building Research, November-December 1964, pp. 39-41.

WATERPROOFING (General)

Gill, G.W.; "Waterproofing Buildings Below Grade," Civil Engineering, January 1959, pp. 3-5.

Maxwell Cook, J.C.; Structural Waterproofing, Butler-Worths, London, 1967.

Protection from Moisture, National Academy of Science, NRC No. 707.

"Subgrade Waterproofing," Building Research, November-December 1964.

UPSIDE-DOWN ROOF SYSTEMS

Baker, M.C., and Hedlin, C.P.; "Protected Membrane Roofs," Canadian Building Digest, Division of Building Research, National Research Council of Ottawa, Canada, No. 150, June 1972.

BITUMINOUS WATERPROOFING

Blackburn, J.B.; "A Study of Bituminous Waterproofing Coatings," Proceedings, American Railway Engineering Association 58, 1957. (Advance Report: Purdue University Civil Engineering Reprint No. 119, 1956.)

Draun Eberhard, I.A.; "Determination of Characteristic Indices of Bituminous Materials in Their Application for Waterproofing of Buildings," Bitumen, Vol. 36, No. 3, May-June 1974, pp. 66-67.

American Society for Testing and Materials, Committee D-8 on Bituminous Materials for Roofing Waterproofing and Related Building or Industrial Uses, Symposium on Bituminous Waterproofing and Materials, San Francisco, California, 1960.

BENTONITE WATERPROOFING

Kanarowski, S.M.; "Evaluation of Bentonite Clay for Waterproofing Foundation Walls Below Grade," Army Construction Engineering Research Laboratory, Champaign, Illinois, 1975. (AD-A011 180).

SHEET MEMBRANE WATERPROOFING

Elastomeric Waterproofing Systems, Office of Building and Facilities, Design and Construction Branch, Department of Heath, Education, and Welfare, Public Heath Service, Division 7, Moisture Protection, Washington, D.C., 1969.

Guide Specifications for Military Construction--Elastomeric Waterproofing, Office of the Chief Engineer, U.S. Army Corps of Engineers, Washington, D.C., 1972.

Rossiter, W.J., Jr., and Mathey, R.G.; Elastomeric Roofing: A Survey, National Bureau of Standards, Technical Note TN-972, July 1978. (Available from NTIS, No. NBS-TN-973.)

LIQUID-APPLIED MEMBRANE WATERPROOFING

Rosen, H.J.; "Liquid Membrane Waterproofing," Progressive Architecture, Vol. 52, No. 7, July 7, 1971, pp. 96.

MOISTURE PROBLEMS IN INSULATION

Dechow, J.J., and Epstein, K.A.; "Laboratory and Field Investigations of Moisture Absorption and Its Effects on Thermal Performance for Various Insulation," ASTM Symposium on Advances in Heat Transmission Measurements, Philadelphia, Pennsylvania, 1977.

Joy, F.A.; "Thermal Conductivity of Insulation Containing Moisture," Thermal Conductivity Measurements and Applications of Thermal Insulation, ASTM Special Technical Publication 217, 1957.

Kaplar, C.W.; "Studies of Moisture Effects on Thermal Conductivity of Rigid Insulation," CRREL International Report No. 207, U.S. Army CRREL, Hanover, New York, 1969.

Schaefer, D.; "Water Absorption of Insulation in Protected Membrane Roofing Systems," CRREL Technical Note, U.S. Army CRREL, Hanover, New York, 1974

APPENDIX A

HEAT FLUX TABLES

This appendix consists of Figure 102 and Tables 21 through 42. Refer to Section 18, Energy Calculations, for instructions.

		DEPT OF INSULATION IN FEET														
		R-Value			Full Berm 3-Story			Full Berm 1-Story			Half Berm		Slab-on Grade	
Table #	Location	0.0	10.4	20.8	0.0	8.0	16.0	0.0	8.0	13.0	0.0	5.0	0.0	4.0
21	Minneapolis	●			●			●			●		●	
22	Boston	●			●			●			●		●	
23	Seattle	●			●			●			●		●	
24	Kansas City	●			●			●			●		●	
25	Albuquerque	●			●			●			●		●	
26	Phoenix	●			●			●			●		●	
27	Miami	●			●			●			●		●	
28	Minneapolis		●			●			●			●		●
29	Boston		●			●			●			●		●
30	Seattle		●			●			●			●		●
31	Kansas City		●			●			●			●		●
32	Albuquerque		●			●			●			●		●
33	Phoenix		●			●			●			●		●
34	Miami		●			●			●			●		●
35	Minneapolis			●		●			●			●		●
36	Boston			●		●			●			●		●
37	Seattle			●		●			●			●		●
38	Kansas City			●		●			●			●		●
39	Minneapolis		●				●			●				
40	Boston		●				●			●				
41	Seattle		●				●			●				
42	Kansas City		●				●			●				

INDEX TO COMPUTER SIMULATIONS

FIGURE 102
7-Zone Climate Classification of the United States and Representative Cities

TABLE 21
Monthly Average and Peak Heat Fluxes for Minneapolis with No Insulation

Minneapolis No Insulation

Full Berm 3-Story

	Slab (Perimeter)						Slab (Interior)						Wall (Level-1)						Wall (Level-2)						Wall (Level-3)					
	AVE HEAT LOSS	PEAK HEAT LOSS	AVE HEAT GAIN	PEAK HEAT GAIN	HRS HT ON	HRS CL ON	AVE HEAT LOSS	PEAK HEAT LOSS	AVE HEAT GAIN	PEAK HEAT GAIN	HRS HT ON	HRS CL ON	AVE HEAT LOSS	PEAK HEAT LOSS	AVE HEAT GAIN	PEAK HEAT GAIN	HRS HT ON	HRS CL ON	AVE HEAT LOSS	PEAK HEAT LOSS	AVE HEAT GAIN	PEAK HEAT GAIN	HRS HT ON	HRS CL ON	AVE HEAT LOSS	PEAK HEAT LOSS	AVE HEAT GAIN	PEAK HEAT GAIN	HRS HT ON	HRS CL ON
JAN	0.63	0.65	0.00	0.00	744	0	0.48	0.49	0.00	0.00	744	0	3.57	3.74	0.00	0.00	744	0	1.01	1.04	0.00	0.00	744	0	1.07	1.09	0.00	0.00	744	0
FEB	0.59	0.61	0.00	0.00	672	0	0.44	0.46	0.00	0.00	672	0	3.67	3.75	0.00	0.00	672	0	1.07	1.09	0.00	0.00	672	0	1.05	1.06	0.00	0.00	672	0
MAR	0.55	0.57	0.00	0.00	744	0	0.41	0.43	0.00	0.00	744	0	3.43	3.58	0.00	0.00	744	0	1.11	1.12	0.00	0.00	744	0	1.03	1.04	0.00	0.00	744	0
APR	0.63	0.72	0.00	0.00	720	0	0.49	0.58	0.00	0.00	720	0	3.10	3.81	0.00	0.00	720	0	1.30	1.63	0.00	0.00	720	0	1.19	1.53	0.00	0.00	720	0
MAY	0.67	0.77	0.00	0.00	744	0	0.54	0.63	0.00	0.00	744	0	2.52	3.30	0.00	0.00	744	0	1.36	1.73	0.00	0.00	744	0	1.28	1.64	0.00	0.00	744	0
JUN	0.71	0.81	0.00	0.00	720	0	0.57	0.67	0.00	0.00	720	0	1.59	2.33	0.00	0.00	720	0	1.36	1.76	0.00	0.00	720	0	1.34	1.71	0.00	0.00	720	0
JUL	0.85	1.02	0.00	0.00	744	0	0.71	0.89	0.00	0.00	744	0	1.09	1.85	0.00	0.00	744	0	1.47	2.25	0.00	0.00	744	0	1.57	2.28	0.00	0.00	744	0
AUG	0.63	0.63	0.00	0.00	744	0	0.49	0.50	0.00	0.00	744	0	0.63	1.00	0.00	0.00	744	0	0.94	0.98	0.00	0.00	744	0	1.14	1.17	0.00	0.00	744	0
SEP	0.38	0.40	0.00	0.00	720	0	0.25	0.28	0.00	0.00	720	0	0.91	1.57	0.05	0.18	558	162	0.50	0.58	0.06	0.12	714	6	0.73	0.81	0.00	0.00	720	0
OCT	0.19	0.23	0.00	0.00	744	0	0.08	0.12	0.03	0.08	677	67	1.66	2.07	0.00	0.00	744	0	0.28	0.41	0.14	0.45	680	64	0.44	0.54	0.09	0.22	730	14
NOV	0.24	0.24	0.00	0.00	720	0	0.13	0.14	0.00	0.00	720	0	2.78	3.62	0.00	0.00	720	0	0.48	0.54	0.00	0.00	720	0	0.56	0.56	0.00	0.00	720	0
DEC	0.24	0.24	0.00	0.00	744	0	0.14	0.14	0.00	0.00	744	0	3.37	3.62	0.00	0.00	744	0	0.62	0.69	0.00	0.00	744	0	0.57	0.57	0.00	0.00	744	0

Full Berm 1-Story

	Slab (Perimeter)						Slab (Interior)						Wall (Level-1)					
	AVE HEAT LOSS	PEAK HEAT LOSS	AVE HEAT GAIN	PEAK HEAT GAIN	HRS HT ON	HRS CL ON	AVE HEAT LOSS	PEAK HEAT LOSS	AVE HEAT GAIN	PEAK HEAT GAIN	HRS HT ON	HRS CL ON	AVE HEAT LOSS	PEAK HEAT LOSS	AVE HEAT GAIN	PEAK HEAT GAIN	HRS HT ON	HRS CL ON
JAN	0.96	0.96	0.00	0.00	744	0	0.78	0.79	0.00	0.00	744	0	4.69	4.86	0.00	0.00	744	0
FEB	0.96	0.96	0.00	0.00	672	0	0.76	0.77	0.00	0.00	672	0	4.82	4.89	0.00	0.00	672	0
MAR	0.96	0.96	0.00	0.00	744	0	0.74	0.75	0.00	0.00	744	0	4.63	4.76	0.00	0.00	744	0
APR	1.06	1.13	0.00	0.00	720	0	0.83	0.92	0.00	0.00	720	0	4.32	5.03	0.00	0.00	720	0
MAY	1.12	1.21	0.00	0.00	744	0	0.89	0.98	0.00	0.00	744	0	3.81	4.52	0.00	0.00	744	0
JUN	1.16	1.26	0.00	0.00	720	0	0.93	1.03	0.00	0.00	720	0	3.01	3.67	0.00	0.00	720	0
JUL	1.30	1.48	0.00	0.00	744	0	1.08	1.25	0.00	0.00	744	0	2.65	3.29	0.00	0.00	744	0
AUG	1.07	1.08	0.00	0.00	744	0	0.86	0.87	0.00	0.00	744	0	2.11	2.41	0.00	0.00	744	0
SEP	0.81	0.83	0.00	0.00	720	0	0.63	0.66	0.00	0.00	720	0	1.93	2.69	0.00	0.00	720	0
OCT	0.61	0.65	0.00	0.00	744	0	0.46	0.51	0.00	0.00	744	0	2.58	2.95	0.00	0.00	744	0
NOV	0.67	0.68	0.00	0.00	720	0	0.52	0.53	0.00	0.00	720	0	3.59	4.35	0.00	0.00	720	0
DEC	0.69	0.70	0.00	0.00	744	0	0.53	0.53	0.00	0.00	744	0	4.15	4.36	0.00	0.00	744	0

Half Berm

	Slab (Perimeter)						Slab (Interior)						Wall (Level-1)					
	AVE HEAT LOSS	PEAK HEAT LOSS	AVE HEAT GAIN	PEAK HEAT GAIN	HRS HT ON	HRS CL ON	AVE HEAT LOSS	PEAK HEAT LOSS	AVE HEAT GAIN	PEAK HEAT GAIN	HRS HT ON	HRS CL ON	AVE HEAT LOSS	PEAK HEAT LOSS	AVE HEAT GAIN	PEAK HEAT GAIN	HRS HT ON	HRS CL ON
JAN	1.28	1.29	0.00	0.00	744	0	0.81	0.81	0.00	0.00	744	0	7.06	7.47	0.00	0.00	744	0
FEB	1.30	1.31	0.00	0.00	672	0	0.80	0.81	0.00	0.00	672	0	7.20	7.41	0.00	0.00	672	0
MAR	1.31	1.31	0.00	0.00	744	0	0.80	0.80	0.00	0.00	744	0	6.70	7.02	0.00	0.00	744	0
APR	1.42	1.49	0.00	0.00	720	0	0.90	0.97	0.00	0.00	720	0	5.49	6.98	0.00	0.00	720	0
MAY	1.47	1.57	0.00	0.00	744	0	0.97	1.05	0.00	0.00	744	0	4.19	5.56	0.00	0.00	744	0
JUN	1.50	1.61	0.00	0.00	720	0	1.02	1.11	0.00	0.00	720	0	2.48	3.64	0.00	0.00	720	0
JUL	1.62	1.81	0.00	0.00	744	0	1.17	1.34	0.00	0.00	744	0	1.65	2.16	0.00	0.00	744	0
AUG	1.38	1.39	0.00	0.00	744	0	0.96	0.97	0.00	0.00	744	0	1.47	2.40	0.00	0.00	744	0
SEP	1.12	1.15	0.00	0.00	720	0	0.72	0.76	0.00	0.00	720	0	2.05	3.80	0.00	0.00	720	0
OCT	0.94	1.00	0.00	0.00	744	0	0.55	0.60	0.00	0.00	744	0	4.02	4.77	0.00	0.00	744	0
NOV	1.03	1.06	0.00	0.00	720	0	0.62	0.63	0.00	0.00	720	0	5.90	7.51	0.00	0.00	720	0
DEC	1.09	1.11	0.00	0.00	744	0	0.63	0.64	0.00	0.00	744	0	6.74	7.47	0.00	0.00	744	0

Slab On Grade

	Slab (Perimeter)						Slab (Interior)					
	AVE HEAT LOSS	PEAK HEAT LOSS	AVE HEAT GAIN	PEAK HEAT GAIN	HRS HT ON	HRS CL ON	AVE HEAT LOSS	PEAK HEAT LOSS	AVE HEAT GAIN	PEAK HEAT GAIN	HRS HT ON	HRS CL ON
JAN	2.01	2.08	0.00	0.00	744	0	0.91	0.91	0.00	0.00	744	0
FEB	2.08	2.11	0.00	0.00	672	0	0.92	0.93	0.00	0.00	672	0
MAR	2.04	2.09	0.00	0.00	744	0	0.93	0.94	0.00	0.00	744	0
APR	1.90	2.19	0.00	0.00	720	0	1.06	1.12	0.00	0.00	720	0
MAY	1.71	2.01	0.00	0.00	744	0	1.14	1.22	0.00	0.00	744	0
JUN	1.42	1.73	0.00	0.00	720	0	1.20	1.29	0.00	0.00	720	0
JUL	1.33	1.54	0.00	0.00	744	0	1.35	1.52	0.00	0.00	744	0
AUG	1.08	1.21	0.00	0.00	744	0	1.11	1.12	0.00	0.00	744	0
SEP	0.95	1.27	0.00	0.00	720	0	0.85	0.87	0.00	0.00	720	0
OCT	1.13	1.31	0.00	0.00	744	0	0.65	0.69	0.00	0.00	744	0
NOV	1.54	1.83	0.00	0.00	720	0	0.70	0.72	0.00	0.00	720	0
DEC	1.77	1.81	0.00	0.00	744	0	0.73	0.74	0.00	0.00	744	0

Heat Flux Expressed in Btu/h²

To obtain W/m² units, multiply Btu/h ft² units by 3 156

TABLE 22

Monthly Average and Peak Heat Fluxes for Boston with No Insulation

Boston No Insulation

Full Berm 3-Story

	Slab (Perimeter)						Slab (Interior)						Wall (Level-1)						Wall (Level-2)						Wall (Level-3)					
	AVE HEAT LOSS	PEAK HEAT LOSS	AVE HEAT GAIN	PEAK HEAT GAIN	HRS HT ON	HRS CL ON	AVE HEAT LOSS	PEAK HEAT LOSS	AVE HEAT GAIN	PEAK HEAT GAIN	HRS HT ON	HRS CL ON	AVE HEAT LOSS	PEAK HEAT LOSS	AVE HEAT GAIN	PEAK HEAT GAIN	HRS HT ON	HRS CL ON	AVE HEAT LOSS	PEAK HEAT LOSS	AVE HEAT GAIN	PEAK HEAT GAIN	HRS HT ON	HRS CL ON	AVE HEAT LOSS	PEAK HEAT LOSS	AVE HEAT GAIN	PEAK HEAT GAIN	HRS HT ON	HRS CL ON
JAN	0 41	0.43	0 00	0 00	744	0	0 31	0 32	0 00	0 00	744	0	2 72	2 83	0 00	0 00	744	0	0 75	0 77	0 00	0 00	744	0	0 74	0 75	0 00	0 00	744	0
FEB	0 39	0 40	0 00	0 00	672	0	0 29	0 30	0 00	0 00	672	0	2 76	2 85	0 00	0 00	672	0	0 79	0 81	0 00	0 00	672	0	0 73	0 74	0 00	0 00	672	0
MAR	0 37	0 38	0 00	0 00	744	0	0 27	0 28	0 00	0 00	744	0	2 60	2 64	0 00	0 00	744	0	0 82	0 82	0 00	0 00	744	0	0 72	0 72	0 00	0 00	744	0
APR	0 46	0 54	0 00	0 00	720	0	0 36	0 44	0 00	0 00	720	0	2 78	3 06	0 00	0 00	720	0	1 01	1 34	0 00	0 00	720	0	0 90	1 23	0 00	0 00	720	0
MAY	0 51	0 60	0 00	0 00	744	0	0 41	0 51	0 00	0 00	744	0	2 51	3 03	0 00	0 00	744	0	1 11	1 46	0 00	0 00	744	0	0 99	1 34	0 00	0 00	744	0
JUN	0 56	0 65	0 00	0 00	720	0	0 46	0 55	0 00	0 00	720	0	1 56	2 17	0 00	0 00	720	0	1 15	1 53	0 00	0 00	720	0	1 07	1 42	0 00	0 00	720	0
JUL	0 71	0 88	0 00	0 00	744	0	0 60	0 78	0 00	0 00	744	0	1 25	2 11	0 00	0 00	744	0	1 30	2 05	0 00	0 00	744	0	1 32	2 01	0 00	0 00	744	0
AUG	0 49	0 50	0 00	0 00	744	0	0 39	0 40	0 00	0 00	744	0	0 70	0 89	0 00	0 00	744	0	0 81	0 83	0 00	0 00	744	0	0 91	0 93	0 00	0 00	744	0
SEP	0 25	0 28	0 00	0 00	720	0	0 16	0 19	0 00	0 00	720	0	0 80	1 32	0 07	0 14	714	6	0 39	0 47	0 10	0 25	702	18	0 51	0 60	0 05	0 12	714	6
OCT	0 08	0 12	0 09	0 27	697	47	0 02	0 04	0 07	0 61	396	348	1 28	1 94	0 00	0 00	744	0	0 20	0 30	0 16	0 55	628	116	0 26	0 35	0 14	0 43	687	57
NOV	0 13	0 14	0 00	0 00	720	0	0 05	0 06	0 00	0 00	720	0	2 18	2 89	0 00	0 00	720	0	0 36	0 41	0 00	0 00	720	0	0 37	0 39	0 00	0 00	720	0
DEC	0 14	0 14	0 00	0 00	744	0	0 07	0 07	0 00	0 00	744	0	2 67	2 91	0 00	0 00	744	0	0 48	0 53	0 00	0 00	744	0	0 39	0 40	0 00	0 00	744	0

Full Berm 1-Story

	Slab (Perimeter)						Slab (Interior)						Wall (Level-1)					
	AVE HEAT LOSS	PEAK HEAT LOSS	AVE HEAT GAIN	PEAK HEAT GAIN	HRS HT ON	HRS CL ON	AVE HEAT LOSS	PEAK HEAT LOSS	AVE HEAT GAIN	PEAK HEAT GAIN	HRS HT ON	HRS CL ON	AVE HEAT LOSS	PEAK HEAT LOSS	AVE HEAT GAIN	PEAK HEAT GAIN	HRS HT ON	HRS CL ON
JAN	0 66	0 67	0 00	0 00	744	0	0 52	0 53	0 00	0 00	744	0	3 49	3 62	0 00	0 00	744	0
FEB	0 67	0 67	0 00	0 00	672	0	0 52	0 52	0 00	0 00	672	0	3 56	3 63	0 00	0 00	672	0
MAR	0 67	0 67	0 00	0 00	744	0	0 51	0 51	0 00	0 00	744	0	3 43	3 48	0 00	0 00	744	0
APR	0 77	0 85	0 00	0 00	720	0	0 60	0 68	0 00	0 00	720	0	3 62	3 90	0 00	0 00	720	0
MAY	0 84	0 93	0 00	0 00	744	0	0 67	0 76	0 00	0 00	744	0	3 41	3 91	0 00	0 00	744	0
JUN	0 90	0 99	0 00	0 00	720	0	0 72	0 81	0 00	0 00	720	0	2 61	3 16	0 00	0 00	720	0
JUL	1 05	1 22	0 00	0 00	744	0	0 87	1 04	0 00	0 00	744	0	2 45	3 19	0 00	0 00	744	0
AUG	0 83	0 84	0 00	0 00	744	0	0 66	0 66	0 00	0 00	744	0	1 83	1 99	0 00	0 00	744	0
SEP	0 58	0 61	0 00	0 00	720	0	0 43	0 47	0 00	0 00	720	0	1 71	2 16	0 00	0 00	720	0
OCT	0 39	0 44	0 00	0 00	744	0	0 26	0 32	0 00	0 00	744	0	1 94	2 55	0 00	0 00	744	0
NOV	0 45	0 47	0 00	0 00	720	0	0 33	0 35	0 00	0 00	720	0	2 74	3 40	0 00	0 00	720	0
DEC	0 48	0 49	0 00	0 00	744	0	0 35	0 36	0 00	0 00	744	0	3 21	3 42	0 00	0 00	744	0

Half Berm

	Slab (Perimeter)						Slab (Interior)						Wall (Level-1)					
	AVE HEAT LOSS	PEAK HEAT LOSS	AVE HEAT GAIN	PEAK HEAT GAIN	HRS HT ON	HRS CL ON	AVE HEAT LOSS	PEAK HEAT LOSS	AVE HEAT GAIN	PEAK HEAT GAIN	HRS HT ON	HRS CL ON	AVE HEAT LOSS	PEAK HEAT LOSS	AVE HEAT GAIN	PEAK HEAT GAIN	HRS HT ON	HRS CL ON
JAN	0 89	0 90	0 00	0 00	744	0	0 55	0 55	0 00	0 00	744	0	5 26	5 53	0 00	0 00	744	0
FEB	0 90	0 91	0 00	0 00	672	0	0 55	0 55	0 00	0 00	672	0	5 32	5 55	0 00	0 00	672	0
MAR	0 91	0 91	0 00	0 00	744	0	0 54	0 55	0 00	0 00	744	0	5 02	5 15	0 00	0 00	744	0
APR	1 03	1 09	0 00	0 00	720	0	0 65	0 72	0 00	0 00	720	0	4 97	5 69	0 00	0 00	720	0
MAY	1 10	1 18	0 00	0 00	744	0	0 72	0 81	0 00	0 00	744	0	4 20	5 32	0 00	0 00	744	0
JUN	1 15	1 25	0 00	0 00	720	0	0 78	0 87	0 00	0 00	720	0	2 34	3 41	0 00	0 00	720	0
JUL	1 28	1 46	0 00	0 00	744	0	0 94	1 11	0 00	0 00	744	0	1 85	2 62	0 00	0 00	744	0
AUG	1 05	1 06	0 00	0 00	744	0	0 73	0 74	0 00	0 00	744	0	1 38	1 92	0 00	0 00	744	0
SEP	0 80	0 83	0 00	0 00	720	0	0 50	0 54	0 00	0 00	720	0	2 04	3 10	0 00	0 00	720	0
OCT	0 63	0 68	0 00	0 00	744	0	0 33	0 39	0 00	0 00	744	0	3 07	4 35	0 00	0 00	744	0
NOV	0 71	0 74	0 00	0 00	720	0	0 41	0 42	0 00	0 00	720	0	4 53	5 94	0 00	0 00	720	0
DEC	0 76	0 78	0 00	0 00	744	0	0 43	0 43	0 00	0 00	744	0	5 24	5 97	0 00	0 00	744	0

Slab On Grade

	Slab (Perimeter)						Slab (Interior)					
	AVE HEAT LOSS	PEAK HEAT LOSS	AVE HEAT GAIN	PEAK HEAT GAIN	HRS HT ON	HRS CL ON	AVE HEAT LOSS	PEAK HEAT LOSS	AVE HEAT GAIN	PEAK HEAT GAIN	HRS HT ON	HRS CL ON
JAN	1 49	1 55	0 00	0 00	744	0	0 63	0 64	0 00	0 00	744	0
FEB	1 53	1 55	0 00	0 00	672	0	0 65	0 65	0 00	0 00	672	0
MAR	1 50	1 53	0 00	0 00	744	0	0 66	0 67	0 00	0 00	744	0
APR	1 57	1 71	0 00	0 00	720	0	0 78	0 85	0 00	0 00	720	0
MAY	1 49	1 71	0 00	0 00	744	0	0 87	0 94	0 00	0 00	744	0
JUN	1 21	1 48	0 00	0 00	720	0	0 93	1 02	0 00	0 00	720	0
JUL	1 19	1 42	0.00	0 00	744	0	1 09	1 26	0 00	0 00	744	0
AUG	0 91	0 99	0 00	0 00	744	0	0 87	0 88	0 00	0 00	744	0
SEP	0 81	0 99	0 00	0 00	720	0	0 62	0 65	0 00	0 00	720	0
OCT	0 85	1 10	0 00	0 00	744	0	0 43	0 47	0 00	0 00	744	0
NOV	1 16	1 43	0 00	0 00	720	0	0 49	0 51	0 00	0 00	720	0
DEC	1 35	1 42	0.00	0 00	744	0	0 52	0 53	0 00	0 00	744	0

Heat Flux Expressed in Btu/h^2

To obtain W/m^2 units, multiply Btu/h ft^2 units by 3 156

TABLE 23

Monthly Average and Peak Heat Fluxes for Seattle with No Insulation

Seattle No Insulation

Full Berm 3-Story

	Slab (Perimeter)						Slab (Interior)						Wall (Level-1)						Wall (Level-2)						Wall (Level-3)					
	AVE HEAT LOSS	PEAK HEAT LOSS	AVE HEAT GAIN	PEAK HEAT GAIN	HRS HT ON	HRS CL ON	AVE HEAT LOSS	PEAK HEAT LOSS	AVE HEAT GAIN	PEAK HEAT GAIN	HRS HT ON	HRS CL ON	AVE HEAT LOSS	PEAK HEAT LOSS	AVE HEAT GAIN	PEAK HEAT GAIN	HRS HT ON	HRS CL ON	AVE HEAT LOSS	PEAK HEAT LOSS	AVE HEAT GAIN	PEAK HEAT GAIN	HRS HT ON	HRS CL ON	AVE HEAT LOSS	PEAK HEAT LOSS	AVE HEAT GAIN	PEAK HEAT GAIN	HRS HT ON	HRS CL ON
JAN	0 37	0 38	0 00	0 00	744	0	0 26	0 27	0 00	0 00	744	0	3 26	3 54	0 00	0 00	744	0	0 77	0 81	0 00	0 00	744	0	0 72	0 72	0 00	0 00	744	0
FEB	0 35	0 36	0 00	0 00	672	0	0 25	0 26	0 00	0 00	672	0	3 28	3 55	0 00	0 00	672	0	0 84	0 87	0 00	0 00	672	0	0 71	0 71	0 00	0 00	672	0
MAR	0 33	0 34	0 00	0 00	744	0	0 24	0 24	0 00	0 00	744	0	3 17	3 53	0 00	0 00	744	0	0 89	0 91	0 00	0 00	744	0	0 71	0 71	0 00	0 00	744	0
APR	0 42	0 50	0 00	0 00	720	0	0 33	0 41	0 00	0 00	720	0	2 91	3 29	0 00	0 00	720	0	1 09	1 43	0 00	0 00	720	0	0 89	1 22	0 00	0 00	720	0
MAY	0 59	0 75	0 00	0 00	744	0	0 49	0 66	0 00	0 00	744	0	2 72	3 68	0 00	0 00	744	0	1 35	2 05	0 00	0 00	744	0	1 18	1 86	0 00	0 00	744	0
JUN	0 61	0 71	0 00	0 00	720	0	0 51	0 61	0 00	0 00	720	0	2 14	2 68	0 00	0 00	720	0	1 30	1 71	0 00	0 00	720	0	1 19	1 56	0 00	0 00	720	0
JUL	0 64	0 74	0 00	0 00	744	0	0 54	0 64	0 00	0 00	744	0	1 87	2 35	0 00	0 00	744	0	1 27	1 68	0 00	0 00	744	0	1 23	1 60	0 00	0 00	744	0
AUG	0 56	0 59	0 00	0 00	744	0	0 46	0 49	0 00	0 00	744	0	1 50	1 62	0 00	0 00	744	0	1 05	1 13	0 00	0 00	744	0	1 08	1 13	0 00	0 00	744	0
SEP	0 30	0 33	0 00	0 00	720	0	0 20	0 23	0 00	0 00	720	0	1 32	1 67	0 00	0 00	720	0	0 57	0 66	0 03	0 05	718	2	0 62	0 71	0 00	0 00	720	0
OCT	0 11	0 15	0 05	0 12	730	14	0 04	0 07	0 06	0 47	561	183	1 61	2 05	0 00	0 00	744	0	0 34	0 46	0 13	0 37	705	39	0 36	0 45	0 12	0 32	714	30
NOV	0 06	0 09	0 04	0 09	697	23	0 00	1 01	0 04	0 40	149	571	2 13	2 47	0 00	0 00	720	0	0 33	0 44	0 03	0 06	715	5	0 29	0 36	0 03	0 06	714	6
DEC	0 10	0 11	0 00	0 00	744	0	0 02	0 03	0 00	0 00	744	0	2 79	3 10	0 00	0 00	744	0	0 50	0 56	0 00	0 00	744	0	0 38	0 39	0 00	0 00	744	0

Full Berm 1-Story

	Slab (Perimeter)						Slab (Interior)						Wall (Level-1)					
	AVE HEAT LOSS	PEAK HEAT LOSS	AVE HEAT GAIN	PEAK HEAT GAIN	HRS HT ON	HRS CL ON	AVE HEAT LOSS	PEAK HEAT LOSS	AVE HEAT GAIN	PEAK HEAT GAIN	HRS HT ON	HRS CL ON	AVE HEAT LOSS	PEAK HEAT LOSS	AVE HEAT GAIN	PEAK HEAT GAIN	HRS HT ON	HRS CL ON
JAN	0 64	0 65	0 00	0 00	744	0	0 49	0 49	0 00	0 00	744	0	3 91	4 20	0 00	0 00	744	0
FEB	0 65	0 66	0 00	0 00	672	0	0 49	0 49	0 00	0 00	672	0	3 94	4 20	0 00	0 00	672	0
MAR	0 66	0 66	0 00	0 00	744	0	0 48	0 49	0 00	0 00	744	0	3 87	4 21	0 00	0 00	744	0
APR	0 77	0 84	0 00	0 00	720	0	0 58	0 66	0 00	0 00	720	0	3 71	4 01	0 00	0 00	720	0
MAY	0 95	1 11	0 00	0 00	744	0	0 75	0 92	0 00	0 00	744	0	3 69	4 51	0 00	0 00	744	0
JUN	0 97	1 07	0 00	0 00	720	0	0 77	0 87	0 00	0 00	720	0	3 23	3 70	0 00	0 00	720	0
JUL	1 01	1 11	0 00	0 00	744	0	0 81	0 91	0 00	0 00	744	0	3 02	3 46	0 00	0 00	744	0
AUG	0 93	0 96	0 00	0 00	744	0	0 73	0 76	0 00	0 00	744	0	2 65	2 78	0 00	0 00	744	0
SEP	0 66	0 68	0 00	0 00	720	0	0 48	0 51	0 00	0 00	720	0	2 29	2 58	0 00	0 00	720	0
OCT	0 46	0 51	0 00	0 00	744	0	0 31	0 36	0 00	0 00	744	0	2 36	2 75	0 00	0 00	744	0
NOV	0 41	0 45	0 00	0 00	720	0	0 27	0 30	0 00	0 00	720	0	2 72	3 02	0 00	0 00	720	0
DEC	0 46	0 48	0 00	0 00	744	0	0 32	0 33	0 00	0 00	744	0	3 32	3 63	0 00	0 00	744	0

Half Berm

	Slab (Perimeter)						Slab (Interior)						Wall (Level-1)					
	AVE HEAT LOSS	PEAK HEAT LOSS	AVE HEAT GAIN	PEAK HEAT GAIN	HRS HT ON	HRS CL ON	AVE HEAT LOSS	PEAK HEAT LOSS	AVE HEAT GAIN	PEAK HEAT GAIN	HRS HT ON	HRS CL ON	AVE HEAT LOSS	PEAK HEAT LOSS	AVE HEAT GAIN	PEAK HEAT GAIN	HRS HT ON	HRS CL ON
JAN	0 87	0 88	0 00	0 00	744	0	0 52	0 52	0 00	0 00	744	0	6 06	6 65	0 00	0 00	744	0
FEB	0 90	0 90	0 00	0 00	672	0	0 53	0 53	0 00	0 00	672	0	5 88	6 66	0 00	0 00	672	0
MAR	0 91	0 91	0 00	0 00	744	0	0 53	0 53	0 00	0 00	744	0	5 60	6 53	0 00	0 00	744	0
APR	1 02	1 09	0 00	0 00	720	0	0 64	0 71	0 00	0 00	720	0	4 93	5 32	0 00	0 00	720	0
MAY	1 20	1 36	0 00	0 00	744	0	0 82	0 98	0 00	0 00	744	0	4 36	5 52	0 00	0 00	744	0
JUN	1 22	1 33	0 00	0 00	720	0	0 85	0 94	0 00	0 00	720	0	3 43	3 97	0 00	0 00	720	0
JUL	1 25	1 35	0 00	0 00	744	0	0 89	0 98	0 00	0 00	744	0	3 05	3 62	0 00	0 00	744	0
AUG	1 16	1 20	0 00	0 00	744	0	0 81	0 84	0 00	0 00	744	0	2 67	2 91	0 00	0 00	744	0
SEP	0 89	0 92	0 00	0 00	720	0	0 56	0 59	0 00	0 00	720	0	2 80	3 41	0 00	0 00	720	0
OCT	0 71	0 76	0 00	0 00	744	0	0 38	0 43	0 00	0 00	744	0	3 54	4 39	0 00	0 00	744	0
NOV	0 67	0 71	0 00	0 00	720	0	0 34	0 38	0 00	0 00	720	0	4 36	4 89	0 00	0 00	720	0
DEC	0 74	0 77	0 00	0 00	744	0	0 40	0 41	0 00	0 00	744	0	5 40	6 35	0 00	0 00	744	0

Slab On Grade

	Slab (Perimeter)						Slab (Interior)					
	AVE HEAT LOSS	PEAK HEAT LOSS	AVE HEAT GAIN	PEAK HEAT GAIN	HRS HT ON	HRS CL ON	AVE HEAT LOSS	PEAK HEAT LOSS	AVE HEAT GAIN	PEAK HEAT GAIN	HRS HT ON	HRS CL ON
JAN	1 60	1 72	0 00	0 00	744	0	0 62	0 63	0 00	0 00	744	0
FEB	1 62	1 72	0 00	0 00	672	0	0 64	0 65	0 00	0 00	672	0
MAR	1 59	1 73	0 00	0 00	744	0	0 66	0 67	0 00	0 00	744	0
APR	1 57	1 65	0 00	0 00	720	0	0 79	0 85	0 00	0 00	720	0
MAY	1 61	1 85	0 00	0 00	744	0	0 98	1 13	0 00	0 00	744	0
JUN	1 45	1 59	0 00	0 00	720	0	1 02	1 11	0 00	0 00	720	0
JUL	1 38	1 55	0 00	0 00	744	0	1 05	1 15	0 00	0 00	744	0
AUG	1 24	1 31	0 00	0 00	744	0	0 98	1 01	0 00	0 00	744	0
SEP	1 04	1 16	0 00	0 00	720	0	0 70	0 73	0 00	0 00	720	0
OCT	1 02	1 18	0 00	0 00	744	0	0 51	0 56	0 00	0 00	744	0
NOV	1 14	1 29	0 00	0 00	720	0	0 46	0 50	0 00	0 00	720	0
DEC	1 39	1 52	0 00	0 00	744	0	0 52	0 54	0 00	0 00	744	0

Heat Flux Expressed in Btu/h^2

To obtain W/m^2 units, multiply Btu/h ft^2 units by 3 156

TABLE 24
Monthly Average and Peak Heat Fluxes for Kansas City with No Insulation

Kansas City No Insulation

Full Berm 3-Story

	Slab (Perimeter)						Slab (Interior)						Wall (Level-1)						Wall (Level-2)						Wall (Level-3)					
	AVE HEAT LOSS	PEAK HEAT LOSS	AVE HEAT GAIN	PEAK HEAT GAIN	HRS HT ON	HRS CL ON	AVE HEAT LOSS	PEAK HEAT LOSS	AVE HEAT GAIN	PEAK HEAT GAIN	HRS HT ON	HRS CL ON	AVE HEAT LOSS	PEAK HEAT LOSS	AVE HEAT GAIN	PEAK HEAT GAIN	HRS HT ON	HRS CL ON	AVE HEAT LOSS	PEAK HEAT LOSS	AVE HEAT GAIN	PEAK HEAT GAIN	HRS HT ON	HRS CL ON	AVE HEAT LOSS	PEAK HEAT LOSS	AVE HEAT GAIN	PEAK HEAT GAIN	HRS HT ON	HRS CL ON
JAN	0 31	0 32	0 00	0 00	744	0	0 22	0 22	0 00	0 00	744	0	3 62	4 47	0 00	0 00	744	0	0 64	0 71	0 00	0 00	744	0	0 57	0 57	0 00	0 00	744	0
FEB	0 30	0 30	0 00	0 00	672	0	0 21	0 21	0 00	0 00	672	0	4 07	4 53	0 00	0 00	672	0	0 77	0 83	0 00	0 00	672	0	0 58	0 58	0 00	0 00	672	0
MAR	0 28	0 29	0 00	0 00	744	0	0 20	0 20	0 00	0 00	744	0	3 43	3 94	0 00	0 00	744	0	0 88	0 90	0 00	0 00	744	0	0 59	0 60	0 00	0 00	744	0
APR	0 38	0 45	0 00	0 00	720	0	0 29	0 37	0 00	0 00	720	0	2 75	3 78	0 00	0 00	720	0	1 09	1 42	0 00	0 00	720	0	0 79	1 11	0 00	0 00	720	0
MAY	0 55	0 71	0 00	0 00	744	0	0 46	0 62	0 00	0 00	744	0	1 78	2 73	0 00	0 00	744	0	1 32	2 05	0 00	0 00	744	0	1 10	1 77	0 00	0 00	744	0
JUN	0 57	0 67	0 00	0 00	720	0	0 48	0 58	0 00	0 00	720	0	1 05	1 73	0 00	0 00	720	0	1 20	1 64	0 00	0 00	720	0	1 11	1 48	0 00	0 00	720	0
JUL	0 60	0 70	0 00	0 00	744	0	0 51	0 61	0 00	0 00	744	0	0 41	0 99	0 00	0 00	744	0	1 11	1 56	0 00	0 00	744	0	1 15	1 52	0 00	0 00	744	0
AUG	0 52	0 55	0 00	0 00	744	0	0 43	0 46	0 00	0 00	744	0	0 31	0 62	0 00	0 00	744	0	0 84	0 94	0 00	0 00	744	0	0 99	1 04	0 00	0 00	744	0
SEP	0 26	0 29	0 00	0 00	720	0	0 17	0 20	0 00	0 00	720	0	0 43	1 13	0 25	0 65	675	45	0 34	0 42	0 11	0 28	697	23	0 52	0 60	0 05	0 09	716	4
OCT	0 08	0 12	0 08	0 25	703	41	0 03	0 04	0 07	0 56	459	285	1 12	1 53	0 00	0 00	744	0	0 15	0 23	0 16	0 61	577	167	0 26	0 34	0 14	0 43	686	58
NOV	0 03	0 06	0 06	0 21	626	94	0 00	0 00	0 06	0 48	0	720	1 78	2 68	0 00	0 00	720	0	0 15	0 24	0 09	0 28	622	98	0 18	0 24	0 06	0 18	684	36
DEC	0 07	0 07	0 00	0 00	744	0	0 01	0 01	0 01	0 01	454	290	3 29	3 73	0 00	0 00	744	0	0 33	0 42	0 00	0 00	744	0	0 26	0 27	0 00	0 00	744	0

Full Berm 1-Story

	Slab (Perimeter)						Slab (Interior)						Wall (Level-1)					
	AVE HEAT LOSS	PEAK HEAT LOSS	AVE HEAT GAIN	PEAK HEAT GAIN	HRS HT ON	HRS CL ON	AVE HEAT LOSS	PEAK HEAT LOSS	AVE HEAT GAIN	PEAK HEAT GAIN	HRS HT ON	HRS CL ON	AVE HEAT LOSS	PEAK HEAT LOSS	AVE HEAT GAIN	PEAK HEAT GAIN	HRS HT ON	HRS CL ON
JAN	0 53	0 54	0 00	0 00	744	0	0 41	0 41	0 00	0 00	744	0	4 06	4 92	0 00	0 00	744	0
FEB	0 55	0 56	0 00	0 00	672	0	0 41	0 41	0 00	0 00	672	0	4 51	4 97	0 00	0 00	672	0
MAR	0 57	0 57	0 00	0 00	744	0	0 41	0 41	0 00	0 00	744	0	3 97	4 44	0 00	0 00	744	0
APR	0 69	0 76	0 00	0 00	720	0	0 51	0 58	0 00	0 00	720	0	3 45	4 38	0 00	0 00	720	0
MAY	0 87	1 03	0 00	0 00	744	0	0 68	0 84	0 00	0 00	744	0	2 73	3 50	0 00	0 00	744	0
JUN	0 89	0 99	0 00	0 00	720	0	0 70	0 80	0 00	0 00	720	0	2 13	2 74	0 00	0 00	720	0
JUL	0 91	1 02	0 00	0 00	744	0	0 74	0 84	0 00	0 00	744	0	1 57	2 09	0 00	0 00	744	0
AUG	0 82	0 86	0 00	0 00	744	0	0 67	0 69	0 00	0 00	744	0	1 41	1 71	0 00	0 00	744	0
SEP	0 55	0 57	0 00	0 00	720	0	0 42	0 45	0 00	0 00	720	0	1 27	1 93	0 00	0 00	720	0
OCT	0 35	0 40	0 00	0 00	744	0	0 24	0 29	0 00	0 00	744	0	1 72	2 09	0 00	0 00	744	0
NOV	0 30	0 34	0 00	0 00	720	0	0 20	0 24	0 00	0 00	720	0	2 23	3 07	0 00	0 00	720	0
DEC	0 36	0 38	0 00	0 00	744	0	0 25	0 27	0 00	0 00	744	0	3 61	4 03	0 00	0 00	744	0

Half Berm

	Slab (Perimeter)						Slab (Interior)						Wall (Level-1)					
	AVE HEAT LOSS	PEAK HEAT LOSS	AVE HEAT GAIN	PEAK HEAT GAIN	HRS HT ON	HRS CL ON	AVE HEAT LOSS	PEAK HEAT LOSS	AVE HEAT GAIN	PEAK HEAT GAIN	HRS HT ON	HRS CL ON	AVE HEAT LOSS	PEAK HEAT LOSS	AVE HEAT GAIN	PEAK HEAT GAIN	HRS HT ON	HRS CL ON
JAN	0 77	0 79	0 00	0 00	744	0	0 43	0 43	0 00	0 00	744	0	6 71	8 71	0 00	0 00	744	0
FEB	0 82	0 83	0 00	0 00	672	0	0 44	0 44	0 00	0 00	672	0	7 36	8 71	0 00	0 00	672	0
MAR	0 84	0 85	0 00	0 00	744	0	0 44	0 45	0 00	0 00	744	0	5 85	7 07	0 00	0 00	744	0
APR	0 96	1 03	0 00	0 00	720	0	0 56	0 63	0 00	0 00	720	0	4 36	6 22	0 00	0 00	720	0
MAY	1 12	1 29	0 00	0 00	744	0	0 74	0 90	0 00	0 00	744	0	2 45	3 64	0 00	0 00	744	0
JUN	1 12	1 23	0 00	0 00	720	0	0 77	0 87	0 00	0 00	720	0	1 39	2 59	0 00	0 00	720	0
JUL	1 12	1 24	0 00	0 00	744	0	0 81	0 91	0 00	0 00	744	0	0 44	1 14	0 02	0 03	731	13
AUG	1 02	1 06	0 00	0 00	744	0	0 74	0 77	0 00	0 00	744	0	0 71	1 55	0 06	0 12	691	53
SEP	0 75	0 78	0 00	0 00	720	0	0 49	0 52	0 00	0 00	720	0	1 32	2 89	0 08	0 16	715	5
OCT	0 57	0 63	0 00	0 00	744	0	0 31	0 36	0 00	0 00	744	0	2 85	3 59	0 00	0 00	744	0
NOV	0 54	0 59	0 00	0 00	720	0	0 27	0 30	0 00	0 00	720	0	3 95	5 74	0 00	0 00	720	0
DEC	0 63	0 67	0 00	0 00	744	0	0 32	0 33	0 00	0 00	744	0	6 61	7 59	0 00	0 00	744	0

Slab On Grade

	Slab (Perimeter)						Slab (Interior)					
	AVE HEAT LOSS	PEAK HEAT LOSS	AVE HEAT GAIN	PEAK HEAT GAIN	HRS HT ON	HRS CL ON	AVE HEAT LOSS	PEAK HEAT LOSS	AVE HEAT GAIN	PEAK HEAT GAIN	HRS HT ON	HRS CL ON
JAN	1 64	2 03	0 00	0 00	744	0	0 50	0 52	0 00	0 00	744	0
FEB	1 79	2 03	0 00	0 00	672	0	0 53	0 55	0 00	0 00	672	0
MAR	1 62	1 82	0 00	0 00	744	0	0 56	0 58	0 00	0 00	744	0
APR	1 44	1 80	0 00	0 00	720	0	0 70	0 76	0 00	0 00	720	0
MAY	1 24	1 53	0 00	0 00	744	0	0 91	1 05	0 00	0 00	744	0
JUN	1 02	1 29	0 00	0 00	720	0	0 94	1 04	0 00	0 00	720	0
JUL	0 83	1 02	0 00	0 00	744	0	0 97	1 07	0 00	0 00	744	0
AUG	0 74	0 89	0 00	0 00	744	0	0 87	0 91	0 00	0 00	744	0
SEP	0 63	0 92	0 00	0 00	720	0	0 59	0 61	0 00	0 00	720	0
OCT	0 74	0 92	0 00	0 00	744	0	0 39	0 43	0 00	0 00	744	0
NOV	0 94	1 26	0 00	0 00	720	0	0 33	0 37	0 00	0 00	720	0
DEC	1 45	1 66	0 00	0 00	744	0	0 39	0 41	0 00	0 00	744	0

Heat Flux Expressed in Btu/h²

To obtain W/m² units, multiply Btu/h ft² units by 3 156

TABLE 25

Monthly Average and Peak Heat Fluxes for Albuquerque with No Insulation

Albuquerque — No Insulation

Full Berm 3-Story

	Slab (Perimeter)						Slab (Interior)						Wall (Level-1)						Wall (Level-2)						Wall (Level-3)					
	AVE HEAT LOSS	PEAK HEAT LOSS	AVE HEAT GAIN	PEAK HEAT GAIN	HRS HT ON	HRS CL ON	AVE HEAT LOSS	PEAK HEAT LOSS	AVE HEAT GAIN	PEAK HEAT GAIN	HRS HT ON	HRS CL ON	AVE HEAT LOSS	PEAK HEAT LOSS	AVE HEAT GAIN	PEAK HEAT GAIN	HRS HT ON	HRS CL ON	AVE HEAT LOSS	PEAK HEAT LOSS	AVE HEAT GAIN	PEAK HEAT GAIN	HRS HT ON	HRS CL ON	AVE HEAT LOSS	PEAK HEAT LOSS	AVE HEAT GAIN	PEAK HEAT GAIN	HRS HT ON	HRS CL ON
JAN	0 23	0 23	0 00	0 00	744	0	0 16	0 16	0 00	0 00	744	0	3 47	3 70	0 00	0 00	744	0	0 55	0 63	0 00	0 00	744	0	0 45	0 45	0 00	0 00	744	0
FEB	0 22	0 23	0 00	0 00	672	0	0 15	0 15	0 00	0 00	672	0	3 33	3 64	0 00	0 00	672	0	0 68	0 72	0 00	0 00	672	0	0 46	0 47	0 00	0 00	672	0
MAR	0 21	0 22	0 00	0 00	744	0	0 14	0 15	0 00	0 00	744	0	3 09	3 68	0 00	0 00	744	0	0 75	0 78	0 00	0 00	744	0	0 48	0 49	0 00	0 00	744	0
APR	0 31	0 39	0 00	0 00	720	0	0 24	0 32	0 00	0 00	720	0	2 37	2 85	0 00	0 00	720	0	0 96	1 30	0 00	0 00	720	0	0 68	1 00	0 00	0 00	720	0
MAY	0 48	0 64	0 00	0 00	744	0	0 41	0 58	0 00	0 00	744	0	1 93	3 11	0 00	0 00	744	0	1 20	1 92	0 00	0 00	744	0	0 99	1 66	0 00	0 00	744	0
JUN	0 51	0 61	0 00	0 00	720	0	0 43	0 54	0 00	0 00	720	0	1 23	1 97	0 00	0 00	720	0	1 12	1 54	0 00	0 00	720	0	1 00	1 37	0 00	0 00	720	0
JUL	0 55	0 65	0 00	0 00	744	0	0 47	0 57	0 00	0 00	744	0	0 53	1 29	0 00	0 00	744	0	1 05	1 48	0 00	0 00	744	0	1 05	1 42	0 00	0 00	744	0
AUG	0 48	0 50	0 00	0 00	744	0	0 39	0 42	0 00	0 00	744	0	0 33	0 47	0 00	0 00	744	0	0 79	0 89	0 00	0 00	744	0	0 90	0 95	0 00	0 00	744	0
SEP	0 22	0 25	0 00	0 00	720	0	0 14	0 17	0 00	0 00	720	0	0 45	0 81	0 18	0 67	614	106	0 31	0 38	0 12	0 32	690	30	0 44	0 52	0 08	0 18	710	10
OCT	0 05	0 08	0 11	0 41	638	106	0 01	0 02	0 08	0 68	227	517	0 87	1 38	0 13	0 26	733	11	0 11	0 18	0 16	0 65	533	211	0 19	0 27	0 15	0 50	652	92
NOV	0 01	0 02	0 07	0 36	388	332	0 00	0 00	0 11	0 59	0	720	1 88	2 52	0 00	0 00	720	0	0 11	0 20	0 09	0 33	558	162	0 12	0 17	0 08	0 25	638	82
DEC	0 03	0 04	0 00	0 00	744	0	0 00	0 00	0 02	0 03	0	744	2 87	3 54	0 00	0 00	744	0	0 28	0 37	0 00	0 00	744	0	0 19	0 21	0 00	0 00	744	0

Full Berm 1-Story

	Slab (Perimeter)						Slab (Interior)						Wall (Level-1)					
	AVE HEAT LOSS	PEAK HEAT LOSS	AVE HEAT GAIN	PEAK HEAT GAIN	HRS HT ON	HRS CL ON	AVE HEAT LOSS	PEAK HEAT LOSS	AVE HEAT GAIN	PEAK HEAT GAIN	HRS HT ON	HRS CL ON	AVE HEAT LOSS	PEAK HEAT LOSS	AVE HEAT GAIN	PEAK HEAT GAIN	HRS HT ON	HRS CL ON
JAN	0 42	0 43	0 00	0 00	744	0	0 31	0 32	0 00	0 00	744	0	3 79	4 03	0 00	0 00	744	0
FEB	0 44	0 45	0 00	0 00	672	0	0 32	0 32	0 00	0 00	672	0	3 70	4 03	0 00	0 00	672	0
MAR	0 46	0 47	0 00	0 00	744	0	0 32	0 32	0 00	0 00	744	0	3 51	4 07	0 00	0 00	744	0
APR	0 58	0 65	0 00	0 00	720	0	0 42	0 50	0 00	0 00	720	0	2 96	3 33	0 00	0 00	720	0
MAY	0 76	0 92	0 00	0 00	744	0	0 60	0 76	0 00	0 00	744	0	2 71	3 74	0 00	0 00	744	0
JUN	0 79	0 89	0 00	0 00	720	0	0 62	0 72	0 00	0 00	720	0	2 15	2 83	0 00	0 00	720	0
JUL	0 82	0 92	0 00	0 00	744	0	0 66	0 76	0 00	0 00	744	0	1 53	2 22	0 00	0 00	744	0
AUG	0 74	0 77	0 00	0 00	744	0	0 59	0 62	0 00	0 00	744	0	1 31	1 42	0 00	0 00	744	0
SEP	0 46	0 49	0 00	0 00	720	0	0 34	0 38	0 00	0 00	720	0	1 13	1 50	0 00	0 00	720	0
OCT	0 27	0 32	0 00	0 00	744	0	0 17	0 22	0 00	0 00	744	0	1 38	1 82	0 00	0 00	744	0
NOV	0 22	0 26	0 00	0 00	720	0	0 13	0 17	0 00	0 00	720	0	2 21	2 80	0 00	0 00	720	0
DEC	0 28	0 30	0 00	0 00	744	0	0 19	0 20	0 00	0 00	744	0	3 12	3 76	0 00	0 00	744	0

Half Berm

	Slab (Perimeter)						Slab (Interior)						Wall (Level-1)					
	AVE HEAT LOSS	PEAK HEAT LOSS	AVE HEAT GAIN	PEAK HEAT GAIN	HRS HT ON	HRS CL ON	AVE HEAT LOSS	PEAK HEAT LOSS	AVE HEAT GAIN	PEAK HEAT GAIN	HRS HT ON	HRS CL ON	AVE HEAT LOSS	PEAK HEAT LOSS	AVE HEAT GAIN	PEAK HEAT GAIN	HRS HT ON	HRS CL ON
JAN	0 64	0 66	0 00	0 00	744	0	0 33	0 34	0 00	0 00	744	0	6 42	7 00	0 00	0 00	744	0
FEB	0 68	0 69	0 00	0 00	672	0	0 34	0 35	0 00	0 00	672	0	5 91	6 73	0 00	0 00	672	0
MAR	0 70	0 71	0 00	0 00	744	0	0 35	0 35	0 00	0 00	744	0	5 27	6 73	0 00	0 00	744	0
APR	0 81	0 89	0 00	0 00	720	0	0 47	0 53	0 00	0 00	720	0	3 76	4 41	0 00	0 00	720	0
MAY	0 99	1 15	0 00	0 00	744	0	0 65	0 81	0 00	0 00	744	0	2 78	4 39	0 00	0 00	744	0
JUN	1 00	1 10	0 00	0 00	720	0	0 69	0 78	0 00	0 00	720	0	1 76	2 80	0 00	0 00	720	0
JUL	1 01	1 12	0 00	0 00	744	0	0 73	0 82	0 00	0 00	744	0	0 54	1 51	0 00	0 00	744	0
AUG	0 91	0 95	0 00	0 00	744	0	0 66	0 68	0 00	0 00	744	0	0 63	1 04	0 00	0 00	744	0
SEP	0 64	0 67	0 00	0 00	720	0	0 40	0 44	0 00	0 00	720	0	1 20	2 00	0 14	0 28	709	11
OCT	0 46	0 51	0 00	0 00	744	0	0 23	0 28	0 00	0 00	744	0	2 30	3 16	0 00	0 00	744	0
NOV	0 43	0 48	0 00	0 00	720	0	0 19	0 23	0 00	0 00	720	0	4 16	5 27	0 00	0 00	720	0
DEC	0 52	0 56	0 00	0 00	744	0	0 25	0 26	0 00	0 00	744	0	5 69	6 99	0 00	0 00	744	0

Slab On Grade

	Slab (Perimeter)						Slab (Interior)					
	AVE HEAT LOSS	PEAK HEAT LOSS	AVE HEAT GAIN	PEAK HEAT GAIN	HRS HT ON	HRS CL ON	AVE HEAT LOSS	PEAK HEAT LOSS	AVE HEAT GAIN	PEAK HEAT GAIN	HRS HT ON	HRS CL ON
JAN	1 49	1 62	0 00	0 00	744	0	0 40	0 42	0 00	0 00	744	0
FEB	1 48	1 63	0 00	0 00	672	0	0 43	0 45	0 00	0 00	672	0
MAR	1 41	1 68	0 00	0 00	744	0	0 46	0 48	0 00	0 00	744	0
APR	1 26	1 43	0 00	0 00	720	0	0 60	0 66	0 00	0 00	720	0
MAY	1 20	1 54	0 00	0 00	744	0	0 80	0 95	0 00	0 00	744	0
JUN	1 01	1 28	0 00	0 00	720	0	0 84	0 93	0 00	0 00	720	0
JUL	0 79	1 00	0 00	0 00	744	0	0 87	0 97	0 00	0 00	744	0
AUG	0 69	0 75	0 00	0 00	744	0	0 79	0 82	0 00	0 00	744	0
SEP	0 56	0 72	0 00	0 00	720	0	0 51	0 53	0 00	0 00	720	0
OCT	0 60	0 80	0 00	0 00	744	0	0 31	0 35	0 00	0 00	744	0
NOV	0 90	1 15	0 00	0 00	720	0	0 26	0 29	0 00	0 00	720	0
DEC	1 26	1 53	0 00	0 00	744	0	0 31	0 33	0 00	0 00	744	0

Heat Flux Expressed in Btu/h²

To obtain W/m² units, multiply Btu/h ft² units by 3 156

TABLE 26

Monthly Average and Peak Heat Fluxes for Phoenix with No Insulation

Phoenix — No Insulation

Full Berm 3-Story

	Slab (Perimeter)						Slab (Interior)						Wall (Level-1)						Wall (Level-2)						Wall (Level-3)					
	AVE HEAT LOSS	PEAK HEAT LOSS	AVE HEAT GAIN	PEAK HEAT GAIN	HRS HT ON	HRS CL ON	AVE HEAT LOSS	PEAK HEAT LOSS	AVE HEAT GAIN	PEAK HEAT GAIN	HRS HT ON	HRS CL ON	AVE HEAT LOSS	PEAK HEAT LOSS	AVE HEAT GAIN	PEAK HEAT GAIN	HRS HT ON	HRS CL ON	AVE HEAT LOSS	PEAK HEAT LOSS	AVE HEAT GAIN	PEAK HEAT GAIN	HRS HT ON	HRS CL ON	AVE HEAT LOSS	PEAK HEAT LOSS	AVE HEAT GAIN	PEAK HEAT GAIN	HRS HT ON	HRS CL ON
JAN	0 00	0 00	0 30	0 78	0	744	0 00	0 00	0 30	0 79	0	744	1 80	1 97	0 00	0 00	744	0	0 00	0 00	0 21	0 64	0	744	0 00	0 00	0 31	0 67	0	744
FEB	0 11	0 30	0 00	0 00	672	0	0 11	0 30	0 00	0 00	672	0	2 10	2 61	0 00	0 00	672	0	0 39	0 97	0 00	0 00	672	0	0 20	0 81	0 00	0 00	672	0
MAR	0 15	0 24	0 00	0 00	744	0	0 15	0 24	0 00	0 00	744	0	1 85	2 35	0 00	0 00	744	0	0 50	0 83	0 00	0 00	744	0	0 28	0 62	0 00	0 00	744	0
APR	0 22	0 30	0 00	0 00	720	0	0 20	0 29	0 00	0 00	720	0	1 22	1 76	0 00	0 00	720	0	0 61	0 96	0 00	0 00	720	0	0 40	0 73	0 00	0 00	720	0
MAY	0 27	0 36	0 00	0 00	744	0	0 26	0 35	0 00	0 00	744	0	0 67	1 33	0 00	0 00	744	0	0 64	1 03	0 00	0 00	744	0	0 50	0 85	0 00	0 00	744	0
JUN	0 22	0 24	0 00	0 00	720	0	0 20	0 22	0 00	0 00	720	0	0 06	0 11	0 36	0 64	44	676	0 46	0 53	0 00	0 00	720	0	0 40	0 43	0 00	0 00	720	0
JUL	0 20	0 21	0 00	0 00	744	0	0 17	0 19	0 00	0 00	744	0	0 00	0 00	0 85	1 20	0	744	0 32	0 39	0 00	0 00	744	0	0 36	0 38	0 00	0 00	744	0
AUG	0 18	0 19	0 00	0 00	744	0	0 15	0 16	0 00	0 00	744	0	0 00	0 00	1 07	1 25	0	744	0 19	0 25	0 00	0 00	744	0	0 32	0 34	0 00	0 00	744	0
SEP	0 16	0 17	0 00	0 00	720	0	0 14	0 14	0 00	0 00	720	0	0 00	0 00	0 80	1 04	0	720	0 09	0 13	0 00	0 00	720	0	0 28	0 30	0 00	0 00	720	0
OCT	0 00	0 00	0 10	0 72	0	744	0 00	0 00	0 14	0 82	0	744	0 05	0 14	0 56	1 54	224	520	0 00	0 00	0 32	0 97	0	744	0 00	0 00	0 13	0 77	0	744
NOV	0 00	0 00	0 19	0 66	0	720	0 00	0 00	0 21	0 69	0	720	0 53	1 12	0 17	0 39	658	62	0 00	0 00	0 31	0 70	0	720	0 00	0 00	0 20	0 54	0	720
DEC	0 00	0 00	0 28	0 76	0	744	0 00	0 00	0 28	0 77	0	744	1 33	2 15	0 00	0 00	744	0	0 00	0 00	0 30	0 71	0	744	0 00	0 00	0 29	0 64	0	744

Full Berm 1-Story

	Slab (Perimeter)						Slab (Interior)						Wall (Level-1)					
	AVE HEAT LOSS	PEAK HEAT LOSS	AVE HEAT GAIN	PEAK HEAT GAIN	HRS HT ON	HRS CL ON	AVE HEAT LOSS	PEAK HEAT LOSS	AVE HEAT GAIN	PEAK HEAT GAIN	HRS HT ON	HRS CL ON	AVE HEAT LOSS	PEAK HEAT LOSS	AVE HEAT GAIN	PEAK HEAT GAIN	HRS HT ON	HRS CL ON
JAN	0 00	0 00	0 24	0 73	0	744	0 00	0 00	0 24	0 73	0	744	1 56	1 72	0 00	0 00	744	0
FEB	0 17	0 32	0 00	0 00	672	0	0 14	0 32	0 00	0 00	672	0	2 03	2 40	0 00	0 00	672	0
MAR	0 23	0 31	0 00	0 00	744	0	0 18	0 27	0 00	0 00	744	0	1 92	2 32	0 00	0 00	744	0
APR	0 31	0 39	0 00	0 00	720	0	0 24	0 33	0 00	0 00	720	0	1 45	1 89	0 00	0 00	720	0
MAY	0 37	0 46	0 00	0 00	744	0	0 30	0 39	0 00	0 00	744	0	1 03	1 60	0 00	0 00	744	0
JUN	0 31	0 33	0 00	0 00	720	0	0 24	0 26	0 00	0 00	720	0	0 18	0 51	0 08	0 19	507	213
JUL	0 27	0 29	0 00	0 00	744	0	0 22	0 23	0 00	0 00	744	0	0 00	0 00	0 40	0 75	0	744
AUG	0 24	0 26	0 00	0 00	744	0	0 20	0 21	0 00	0 00	744	0	0 00	0 00	0 64	0 80	0	744
SEP	0 21	0 23	0 00	0 00	720	0	0 19	0 19	0 00	0 00	720	0	0 00	0 00	0 45	0 66	0	720
OCT	0 01	0 02	0 08	0 57	323	421	0 01	0 02	0 09	0 63	290	454	0 11	0 21	0 46	1 23	295	449
NOV	0 00	0 00	0 14	0 58	0	720	0 00	0 00	0 13	0 58	0	720	0 49	1 05	0 14	0 33	662	58
DEC	0 00	0 00	0 22	0 70	0	744	0 00	0 00	0 20	0 69	0	744	1 18	1 95	0 00	0 00	744	0

Half Berm

	Slab (Perimeter)						Slab (Interior)						Wall (Level-1)					
	AVE HEAT LOSS	PEAK HEAT LOSS	AVE HEAT GAIN	PEAK HEAT GAIN	HRS HT ON	HRS CL ON	AVE HEAT LOSS	PEAK HEAT LOSS	AVE HEAT GAIN	PEAK HEAT GAIN	HRS HT ON	HRS CL ON	AVE HEAT LOSS	PEAK HEAT LOSS	AVE HEAT GAIN	PEAK HEAT GAIN	HRS HT ON	HRS CL ON
JAN	0 03	0 06	0 11	0 56	316	428	0 00	0 00	0 24	0 73	0	744	3 52	4 27	0 00	0 00	744	0
FEB	0 27	0 38	0 00	0 00	672	0	0 15	0 32	0 00	0 00	672	0	3 53	4 07	0 00	0 00	672	0
MAR	0 35	0 42	0 00	0 00	744	0	0 19	0 28	0 00	0 00	744	0	2 91	3 51	0 00	0 00	744	0
APR	0 42	0 50	0 00	0 00	720	0	0 26	0 34	0 00	0 00	720	0	1 68	2 39	0 00	0 00	720	0
MAY	0 47	0 56	0 00	0 00	744	0	0 32	0 41	0 00	0 00	744	0	0 80	1 51	0 23	0 37	669	75
JUN	0 39	0 43	0 00	0 00	720	0	0 27	0 28	0 00	0 00	720	0	0 00	0 00	1 02	1 48	0	720
JUL	0 34	0 36	0 00	0 00	744	0	0 24	0 25	0 00	0 00	744	0	0 00	0 00	1 72	2 42	0	744
AUG	0 29	0 31	0 00	0 00	744	0	0 23	0 23	0 00	0 00	744	0	0 00	0 00	1 85	2 45	0	744
SEP	0 26	0 27	0 00	0 00	720	0	0 21	0 22	0 00	0 00	720	0	0 00	0 00	1 18	1 76	0	720
OCT	0 05	0 08	0 10	0 41	631	113	0 02	0 03	0 10	0 56	434	310	0 60	0 84	0 54	1 57	369	375
NOV	0 05	0 09	0 08	0 38	454	266	0 00	0 00	0 11	0 54	0	720	1 51	2 72	0 00	0 00	720	0
DEC	0 04	0 07	0 11	0 55	306	438	0 00	0 00	0 19	0 68	0	744	3 04	4 58	0 00	0 00	744	0

Slab On Grade

	Slab (Perimeter)						Slab (Interior)					
	AVE HEAT LOSS	PEAK HEAT LOSS	AVE HEAT GAIN	PEAK HEAT GAIN	HRS HT ON	HRS CL ON	AVE HEAT LOSS	PEAK HEAT LOSS	AVE HEAT GAIN	PEAK HEAT GAIN	HRS HT ON	HRS CL ON
JAN	0 51	0 58	0 00	0 00	744	0	0 00	0 00	0 13	0 23	0	744
FEB	0 78	0 90	0 00	0 00	672	0	0 14	0 26	0 00	0 00	672	0
MAR	0 77	0 92	0 00	0 00	744	0	0 23	0 30	0 00	0 00	744	0
APR	0 62	0 82	0 00	0 00	720	0	0 32	0 39	0 00	0 00	720	0
MAY	0 48	0 66	0 00	0 00	744	0	0 39	0 47	0 00	0 00	744	0
JUN	0 15	0 31	0 00	0 00	720	0	0 34	0 36	0 00	0 00	720	0
JUL	0 03	0 08	0 07	0 16	317	427	0 30	0 32	0 00	0 00	744	0
AUG	0 00	0 00	0 14	0 54	0	744	0 35	0 95	0 00	0 00	744	0
SEP	0 08	0 20	0 30	0 60	108	612	0 84	0 93	0 00	0 00	720	0
OCT	0 23	0 48	0 39	1 36	375	369	0 03	0 05	0 09	0 51	459	285
NOV	0 38	0 71	0 14	0 38	690	30	0 00	0 00	0 11	0 59	0	720
DEC	0 51	0 92	0 00	0 00	744	0	0 00	0 00	0 13	0 71	0	744

Heat Flux Expressed in Btu/h²

To obtain W/m² units, multiply Btu/h ft² units by 3 156

TABLE 27
Monthly Average and Peak Heat Fluxes for Miami with No Insulation

Miami No Insulation

Full Berm 3-Story

	Slab (Perimeter)						Slab (Interior)						Wall (Level-1)						Wall (Level-2)						Wall (Level-3)					
	AVE HEAT LOSS	PEAK HEAT LOSS	AVE HEAT GAIN	PEAK HEAT GAIN	HRS HT ON	HRS CL ON	AVE HEAT LOSS	PEAK HEAT LOSS	AVE HEAT GAIN	PEAK HEAT GAIN	HRS HT ON	HRS CL ON	AVE HEAT LOSS	PEAK HEAT LOSS	AVE HEAT GAIN	PEAK HEAT GAIN	HRS HT ON	HRS CL ON	AVE HEAT LOSS	PEAK HEAT LOSS	AVE HEAT GAIN	PEAK HEAT GAIN	HRS HT ON	HRS CL ON	AVE HEAT LOSS	PEAK HEAT LOSS	AVE HEAT GAIN	PEAK HEAT GAIN	HRS HT ON	HRS CL ON
JAN	0 00	0 00	0 41	0 89	0	744	0 00	0 00	0 36	0 85	0	744	0 22	0 48	0 05	0 26	637	107	0 00	0 00	0 40	0 80	0	744	0 00	0 00	0 46	0 82	0	744
FEB	0 03	0 09	0 12	0 15	72	600	0 04	0 11	0 08	0 10	115	557	0 42	0 78	0 00	0 00	672	0	0 07	0 23	0 07	0 09	78	594	0 05	0 14	0 16	0 18	25	647
MAR	0 03	0 14	0 01	0 02	588	156	0 05	0 15	0 00	0 00	744	0	0 40	1 16	0 00	0 00	744	0	0 10	0 43	0 00	0 00	744	0	0 07	0 33	0 03	0 05	265	479
APR	0 09	0 18	0 00	0 00	720	0	0 11	0 19	0 00	0 00	720	0	0 29	0 88	0 00	0 00	720	0	0 21	0 56	0 00	0 00	720	0	0 13	0 47	0 00	0 00	720	0
MAY	0 16	0 24	0 00	0 00	744	0	0 17	0 26	0 00	0 00	744	0	0 14	0 59	0 07	0 15	494	250	0 28	0 65	0 00	0 00	744	0	0 24	0 58	0 00	0 00	744	0
JUN	0 11	0 13	0 00	0 00	720	0	0 12	0 13	0 00	0 00	720	0	0 00	0 00	0 31	0 44	0	720	0 15	0 19	0 00	0 00	720	0	0 14	0 17	0 00	0 00	720	0
JUL	0 09	0 10	0 00	0 00	744	0	0 10	0 11	0 00	0 00	744	0	0 00	0 00	0 51	0 60	0	744	0 08	0 11	0 00	0 00	744	0	0 12	0 13	0 00	0 00	744	0
AUG	0 08	0 09	0 00	0 00	744	0	0 08	0 09	0 00	0 00	744	0	0 00	0 00	0 62	0 67	0	744	0 02	0 05	0 00	0 00	709	35	0 09	0 10	0 00	0 00	744	0
SEP	0 07	0 08	0 00	0 00	720	0	0 07	0 08	0 00	0 00	720	0	0 00	0 00	0 64	0 71	0	720	0 00	0 00	0 02	0 04	0	720	0 08	0 09	0 00	0 00	720	0
OCT	0 00	0 00	0 07	0 40	0	744	0 00	0 00	0 07	0 40	0	744	0 00	0 00	0 59	1 13	0	744	0 00	0 00	0 23	0 55	0	744	0 00	0 00	0 13	0 45	0	744
NOV	0 00	0 00	0 21	0 67	0	720	0 00	0 00	0 19	0 66	0	720	0 06	0 12	0 32	0 76	63	657	0 00	0 00	0 33	0 69	0	720	0 00	0 00	0 25	0 59	0	720
DEC	0 00	0 00	0 30	0 78	0	744	0 00	0 00	0 27	0 76	0	744	0 37	0 69	0 23	0 40	621	123	0 00	0 00	0 38	0 76	0	744	0 00	0 00	0 35	0 70	0	744

Full Berm 1-Story

	Slab (Perimeter)						Slab (Interior)						Wall (Level-1)					
	AVE HEAT LOSS	PEAK HEAT LOSS	AVE HEAT GAIN	PEAK HEAT GAIN	HRS HT ON	HRS CL ON	AVE HEAT LOSS	PEAK HEAT LOSS	AVE HEAT GAIN	PEAK HEAT GAIN	HRS HT ON	HRS CL ON	AVE HEAT LOSS	PEAK HEAT LOSS	AVE HEAT GAIN	PEAK HEAT GAIN	HRS HT ON	HRS CL ON
JAN	0.00	0.00	0 45	0 94	0	744	0 00	0 00	0 39	0 88	0	744	0 10	0 21	0 15	0 50	177	567
FEB	0 03	0 08	0 15	0 17	49	623	0 04	0 10	0 11	0 14	84	588	0 24	0 54	0 00	0 00	672	0
MAR	0.04	0 13	0 03	0 04	391	353	0 04	0 14	0 01	0 03	510	234	0 29	0 99	0 00	0 00	744	0
APR	0.08	0 17	0 00	0 00	720	0	0 08	0 17	0 00	0 00	720	0	0 27	0 79	0 00	0 00	720	0
MAY	0.14	0 23	0 00	0 00	744	0	0 14	0 23	0 00	0 00	744	0	0 16	0 59	0 02	0 05	707	37
JUN	0.09	0 11	0 00	0 00	720	0	0 09	0 10	0 00	0 00	720	0	0 00	0 00	0 20	0 33	0	720
JUL	0 07	0 08	0 00	0 00	744	0	0 06	0 07	0 00	0 00	744	0	0 00	0 00	0 40	0 50	0	744
AUG	0 05	0 06	0 00	0 00	744	0	0 05	0 06	0 00	0 00	744	0	0 00	0 00	0 53	0 58	0	744
SEP	0 03	0 04	0 00	0 00	720	0	0 04	0 05	0 00	0 00	720	0	0 00	0 00	0 56	0 63	0	720
OCT	0 00	0 00	0 15	0 54	0	744	0 00	0 00	0 11	0 50	0	744	0 00	0 00	0 59	1 06	0	744
NOV	0 00	0.00	0 28	0 74	0	720	0 00	0 00	0 22	0 70	0	720	0 00	0 00	0 38	0 79	1	719
DEC	0 00	0 00	0 36	0 84	0	744	0 00	0 00	0 30	0 79	0	744	0 19	0 48	0 25	0 52	546	198

Half Berm

	Slab (Perimeter)						Slab (Interior)						Wall (Level-1)					
	AVE HEAT LOSS	PEAK HEAT LOSS	AVE HEAT GAIN	PEAK HEAT GAIN	HRS HT ON	HRS CL ON	AVE HEAT LOSS	PEAK HEAT LOSS	AVE HEAT GAIN	PEAK HEAT GAIN	HRS HT ON	HRS CL ON	AVE HEAT LOSS	PEAK HEAT LOSS	AVE HEAT GAIN	PEAK HEAT GAIN	HRS HT ON	HRS CL ON
JAN	0 00	0 00	0 40	0 88	0	744	0 00	0 00	0 40	0 89	0	744	0 62	1 33	0 07	0 12	621	123
FEB	0 03	0 09	0 09	0 12	86	586	0 03	0 09	0 11	0 14	81	591	0 72	1 41	0 00	0 00	672	0
MAR	0 05	0 15	0 00	0 00	744	0	0 04	0 14	0 01	0 04	540	204	0 59	1 89	0 05	0 10	661	83
APR	0 11	0 19	0 00	0 00	720	0	0 08	0 17	0 00	0 00	720	0	0 34	1 12	0 13	0 27	579	141
MAY	0 17	0 26	0 00	0 00	744	0	0 14	0 23	0 00	0 00	744	0	0 10	0 44	0 25	0 49	184	560
JUN	0 12	0 14	0 00	0 00	720	0	0 09	0 11	0 00	0 00	720	0	0 00	0 00	0 72	0 96	0	720
JUL	0 08	0 10	0 00	0 00	744	0	0 07	0 08	0 00	0 00	744	0	0 00	0 00	0 99	1 15	0	744
AUG	0 06	0 07	0 00	0 00	744	0	0 05	0 06	0 00	0 00	744	0	0 00	0 00	1 12	1 23	0	744
SEP	0.04	0 05	0 00	0 00	720	0	0 04	0 05	0 00	0 00	720	0	0 00	0 00	1 10	1 29	0	720
OCT	0 00	0 00	0 13	0 52	0	744	0 00	0 00	0 12	0 51	0	744	0 00	0 00	0 75	1 52	0	744
NOV	0 00	0 00	0 25	0 72	0	720	0 00	0 00	0 24	0 71	0	720	0 35	0 65	0 25	0 63	139	581
DEC	0 00	0 00	0 32	0 81	0	744	0 00	0 00	0 32	0 80	0	744	0 95	1 94	0 00	0 01	737	7

Slab On Grade

	Slab (Perimeter)						Slab (Interior)					
	AVE HEAT LOSS	PEAK HEAT LOSS	AVE HEAT GAIN	PEAK HEAT GAIN	HRS HT ON	HRS CL ON	AVE HEAT LOSS	PEAK HEAT LOSS	AVE HEAT GAIN	PEAK HEAT GAIN	HRS HT ON	HRS CL ON
JAN	0 21	0 60	0 18	0 72	220	524	0 00	0 00	0 44	0 92	0	744
FEB	0 25	0 83	0 05	0 15	598	74	0 03	0 09	0 11	0 16	64	608
MAR	0 11	0 33	0 00	0 00	744	0	0 04	0 14	0 01	0 01	517	227
APR	0 12	0 28	0 00	0 00	720	0	0 09	0 17	0 00	0 00	720	0
MAY	0 10	0 21	0 00	0 00	744	0	0 15	0 24	0 00	0 00	744	0
JUN	0 02	0 03	0 05	0 09	186	534	0 10	0 12	0 00	0 00	720	0
JUL	0 00	0 00	0 37	0 51	0	744	0 23	0 31	0 00	0 00	744	0
AUG	0 00	0 00	0 44	0 56	0	744	0 07	0 10	0 00	0 00	744	0
SEP	0 00	0 00	0 45	0 59	0	720	0 04	0 06	0 00	0 00	720	0
OCT	0 00	0 00	0 45	1 08	0	744	0 00	0 00	0 16	0 61	0	744
NOV	0 10	0 26	0 33	0 81	88	632	0 00	0 00	0 28	0 74	0	720
DEC	0 28	0 89	0 26	0 52	570	174	0 00	0 00	0 36	0 84	0	744

Heat Flux Expressed in Btu/h^2

To obtain W/m^2 units, multiply Btu/h ft^2 units by 3 156

TABLE 28

Monthly Average and Peak Heat Fluxes for Minneapolis with R-10.4 Insulation

Minneapolis R-10.4 Insulation

Full Berm 3-Story

	Slab (Perimeter)						Slab (Interior)						Wall (Level-1)						Wall (Level-2)						Wall (Level-3)					
	AVE HEAT LOSS	PEAK HEAT LOSS	AVE HEAT GAIN	PEAK HEAT GAIN	HRS HT ON	HRS CL ON	AVE HEAT LOSS	PEAK HEAT LOSS	AVE HEAT GAIN	PEAK HEAT GAIN	HRS HT ON	HRS CL ON	AVE HEAT LOSS	PEAK HEAT LOSS	AVE HEAT GAIN	PEAK HEAT GAIN	HRS HT ON	HRS CL ON	AVE HEAT LOSS	PEAK HEAT LOSS	AVE HEAT GAIN	PEAK HEAT GAIN	HRS HT ON	HRS CL ON	AVE HEAT LOSS	PEAK HEAT LOSS	AVE HEAT GAIN	PEAK HEAT GAIN	HRS HT ON	HRS CL ON
JAN	0 63	0 65	0 00	0 00	744	0	0 48	0 49	0 00	0 00	744	0	1 86	1 95	0 00	0.00	744	0	1 08	1 12	0 00	0 00	744	0	1 09	1 10	0 00	0 00	744	0
FEB	0 59	0 61	0 00	0 00	672	0	0 44	0 46	0 00	0 00	672	0	1 98	1 99	0 00	0 00	672	0	1 16	1 19	0 00	0 00	672	0	1 06	1 08	0 00	0 00	672	0
MAR	0 56	0 57	0 00	0 00	744	0	0 41	0 43	0 00	0 00	744	0	1 92	1 96	0 00	0 00	744	0	1 21	1 22	0 00	0 00	744	0	1 04	1 05	0 00	0 00	744	0
APR	0 63	0 72	0 00	0 00	720	0	0 49	0 58	0 00	0 00	720	0	1 83	2 38	0 00	0 00	720	0	1 40	1 74	0 00	0 00	720	0	1 22	1 55	0 00	0 00	720	0
MAY	0 67	0 77	0 00	0 00	744	0	0 54	0 63	0 00	0 00	744	0	1 58	2 15	0 00	0 00	744	0	1 46	1 84	0 00	0 00	744	0	1 30	1 66	0 00	0 00	744	0
JUN	0 71	0 81	0 00	0 00	720	0	0 57	0 67	0 00	0 00	720	0	1 15	1 76	0 00	0 00	720	0	1 45	1 86	0 00	0 00	720	0	1 37	1 73	0 00	0 00	720	0
JUL	0.85	1 03	0.00	0 00	744	0	0 71	0 89	0 00	0 00	744	0	0 90	1 84	0 00	0 00	744	0	1 54	2 33	0 00	0 00	744	0	1 60	2 30	0 00	0 00	744	0
AUG	0 63	0 64	0 00	0 00	744	0	0 49	0 50	0 00	0 00	744	0	0 45	0 53	0 00	0 00	744	0	1 00	1 03	0 00	0 00	744	0	1 17	1 19	0 00	0 00	744	0
SEP	0 38	0 41	0 00	0 00	720	0	0 25	0 28	0 00	0 00	720	0	0 35	0 64	0 17	0 50	657	63	0 54	0 62	0 04	0 07	717	3	0 75	0 83	0 00	0 00	720	0
OCT	0 19	0 23	0 00	0 00	744	0	0 08	0 12	0 03	0 08	678	66	0 68	0 89	0 15	0 39	717	27	0 31	0 45	0 14	0 41	693	51	0 46	0 56	0 08	0 20	732	12
NOV	0 24	0 25	0 00	0 00	720	0	0 13	0 14	0 00	0 00	720	0	1 21	1 57	0 00	0 00	720	0	0 52	0 59	0 00	0 00	720	0	0 58	0 58	0 00	0 00	720	0
DEC	0 25	0 25	0 00	0 00	744	0	0 14	0 14	0 00	0 00	744	0	1 66	1 71	0 00	0 00	744	0	0 68	0 76	0 00	0 00	744	0	0 58	0 59	0 00	0 00	744	0

Full Berm 1-Story

	Slab (Perimeter)						Slab (Interior)						Wall (Level-1)					
	AVE HEAT LOSS	PEAK HEAT LOSS	AVE HEAT GAIN	PEAK HEAT GAIN	HRS HT ON	HRS CL ON	AVE HEAT LOSS	PEAK HEAT LOSS	AVE HEAT GAIN	PEAK HEAT GAIN	HRS HT ON	HRS CL ON	AVE HEAT LOSS	PEAK HEAT LOSS	AVE HEAT GAIN	PEAK HEAT GAIN	HRS HT ON	HRS CL ON
JAN	0 98	0 98	0 00	0 00	744	0	0 78	0 79	0 00	0 00	744	0	3 12	3 23	0 00	0 00	744	0
FEB	0 98	0 98	0 00	0 00	672	0	0 77	0 77	0 00	0 00	672	0	3 27	3 29	0 00	0.00	672	0
MAR	0 98	0 98	0 00	0 00	744	0	0 75	0 76	0 00	0 00	744	0	3 27	3 29	0 00	0 00	744	0
APR	1 09	1 16	0 00	0 00	720	0	0 84	0 92	0 00	0 00	720	0	3 26	3 75	0 00	0 00	720	0
MAY	1 15	1 24	0 00	0 00	744	0	0 90	0 99	0 00	0 00	744	0	3 08	3 59	0 00	0.00	744	0
JUN	1 19	1 29	0 00	0 00	720	0	0 94	1 04	0 00	0 00	720	0	2 74	3 28	0 00	0 00	720	0
JUL	1 33	1 51	0 00	0 00	744	0	1 09	1 26	0 00	0 00	744	0	2 60	3 43	0 00	0 00	744	0
AUG	1 09	1 10	0 00	0 00	744	0	0 87	0 88	0 00	0 00	744	0	2 02	2 08	0 00	0 00	744	0
SEP	0 83	0 85	0 00	0 00	720	0	0 64	0 67	0 00	0 00	720	0	1 61	1 85	0 00	0 00	720	0
OCT	0 63	0 67	0 00	0 00	744	0	0 47	0 51	0 00	0 00	744	0	1 67	1 86	0 00	0 00	744	0
NOV	0 69	0 70	0 00	0 00	720	0	0 53	0 54	0 00	0 00	720	0	2 15	2 49	0 00	0 00	720	0
DEC	0 71	0 72	0 00	0 00	744	0	0 54	0 54	0 00	0 00	744	0	2 59	2 68	0 00	0 00	744	0

Half Berm

	Slab (Perimeter)						Slab (Interior)						Wall (Level-1)					
	AVE HEAT LOSS	PEAK HEAT LOSS	AVE HEAT GAIN	PEAK HEAT GAIN	HRS HT ON	HRS CL ON	AVE HEAT LOSS	PEAK HEAT LOSS	AVE HEAT GAIN	PEAK HEAT GAIN	HRS HT ON	HRS CL ON	AVE HEAT LOSS	PEAK HEAT LOSS	AVE HEAT GAIN	PEAK HEAT GAIN	HRS HT ON	HRS CL ON
JAN	1 33	1 34	0 00	0 00	744	0	0 82	0 82	0 00	0 00	744	0	2 11	2 22	0 00	0 00	744	0
FEB	1 35	1 36	0 00	0 00	672	0	0 81	0 82	0 00	0 00	672	0	2 19	2 23	0 00	0 00	672	0
MAR	1 37	1 37	0 00	0 00	744	0	0 81	0 81	0 00	0 00	744	0	2 06	2 13	0 00	0 00	744	0
APR	1 48	1 55	0 00	0 00	720	0	0 91	0 99	0 00	0 00	720	0	1 72	2 50	0 00	0 00	720	0
MAY	1 53	1 62	0 00	0 00	744	0	0 98	1 07	0 00	0 00	744	0	1 33	1 97	0 00	0 00	744	0
JUN	1 55	1 66	0 00	0 00	720	0	1 04	1 13	0 00	0 00	720	0	0 83	1 41	0 00	0 00	720	0
JUL	1 67	1 86	0 00	0 00	744	0	1 19	1 36	0 00	0 00	744	0	0 59	1 46	0 00	0 00	744	0
AUG	1 42	1 43	0 00	0 00	744	0	0 98	0 98	0 00	0 00	744	0	0 42	0 65	0 04	0 08	736	8
SEP	1 15	1 18	0 00	0 00	720	0	0 74	0 77	0 00	0 00	720	0	0 56	1 02	0 14	0 39	679	41
OCT	0.98	1 04	0.00	0 00	744	0	0 57	0 62	0 00	0 00	744	0	1 10	1 30	0 00	0 00	743	1
NOV	1 07	1 10	0 00	0 00	720	0	0 63	0 64	0 00	0 00	720	0	1 64	2 11	0 00	0 00	720	0
DEC	1 14	1 16	0 00	0 00	744	0	0 65	0 65	0 00	0 00	744	0	2 00	2 12	0 00	0 00	744	0

Slab On Grade

	Slab (Perimeter)						Slab (Interior)					
	AVE HEAT LOSS	PEAK HEAT LOSS	AVE HEAT GAIN	PEAK HEAT GAIN	HRS HT ON	HRS CL ON	AVE HEAT LOSS	PEAK HEAT LOSS	AVE HEAT GAIN	PEAK HEAT GAIN	HRS HT ON	HRS CL ON
JAN	1 38	1.42	0.00	0 00	744	0	0 88	0 88	0 00	0 00	744	0
FEB	1 44	1 45	0 00	0 00	672	0	0 89	0 89	0 00	0 00	672	0
MAR	1 46	1 46	0 00	0 00	744	0	0 89	0 90	0 00	0 00	744	0
APR	1 51	1 63	0 00	0 00	720	0	1 01	1 08	0 00	0 00	720	0
MAY	1 48	1 62	0 00	0 00	744	0	1 09	1 17	0 00	0 00	744	0
JUN	1 40	1 56	0 00	0 00	720	0	1 16	1 24	0 00	0 00	720	0
JUL	1 42	1 63	0 00	0 00	744	0	1 31	1 48	0 00	0 00	744	0
AUG	1 12	1 15	0 00	0 00	744	0	1 09	1 09	0 00	0 00	744	0
SEP	0.87	0 95	0 00	0 00	720	0	0 83	0 85	0 00	0 00	720	0
OCT	0 80	0 88	0 00	0 00	744	0	0 63	0 67	0 00	0 00	744	0
NOV	0 99	1 10	0 00	0 00	720	0	0 69	0 70	0 00	0 00	720	0
DEC	1.15	1.19	0 00	0 00	744	0	0 70	0 71	0 00	0 00	744	0

INSULATION DEPTH

Full Berm, 3 Story	8 0 ft
Full Berm, 1 Story	8 0 ft
Half Berm	5 0 ft
Slab on Grade	4 0 ft

Heat Flux Expressed in Btu/h ft²

To obtain W/m² units, multiply Btu/h ft² units by 3 156

TABLE 29

Monthly Average and Peak Heat Fluxes for Boston with R-10.4 Insulation

Boston R-10.4 Insulation

Full Berm 3-Story

	Slab (Perimeter) AVE HEAT LOSS	PEAK HEAT LOSS	AVE HEAT GAIN	PEAK HEAT GAIN	HRS HT ON	HRS CL ON	Slab (Interior) AVE HEAT LOSS	PEAK HEAT LOSS	AVE HEAT GAIN	PEAK HEAT GAIN	HRS HT ON	HRS CL ON
JAN	0 41	0 43	0 00	0 00	744	0	0 31	0 32	0 00	0 00	744	0
FEB	0 39	0 40	0 00	0 00	672	0	0 29	0 30	0 00	0 00	672	0
MAR	0 37	0 38	0 00	0 00	744	0	0 27	0 28	0 00	0 00	744	0
APR	0 46	0 54	0 00	0 00	720	0	0 36	0 44	0 00	0 00	720	0
MAY	0 51	0 60	0 00	0 00	744	0	0 41	0 51	0 00	0 00	744	0
JUN	0 56	0 65	0 00	0 00	720	0	0 46	0 56	0 00	0 00	720	0
JUL	0 71	0 88	0 00	0 00	744	0	0 60	0 78	0 00	0 00	744	0
AUG	0 50	0 50	0 00	0 00	744	0	0 39	0 40	0 00	0 00	744	0
SEP	0 26	0 29	0 00	0 00	720	0	0 16	0 19	0 00	0 00	720	0
OCT	0 08	0 12	0 09	0 27	699	45	0 02	0 04	0 07	0 61	400	344
NOV	0 13	0 14	0 00	0 00	720	0	0 05	0 06	0 00	0 00	720	0
DEC	0 14	0 15	0 00	0 00	744	0	0 07	0 07	0 00	0 00	744	0

	Wall (Level-1) AVE HEAT LOSS	PEAK HEAT LOSS	AVE HEAT GAIN	PEAK HEAT GAIN	HRS HT ON	HRS CL ON	Wall (Level-2) AVE HEAT LOSS	PEAK HEAT LOSS	AVE HEAT GAIN	PEAK HEAT GAIN	HRS HT ON	HRS CL ON	Wall (Level-3) AVE HEAT LOSS	PEAK HEAT LOSS	AVE HEAT GAIN	PEAK HEAT GAIN	HRS HT ON	HRS CL ON
JAN	1 43	1 48	0 00	0 00	744	0	0 80	0 84	0 00	0 00	744	0	0 75	0 76	0 00	0 00	744	0
FEB	1 49	1 50	0 00	0 00	672	0	0 86	0 88	0 00	0 00	672	0	0 74	0 75	0 00	0 00	672	0
MAR	1 45	1 47	0 00	0 00	744	0	0 90	0 91	0 00	0 00	744	0	0 73	0 74	0 00	0 00	744	0
APR	1 55	1 94	0 00	0 00	720	0	1 09	1 42	0 00	0 00	720	0	0 91	1 24	0 00	0 00	720	0
MAY	1 48	1 95	0 00	0 00	744	0	1 19	1 55	0 00	0 00	744	0	1 01	1 36	0 00	0 00	744	0
JUN	1 09	1 70	0 00	0 00	720	0	1 22	1 62	0 00	0 00	720	0	1 09	1 45	0 00	0 00	720	0
JUL	0 92	1 86	0 00	0 00	744	0	1 36	2 12	0 00	0 00	744	0	1 34	2 03	0 00	0 00	744	0
AUG	0 49	0 53	0 00	0 00	744	0	0 86	0 89	0 00	0 00	744	0	0 93	0 95	0 00	0 00	744	0
SEP	0 38	0 59	0 18	0 52	666	54	0 43	0 51	0 08	0 20	707	13	0 53	0 62	0 05	0 10	716	4
OCT	0 53	0 79	0 17	0 44	706	38	0 22	0 33	0 15	0 52	650	94	0 28	0 37	0 14	0 41	693	51
NOV	0 97	1 24	0 00	0 00	720	0	0 40	0 46	0 00	0 00	720	0	0 39	0 40	0 00	0 00	720	0
DEC	1 31	1 34	0 00	0 00	744	0	0 52	0 59	0 00	0 00	744	0	0 41	0 41	0 00	0 00	744	0

Full Berm 1-Story

	Slab (Perimeter) AVE HEAT LOSS	PEAK HEAT LOSS	AVE HEAT GAIN	PEAK HEAT GAIN	HRS HT ON	HRS CL ON	Slab (Interior) AVE HEAT LOSS	PEAK HEAT LOSS	AVE HEAT GAIN	PEAK HEAT GAIN	HRS HT ON	HRS CL ON	Wall (Level-1) AVE HEAT LOSS	PEAK HEAT LOSS	AVE HEAT GAIN	PEAK HEAT GAIN	HRS HT ON	HRS CL ON
JAN	0 68	0 68	0 00	0 00	744	0	0 53	0 53	0 00	0 00	744	0	2 31	2 38	0 00	0 00	744	0
FEB	0 68	0 69	0 00	0 00	672	0	0 52	0 52	0 00	0 00	672	0	2 40	2 41	0 00	0 00	672	0
MAR	0 69	0 69	0 00	0 00	744	0	0 51	0 52	0 00	0 00	744	0	2 40	2 41	0 00	0 00	744	0
APR	0 79	0 87	0 00	0 00	720	0	0 61	0 69	0 00	0 00	720	0	2 57	2 91	0 00	0 00	720	0
MAY	0 87	0 95	0 00	0 00	744	0	0 67	0 76	0 00	0 00	744	0	2 57	2 97	0 00	0 00	744	0
JUN	0 92	1 01	0 00	0 00	720	0	0 72	0 82	0 00	0 00	720	0	2 29	2 82	0 00	0 00	720	0
JUL	1 07	1 24	0 00	0 00	744	0	0 87	1 04	0 00	0 00	744	0	2 26	3 07	0 00	0 00	744	0
AUG	0 85	0 86	0 00	0 00	744	0	0 67	0 67	0 00	0 00	744	0	1 70	1 77	0 00	0 00	744	0
SEP	0 60	0 62	0 00	0 00	720	0	0 44	0 47	0 00	0 00	720	0	1 32	1 50	0 00	0 00	720	0
OCT	0 41	0 45	0 00	0 00	744	0	0 27	0 32	0 00	0 00	744	0	1 23	1 48	0 00	0 00	744	0
NOV	0 47	0 48	0 00	0 00	720	0	0 34	0 35	0 00	0 00	720	0	1 64	1 89	0 00	0 00	720	0
DEC	0 50	0 51	0 00	0 00	744	0	0 36	0 37	0 00	0 00	744	0	1 97	2 04	0 00	0 00	744	0

Half Berm

	Slab (Perimeter) AVE HEAT LOSS	PEAK HEAT LOSS	AVE HEAT GAIN	PEAK HEAT GAIN	HRS HT ON	HRS CL ON	Slab (Interior) AVE HEAT LOSS	PEAK HEAT LOSS	AVE HEAT GAIN	PEAK HEAT GAIN	HRS HT ON	HRS CL ON	Wall (Level-1) AVE HEAT LOSS	PEAK HEAT LOSS	AVE HEAT GAIN	PEAK HEAT GAIN	HRS HT ON	HRS CL ON
JAN	0 92	0 93	0 00	0 00	744	0	0 55	0 55	0 00	0 00	744	0	1 57	1 64	0 00	0 00	744	0
FEB	0 94	0 95	0 00	0 00	672	0	0 55	0 55	0 00	0 00	672	0	1 61	1 66	0 00	0 00	672	0
MAR	0 95	0 96	0 00	0 00	744	0	0 55	0 55	0 00	0 00	744	0	1 53	1 56	0 00	0 00	744	0
APR	1 07	1 14	0 00	0 00	720	0	0 66	0 73	0 00	0 00	720	0	1 52	2 02	0 00	0 00	720	0
MAY	1 14	1 23	0 00	0 00	744	0	0 73	0 82	0 00	0 00	744	0	1 31	1 85	0 00	0 00	744	0
JUN	1 19	1 29	0 00	0 00	720	0	0 79	0 88	0 00	0 00	720	0	0 77	1 37	0 00	0 00	720	0
JUL	1 32	1 50	0 00	0 00	744	0	0 95	1 12	0 00	0 00	744	0	0 64	1 56	0 00	0 00	744	0
AUG	1 08	1 09	0 00	0 00	744	0	0 75	0 75	0 00	0 00	744	0	0 39	0 53	0 02	0 03	742	2
SEP	0 83	0 86	0 00	0 00	720	0	0 51	0 55	0 00	0 00	720	0	0 56	0 85	0 19	0 50	678	42
OCT	0 65	0 71	0 00	0 00	744	0	0 35	0 40	0 00	0 00	744	0	0 82	1 16	0 08	0 18	735	9
NOV	0 74	0 77	0 00	0 00	720	0	0 42	0 43	0 00	0 00	720	0	1 28	1 63	0 00	0 00	720	0
DEC	0 79	0 82	0 00	0 00	744	0	0 44	0 45	0 00	0 00	744	0	1 55	1 67	0 00	0 00	744	0

Slab On Grade

	Slab (Perimeter) AVE HEAT LOSS	PEAK HEAT LOSS	AVE HEAT GAIN	PEAK HEAT GAIN	HRS HT ON	HRS CL ON	Slab (Interior) AVE HEAT LOSS	PEAK HEAT LOSS	AVE HEAT GAIN	PEAK HEAT GAIN	HRS HT ON	HRS CL ON
JAN	1 01	1 03	0 00	0 00	744	0	0 61	0 61	0 00	0 00	744	0
FEB	1 05	1 05	0 00	0 00	672	0	0 62	0 62	0 00	0 00	672	0
MAR	1 06	1 06	0 00	0 00	744	0	0 63	0 63	0 00	0 00	744	0
APR	1 16	1 24	0 00	0 00	720	0	0 75	0 81	0 00	0 00	720	0
MAY	1 20	1 30	0 00	0 00	744	0	0 83	0 91	0 00	0 00	744	0
JUN	1 14	1 28	0 00	0 00	720	0	0 90	0 98	0 00	0 00	720	0
JUL	1 20	1 41	0 00	0 00	744	0	1 06	1 22	0 00	0 00	744	0
AUG	0 92	0 94	0 00	0 00	744	0	0 84	0 85	0 00	0 00	744	0
SEP	0 68	0 74	0 00	0 00	720	0	0 60	0 63	0 00	0 00	720	0
OCT	0 57	0 66	0 00	0 00	744	0	0 41	0 46	0 00	0 00	744	0
NOV	0 73	0 82	0 00	0 00	720	0	0 47	0 49	0 00	0 00	720	0
DEC	0 86	0 89	0 00	0 00	744	0	0 50	0 50	0 00	0 00	744	0

INSULATION DEPTH

Full Berm, 3 Story	8 0 ft
Full Berm, 1 Story	8 0 ft
Half Berm	5 0 ft
Slab on Grade	4 0 ft

Heat Flux Expressed in Btu/h ft²

To obtain W/m² units, multiply Btu/h ft² units by 3 156

TABLE 30

Monthly Average and Peak Heat Fluxes for Seattle with R-10.4 Insulation

Seattle R-10.4 Insulation

Full Berm 3-Story

	Slab (Perimeter)						Slab (Interior)						Wall (Level-1)						Wall (Level-2)						Wall (Level-3)					
	AVE HEAT LOSS	PEAK HEAT LOSS	AVE HEAT GAIN	PEAK HEAT GAIN	HRS HT ON	HRS CL ON	AVE HEAT LOSS	PEAK HEAT LOSS	AVE HEAT GAIN	PEAK HEAT GAIN	HRS HT ON	HRS CL ON	AVE HEAT LOSS	PEAK HEAT LOSS	AVE HEAT GAIN	PEAK HEAT GAIN	HRS HT ON	HRS CL ON	AVE HEAT LOSS	PEAK HEAT LOSS	AVE HEAT GAIN	PEAK HEAT GAIN	HRS HT ON	HRS CL ON	AVE HEAT LOSS	PEAK HEAT LOSS	AVE HEAT GAIN	PEAK HEAT GAIN	HRS HT ON	HRS CL ON
JAN	0 37	0 38	0 00	0 00	744	0	0 26	0 27	0 00	0 00	744	0	1 60	1 71	0 00	0 00	744	0	0 82	0 87	0 00	0 00	744	0	0 73	0 73	0 00	0 00	744	0
FEB	0 35	0 36	0 00	0 00	672	0	0 25	0 26	0 00	0 00	672	0	1 69	1 74	0 00	0 00	672	0	0 91	0 94	0 00	0 00	672	0	0 72	0 73	0 00	0 00	672	0
MAR	0 33	0 34	0 00	0 00	744	0	0 24	0 24	0 00	0 00	744	0	1 68	1 74	0 00	0 00	744	0	0 96	0 98	0 00	0 00	744	0	0 72	0 72	0 00	0 00	744	0
APR	0 43	0 51	0 00	0 00	720	0	0 33	0 41	0 00	0 00	720	0	1 63	2 09	0 00	0 00	720	0	1 16	1 50	0 00	0 00	720	0	0 91	1 23	0 00	0 00	720	0
MAY	0 59	0 76	0 00	0 00	744	0	0 49	0 66	0 00	0 00	744	0	1 63	2 53	0 00	0 00	744	0	1 42	2 12	0 00	0 00	744	0	1 20	1 87	0 00	0 00	744	0
JUN	0 61	0 71	0 00	0 00	720	0	0 51	0 61	0 00	0 00	720	0	1 33	1 86	0 00	0 00	720	0	1 36	1 77	0 00	0 00	720	0	1 20	1 57	0 00	0 00	720	0
JUL	0 64	0 74	0 00	0 00	744	0	0 54	0 64	0 00	0 00	744	0	1 18	1 66	0 00	0 00	744	0	1 33	1 74	0 00	0 00	744	0	1 24	1 62	0 00	0 00	744	0
AUG	0 57	0 59	0 00	0 00	744	0	0 46	0 49	0 00	0 00	744	0	0 91	0 96	0 00	0 00	744	0	1 11	1 19	0 00	0 00	744	0	1 09	1 14	0 00	0 00	744	0
SEP	0 30	0 33	0 00	0 00	720	0	0 20	0 23	0 00	0 00	720	0	0 66	0 83	0 08	0 16	713	7	0 62	0 70	0 00	0 00	720	0	0 63	0 72	0 00	0 00	720	0
OCT	0 12	0 16	0 05	0 12	731	13	0 04	0 07	0 06	0 47	562	182	0 72	0 95	0 08	0 20	733	11	0 38	0 50	0 12	0 33	714	30	0 37	0 47	0 11	0 31	717	27
NOV	0 06	0 09	0 03	0 09	699	21	0 00	0 01	0 04	0 40	157	563	0 97	1 14	0 00	0 00	720	0	0 37	0 49	0 02	0 02	719	1	0 30	0 37	0 03	0 05	716	4
DEC	0 10	0 11	0 00	0 00	744	0	0 02	0 03	0 00	0 00	744	0	1 32	1 45	0 00	0 00	744	0	0 55	0 61	0 00	0 00	744	0	0 39	0 40	0 00	0 00	744	0

Full Berm 1-Story

	Slab (Perimeter)						Slab (Interior)						Wall (Level-1)					
	AVE HEAT LOSS	PEAK HEAT LOSS	AVE HEAT GAIN	PEAK HEAT GAIN	HRS HT ON	HRS CL ON	AVE HEAT LOSS	PEAK HEAT LOSS	AVE HEAT GAIN	PEAK HEAT GAIN	HRS HT ON	HRS CL ON	AVE HEAT LOSS	PEAK HEAT LOSS	AVE HEAT GAIN	PEAK HEAT GAIN	HRS HT ON	HRS CL ON
JAN	0 66	0 66	0 00	0 00	744	0	0 49	0 49	0 00	0 00	744	0	2 43	2 55	0 00	0 00	744	0
FEB	0 67	0 67	0 00	0 00	672	0	0 49	0 49	0 00	0 00	672	0	2 55	2 58	0 00	0 00	672	0
MAR	0 68	0 68	0 00	0 00	744	0	0 49	0 49	0 00	0 00	744	0	2 56	2 62	0 00	0 00	744	0
APR	0 79	0 86	0 00	0 00	720	0	0 59	0 66	0 00	0 00	720	0	2 63	3 00	0 00	0 00	720	0
MAY	0 97	1 13	0 00	0 00	744	0	0 76	0 92	0 00	0 00	744	0	2 79	3 54	0 00	0 00	744	0
JUN	1 00	1 09	0 00	0 00	720	0	0 78	0 88	0 00	0 00	720	0	2 59	3 05	0 00	0 00	720	0
JUL	1 03	1 13	0 00	0 00	744	0	0 81	0 91	0 00	0 00	744	u	2 48	2 92	0 00	0 00	744	0
AUG	0 94	0 98	0 00	0 00	744	0	0 74	0 77	0 00	0 00	744	0	2 19	2 26	0 00	0 00	744	0
SEP	0 67	0 70	0 00	0 00	720	0	0 49	0 52	0 00	0 00	720	0	1 72	1 84	0 00	0 00	720	0
OCT	0 48	0 52	0 00	0 00	744	0	0 31	0 36	0 00	0 00	744	0	1 54	1 74	0 00	0 00	744	0
NOV	0 43	0 46	0 00	0 00	720	0	0 27	0 31	0 00	0 00	720	0	1 67	1 83	0 00	0 00	720	0
DEC	0 48	0 50	0 00	0 00	744	0	0 33	0 34	0 00	0 00	744	0	2 01	2 14	0 00	0 00	744	0

Half Berm

	Slab (Perimeter)						Slab (Interior)						Wall (Level-1)					
	AVE HEAT LOSS	PEAK HEAT LOSS	AVE HEAT GAIN	PEAK HEAT GAIN	HRS HT ON	HRS CL ON	AVE HEAT LOSS	PEAK HEAT LOSS	AVE HEAT GAIN	PEAK HEAT GAIN	HRS HT ON	HRS CL ON	AVE HEAT LOSS	PEAK HEAT LOSS	AVE HEAT GAIN	PEAK HEAT GAIN	HRS HT ON	HRS CL ON
JAN	0 90	0 92	0 00	0 00	744	0	0 53	0 53	0 00	0 00	744	0	1 75	1 87	0 00	0 00	744	0
FEB	0 93	0 94	0.00	0 00	672	0	0 53	0 54	0 00	0 00	672	0	1 72	1 88	0 00	0 00	672	0
MAR	0 95	0 95	0 00	0 00	744	0	0 54	0 54	0 00	0 00	744	0	1 65	1 85	0 00	0 00	744	0
APR	1 06	1 13	0 00	0 00	720	0	0 65	0 72	0 00	0 00	720	0	1 49	1 97	0 00	0 00	720	0
MAY	1 24	1 40	0 00	0 00	744	0	0 83	0 99	0 00	0 00	744	0	1 37	2 36	0 00	0 00	744	0
JUN	1 26	1 36	0 00	0 00	720	0	0 86	0 95	0 00	0 00	720	0	1 07	1 59	0 00	0 00	720	0
JUL	1 29	1 39	0 00	0 00	744	0	0 90	0 99	0 00	0 00	744	0	0 96	1 45	0 00	0 00	744	0
AUG	1 20	1 23	0 00	0 00	744	0	0 83	0 85	0 00	0 00	744	0	0 79	0 84	0 00	0 00	744	0
SEP	0 92	0 95	0 00	0 00	720	0	0 57	0 60	0 00	0 00	720	0	0 76	0 96	0 10	0 23	708	12
OCT	0 74	0 79	0 00	0 00	744	0	0 39	0 45	0 00	0 00	744	0	0 96	1 22	0 04	0 07	741	3
NOV	0 70	0 75	0 00	0 00	720	0	0 36	0 39	0 00	0 00	720	0	1 22	1 37	0 00	0 00	720	0
DEC	0 78	0 80	0 00	0 00	744	0	0 41	0 42	0 00	0 00	744	0	1 55	1 72	0 00	0 00	744	0

Slab On Grade

	Slab (Perimeter)						Slab (Interior)					
	AVE HEAT LOSS	PEAK HEAT LOSS	AVE HEAT GAIN	PEAK HEAT GAIN	HRS HT ON	HRS CL ON	AVE HEAT LOSS	PEAK HEAT LOSS	AVE HEAT GAIN	PEAK HEAT GAIN	HRS HT ON	HRS CL ON
JAN	1 03	1 08	0 00	0 00	744	0	0 60	0 60	0 00	0 00	744	0
FEB	1 08	1 10	0 00	0 00	672	0	0 61	0 62	0 00	0 00	672	0
MAR	1 10	1 11	0 00	0 00	744	0	0 62	0 63	0 00	0 00	744	0
APR	1 18	1 26	0 00	0 00	720	0	0 74	0 81	0 00	0 00	720	0
MAY	1 32	1 50	0 00	0 00	744	0	0 94	1 09	0 00	0 00	744	0
JUN	1 27	1 39	0 00	0 00	720	0	0 97	1 07	0 00	0 00	720	0
JUL	1 25	1 37	0 00	0 00	744	0	1 02	1 11	0 00	0 00	744	0
AUG	1 12	1 16	0 00	0 00	744	0	0 94	0 97	0 00	0 00	744	0
SEP	0 85	0.89	0 00	0 00	720	0	0 67	0 70	0 00	0 00	720	0
OCT	0 71	0 79	0 00	0 00	744	0	0 48	0 53	0 00	0 00	744	0
NOV	0 73	0 80	0 00	0 00	720	0	0 43	0 47	0 00	0 00	720	0
DEC	0 87	0 92	0 00	0 00	744	0	0 49	0 50	0 00	0 00	744	0

INSULATION DEPTH

Full Berm, 3 Story	8 0 ft
Full Berm, 1 Story	8 0 ft
Half Berm	5 0 ft
Slab on Grade	4 0 ft

Heat Flux Expressed in Btu/h ft²

To obtain W/m² units, multiply Btu/h ft² units by 3 156

TABLE 31

Monthly Average and Peak Heat Fluxes for Kansas City with R-10.4 Insulation

Kansas City R-10.4 Insulation

Full Berm 3-Story

	Slab (Perimeter)						Slab (Interior)						Wall (Level-1)						Wall (Level-2)						Wall (Level-3)					
	AVE HEAT LOSS	PEAK HEAT LOSS	AVE HEAT GAIN	PEAK HEAT GAIN	HRS HT ON	HRS CL ON	AVE HEAT LOSS	PEAK HEAT LOSS	AVE HEAT GAIN	PEAK HEAT GAIN	HRS HT ON	HRS CL ON	AVE HEAT LOSS	PEAK HEAT LOSS	AVE HEAT GAIN	PEAK HEAT GAIN	HRS HT ON	HRS CL ON	AVE HEAT LOSS	PEAK HEAT LOSS	AVE HEAT GAIN	PEAK HEAT GAIN	HRS HT ON	HRS CL ON	AVE HEAT LOSS	PEAK HEAT LOSS	AVE HEAT GAIN	PEAK HEAT GAIN	HRS HT ON	HRS CL ON
JAN	0 31	0 32	0 00	0 00	744	0	0 22	0 22	0 00	0 00	744	0	1 75	1 97	0 00	0 00	744	0	0 69	0 77	0 00	0 00	744	0	0 58	0 58	0 00	0 00	744	0
FEB	0 30	0 30	0 00	0 00	672	0	0 21	0 21	0 00	0 00	672	0	2 02	2 10	0 00	0 00	672	0	0 84	0 90	0 00	0 00	672	0	0 59	0 59	0 00	0 00	672	0
MAR	0 28	0 29	0 00	0 00	744	0	0 20	0 20	0 00	0 00	744	0	1 85	1 98	0 00	0 00	744	0	0 95	0 98	0 00	0 00	744	0	0 60	0 61	0 00	0 00	744	0
APR	0 38	0 46	0 00	0 00	720	0	0 29	0 37	0 00	0 00	720	0	1 65	2 24	0 00	0 00	720	0	1 17	1 50	0 00	0 00	720	0	0 81	1 13	0 00	0 00	720	0
MAY	0 55	0 71	0 00	0 00	744	0	0 46	0 63	0 00	0 00	744	0	1 27	2 29	0 00	0 00	744	0	1 39	2 12	0 00	0 00	744	0	1 11	1 78	0 00	0 00	744	0
JUN	0 57	0 67	0 00	0 00	720	0	0 48	0 58	0 00	0 00	720	0	0 85	1 44	0 00	0 00	720	0	1 26	1 70	0 00	0 00	720	0	1 12	1 49	0 00	0 00	720	0
JUL	0 60	0 70	0 00	0 00	744	0	0 51	0 61	0 00	0 00	744	0	0 48	1 06	0 00	0 00	744	0	1 15	1 61	0 00	0 00	744	0	1 16	1 54	0 00	0 00	744	0
AUG	0 53	0 55	0 00	0 00	744	0	0 43	0 46	0 00	0 00	744	0	0 29	0 37	0 00	0 00	744	0	0 87	0 97	0 00	0 00	744	0	1 00	1 06	0 00	0 00	744	0
SEP	0 26	0 29	0 00	0 00	720	0	0 17	0 20	0 00	0 00	720	0	0 20	0 39	0 24	0 74	613	107	0 36	0 44	0 10	0 26	701	19	0 53	0 61	0 05	0 08	717	3
OCT	0 09	0 12	0 08	0 25	705	39	0 03	0 04	0 07	0 56	461	283	0 44	0 60	0 24	0 63	680	64	0 16	0 26	0 16	0 59	595	149	0 26	0 35	0 14	0 42	690	54
NOV	0 04	0 06	0 06	0 21	630	90	0 00	0 00	0 06	0 48	0	720	0 70	1 06	0 00	0 00	720	0	0 17	0 27	0 08	0 26	642	78	0 19	0 25	0 06	0 17	688	32
DEC	0 07	0 08	0 00	0 00	744	0	0 01	0 01	0 01	0 01	469	275	1 42	1 64	0 00	0 00	744	0	0 36	0 46	0 00	0 00	744	0	0 27	0 28	0 00	0 00	744	0

Full Berm 1-Story

	Slab (Perimeter)						Slab (Interior)						Wall (Level-1)					
	AVE HEAT LOSS	PEAK HEAT LOSS	AVE HEAT GAIN	PEAK HEAT GAIN	HRS HT ON	HRS CL ON	AVE HEAT LOSS	PEAK HEAT LOSS	AVE HEAT GAIN	PEAK HEAT GAIN	HRS HT ON	HRS CL ON	AVE HEAT LOSS	PEAK HEAT LOSS	AVE HEAT GAIN	PEAK HEAT GAIN	HRS HT ON	HRS CL ON
JAN	0 54	0 56	0 00	0 00	744	0	0 41	0 41	0 00	0 00	744	0	2 38	2 62	0 00	0 00	744	0
FEB	0 57	0 58	0 00	0 00	672	0	0 41	0 41	0 00	0 00	672	0	2 69	2 76	0 00	0 00	672	0
MAR	0 59	0 59	0 00	0 00	744	0	0 41	0 41	0 00	0 00	744	0	2 60	2 70	0 00	0 00	744	0
APR	0 71	0 78	0 00	0 00	720	0	0 51	0 59	0 00	0 00	720	0	2 54	3 03	0 00	0 00	720	0
MAY	0 89	1 05	0 00	0 00	744	0	0 69	0 85	0 00	0 00	744	0	2 38	3 23	0 00	0 00	744	0
JUN	0.91	1 01	0 00	0 00	720	0	0 71	0 81	0 00	0 00	720	0	2 04	2 58	0 00	0 00	720	0
JUL	0 93	1 04	0 00	0 00	744	0	0 74	0 84	0 00	0 00	744	0	1 72	2 25	0 00	0 00	744	0
AUG	0 84	0 88	0 00	0 00	744	0	0 67	0 70	0 00	0 00	744	0	1 45	1 58	0 00	0 00	744	0
SEP	0 56	0 58	0 00	0 00	720	0	0 42	0 45	0 00	0 00	720	0	1 05	1 25	0 00	0 00	720	0
OCT	0 37	0 41	0 00	0 00	744	0	0 25	0 30	0 00	0 00	744	0	1 04	1 21	0 00	0 00	744	0
NOV	0 32	0 35	0 00	0 00	720	0	0 21	0 24	0 00	0 00	720	0	1 23	1 56	0 00	0 00	720	0
DEC	0 37	0 39	0 00	0 00	744	0	0 26	0 27	0 00	0 00	744	0	1 89	2 11	0 00	0 00	744	0

Half Berm

	Slab (Perimeter)						Slab (Interior)						Wall (Level-1)					
	AVE HEAT LOSS	PEAK HEAT LOSS	AVE HEAT GAIN	PEAK HEAT GAIN	HRS HT ON	HRS CL ON	AVE HEAT LOSS	PEAK HEAT LOSS	AVE HEAT GAIN	PEAK HEAT GAIN	HRS HT ON	HRS CL ON	AVE HEAT LOSS	PEAK HEAT LOSS	AVE HEAT GAIN	PEAK HEAT GAIN	HRS HT ON	HRS CL ON
JAN	0 80	0 83	0 00	0 00	744	0	0 43	0 44	0 00	0 00	744	0	1 92	2 28	0 00	0 00	744	0
FEB	0 85	0 87	0 00	0 00	672	0	0 44	0 45	0 00	0 00	672	0	2 16	2 33	0 00	0 00	672	0
MAR	0 88	0 89	0 00	0 00	744	0	0 45	0 45	0 00	0 00	744	0	1 73	1 98	0 00	0 00	744	0
APR	1 00	1 07	0 00	0 00	720	0	0 56	0 63	0 00	0 00	720	0	1 37	2 09	0 00	0 00	720	0
MAY	1 16	1 33	0 00	0 00	744	0	0 75	0 91	0 00	0 00	744	0	0 83	1 81	0 00	0 00	744	0
JUN	1 15	1 26	0 00	0 00	720	0	0 78	0 88	0 00	0 00	720	0	0 50	1 10	0 00	0 00	720	0
JUL	1 15	1 27	0 00	0 00	744	0	0 82	0 92	0 00	0 00	744	0	0 19	0 71	0 00	0 00	744	0
AUG	1 04	1 08	0 00	0 00	744	0	0 75	0 78	0 00	0 00	744	0	0 21	0 38	0 00	0 00	744	0
SEP	0 77	0 80	0 00	0 00	720	0	0 50	0 53	0 00	0 00	720	0	0 36	0 71	0 31	0 77	653	67
OCT	0 59	0 65	0 00	0 00	744	0	0 32	0 37	0 00	0 00	744	0	0 78	0 97	0 13	0 32	725	19
NOV	0 56	0 61	0 00	0 00	720	0	0 28	0 31	0 00	0 00	720	0	1 05	1 57	0 00	0 00	720	0
DEC	0 66	0 70	0 00	0 00	744	0	0 33	0 34	0 00	0 00	744	0	1 88	2 07	0 00	0 00	744	0

Slab On Grade

	Slab (Perimeter)						Slab (Interior)					
	AVE HEAT LOSS	PEAK HEAT LOSS	AVE HEAT GAIN	PEAK HEAT GAIN	HRS HT ON	HRS CL ON	AVE HEAT LOSS	PEAK HEAT LOSS	AVE HEAT GAIN	PEAK HEAT GAIN	HRS HT ON	HRS CL ON
JAN	0.97	1.05	0 00	0 00	744	0	0 48	0 49	0 00	0 00	744	0
FEB	1 08	1 10	0 00	0 00	672	0	0 50	0 51	0 00	0 00	672	0
MAR	1 08	1 10	0 00	0 00	744	0	0 52	0 54	0 00	0 00	744	0
APR	1 12	1 24	0 00	0 00	720	0	0 66	0.72	0 00	0 00	720	0
MAY	1 16	1.38	0 00	0 00	744	0	0 86	1 01	0 00	0 00	744	0
JUN	1 06	1 21	0 00	0 00	720	0	0 90	0 99	0 00	0.00	720	0
JUL	0 97	1.12	0 00	0 00	744	0	0 93	1 03	0 00	0 00	744	0
AUG	0 83	0.89	0.00	0 00	744	0	0 85	0 88	0.00	0 00	744	0
SEP	0.58	0 65	0 00	0.00	720	0	0 57	0.59	0 00	0 00	720	0
OCT	0 48	0.55	0 00	0.00	744	0	0 37	0 42	0 00	0 00	744	0
NOV	0 53	0.65	0.00	0 00	720	0	0 32	0.35	0 00	0.00	720	0
DEC	0 77	0.85	0.00	0 00	744	0	0 37	0 39	0 00	0 00	744	0

INSULATION DEPTH

Full Berm, 3 Story	8 0 ft
Full Berm, 1 Story	8 0 ft
Half Berm	5 0 ft
Slab on Grade	4 0 ft

Heat Flux Expressed in Btu/h ft²

To obtain W/m² units, multiply Btu/h ft² units by 3 156

TABLE 32

Monthly Average and Peak Heat Fluxes for Albuquerque with R-10.4 Insulation

Albuquerque R-10.4 Insulation

Full Berm 3-Story

	Slab (Perimeter)						Slab (Interior)						Wall (Level-1)						Wall (Level-2)						Wall (Level-3)					
	AVE HEAT LOSS	PEAK HEAT LOSS	AVE HEAT GAIN	PEAK HEAT GAIN	HRS HT ON	HRS CL ON	AVE HEAT LOSS	PEAK HEAT LOSS	AVE HEAT GAIN	PEAK HEAT GAIN	HRS HT ON	HRS CL ON	AVE HEAT LOSS	PEAK HEAT LOSS	AVE HEAT GAIN	PEAK HEAT GAIN	HRS HT ON	HRS CL ON	AVE HEAT LOSS	PEAK HEAT LOSS	AVE HEAT GAIN	PEAK HEAT GAIN	HRS HT ON	HRS CL ON	AVE HEAT LOSS	PEAK HEAT LOSS	AVE HEAT GAIN	PEAK HEAT GAIN	HRS HT ON	HRS CL ON
JAN	0 23	0 23	0 00	0 00	744	0	0 16	0 16	0.00	0 00	744	0	1 67	1 74	0 00	0 00	744	0	0 60	0 68	0 00	0 00	744	0	0 45	0 46	0 00	0 00	744	0
FEB	0 22	0 23	0 00	0 00	672	0	0 15	0 15	0 00	0 00	672	0	1 69	1 76	0 00	0 00	672	0	0 74	0 78	0 00	0 00	672	0	0 47	0 48	0 00	0 00	672	0
MAR	0 21	0 22	0 00	0 00	744	0	0 14	0 15	0 00	0 00	744	0	1 66	1 79	0 00	0 00	744	0	0 82	0 85	0 00	0 00	744	0	0 49	0 50	0 00	0 00	744	0
APR	0 31	0 39	0 00	0 00	720	0	0 24	0 32	0 00	0 00	720	0	1 41	1 96	0 00	0 00	720	0	1 03	1 37	0 00	0 00	720	0	0 69	1 01	0 00	0 00	720	0
MAY	0 49	0 64	0 00	0 00	744	0	0 41	0 58	0 00	0 00	744	0	1 28	2 26	0 00	0 00	744	0	1 26	1 98	0 00	0 00	744	0	1 00	1 67	0 00	0 00	744	0
JUN	0 51	0 61	0 00	0 00	720	0	0 43	0 54	0 00	0 00	720	0	0 88	1 44	0 00	0 00	720	0	1 17	1 60	0 00	0 00	720	0	1 02	1 38	0 00	0 00	720	0
JUL	0 55	0 65	0 00	0 00	744	0	0 47	0 57	0 00	0 00	744	0	0 55	1 18	0 00	0 00	744	0	1 09	1 53	0 00	0 00	744	0	1 06	1 43	0 00	0 00	744	0
AUG	0 48	0 50	0 00	0 00	744	0	0 39	0 42	0 00	0 00	744	0	0 31	0 35	0 00	0 00	744	0	0 82	0 93	0 00	0 00	744	0	0 91	0 96	0 00	0 00	744	0
SEP	0 22	0 25	0 00	0 00	720	0	0 14	0 17	0 00	0 00	720	0	0 20	0 34	0 22	0 74	589	131	0 33	0 40	0 11	0 29	695	25	0 45	0 53	0 08	0 16	711	9
OCT	0 05	0 08	0 11	0 41	641	103	0 01	0 02	0 08	0 68	231	513	0 34	0 56	0 23	0 69	655	89	0 13	0 20	0 16	0 63	557	187	0 20	0 28	0 15	0 49	658	86
NOV	0 01	0 02	0 07	0 35	400	320	0 00	0 00	0 11	0 58	0	720	0 72	1 05	0 00	0 00	720	0	0 13	0.22	0 09	0 31	587	133	0 13	0 18	0 08	0 24	646	74
DEC	0 03	0 04	0 00	0 00	744	0	0 00	0 00	0 02	0 03	0	744	1 26	1 54	0 00	0 00	744	0	0 32	0.41	0 00	0 00	744	0	0 20	0 22	0 00	0 00	744	0

Full Berm 1-Story

	Slab (Perimeter)						Slab (Interior)						Wall (Level-1)					
	AVE HEAT LOSS	PEAK HEAT LOSS	AVE HEAT GAIN	PEAK HEAT GAIN	HRS HT ON	HRS CL ON	AVE HEAT LOSS	PEAK HEAT LOSS	AVE HEAT GAIN	PEAK HEAT GAIN	HRS HT ON	HRS CL ON	AVE HEAT LOSS	PEAK HEAT LOSS	AVE HEAT GAIN	PEAK HEAT GAIN	HRS HT ON	HRS CL ON
JAN	0 43	0 45	0 00	0 00	744	0	0 31	0 32	0 00	0 00	744	0	2 16	2 25	0 00	0 00	744	0
FEB	0 46	0 47	0 00	0 00	672	0	0 32	0 32	0 00	0 00	672	0	2 25	2 34	0 00	0 00	672	0
MAR	0 48	0 48	0 00	0 00	744	0	0 32	0 32	0 00	0 00	744	0	2 27	2 37	0 00	0 00	744	0
APR	0 60	0 66	0 00	0 00	720	0	0 43	0 50	0 00	0 00	720	0	2 18	2 61	0 00	0 00	720	0
MAY	0 78	0 94	0 00	0 00	744	0	0 60	0 76	0 00	0 00	744	0	2 22	3 05	0 00	0 00	744	0
JUN	0 81	0 91	0 00	0 00	720	0	0 63	0 73	0 00	0 00	720	0	1 93	2 43	0 00	0 00	720	0
JUL	0 84	0 94	0 00	0 00	744	0	0 66	0 76	0 00	0 00	744	0	1 64	2 21	0 00	0 00	744	0
AUG	0 75	0 78	0 00	0 00	744	0	0 60	0 62	0 00	0 00	744	0	1 34	1 43	0 00	0 00	744	0
SEP	0 48	0 50	0 00	0 00	720	0	0 35	0 38	0 00	0 00	720	0	0 93	1 08	0 00	0 00	720	0
OCT	0 28	0 33	0 00	0 00	744	0	0 18	0 23	0 00	0 00	744	0	0 83	1 07	0 00	0 00	744	0
NOV	0 23	0 27	0 00	0 00	720	0	0 14	0 18	0 00	0 00	720	0	1 13	1 43	0 00	0 00	720	0
DEC	0 29	0 31	0 00	0 00	744	0	0 19	0 20	0 00	0 00	744	0	1 65	1 92	0 00	0 00	744	0

Half Berm

	Slab (Perimeter)						Slab (Interior)						Wall (Level-1)					
	AVE HEAT LOSS	PEAK HEAT LOSS	AVE HEAT GAIN	PEAK HEAT GAIN	HRS HT ON	HRS CL ON	AVE HEAT LOSS	PEAK HEAT LOSS	AVE HEAT GAIN	PEAK HEAT GAIN	HRS HT ON	HRS CL ON	AVE HEAT LOSS	PEAK HEAT LOSS	AVE HEAT GAIN	PEAK HEAT GAIN	HRS HT ON	HRS CL ON
JAN	0 67	0 70	0 00	0 00	744	0	0 34	0 34	0 00	0 00	744	0	1 86	1 96	0 00	0 00	744	0
FEB	0 71	0 72	0 00	0 00	672	0	0 35	0 35	0 00	0 00	672	0	1 72	1 90	0 00	0 00	672	0
MAR	0 74	0 74	0 00	0 00	744	0	0 36	0 36	0 00	0 00	744	0	1 58	1 89	0 00	0 00	744	0
APR	0 85	0 92	0 00	0 00	720	0	0 47	0 54	0 00	0 00	720	0	1 16	1 68	0 00	0 00	720	0
MAY	1 02	1 18	0 00	0 00	744	0	0 66	0 81	0 00	0 00	744	0	0 93	2 06	0 00	0 00	744	0
JUN	1 03	1 13	0 00	0 00	720	0	0 69	0 79	0 00	0 00	720	0	0 58	1 10	0 00	0 00	720	0
JUL	1 04	1 15	0 00	0 00	744	0	0 74	0 83	0 00	0 00	744	0	0 24	0 91	0 00	0 00	744	0
AUG	0 93	0 97	0 00	0 00	744	0	0 67	0 69	0 00	0 00	744	0	0 20	0 29	0 00	0 00	744	0
SEP	0 66	0 69	0 00	0 00	720	0	0 41	0 45	0 00	0 00	720	0	0 35	0 53	0 26	0 77	625	95
OCT	0 48	0 53	0 00	0 00	744	0	0 24	0 29	0 00	0 00	744	0	0 62	0 86	0 20	0 50	707	37
NOV	0 45	0 51	0 00	0 00	720	0	0 20	0 24	0 00	0 00	720	0	1 12	1 45	0 00	0 00	720	0
DEC	0 55	0 59	0 00	0 00	744	0	0 25	0 27	0 00	0 00	744	0	1 61	1 92	0 00	0 00	744	0

Slab On Grade

	Slab (Perimeter)						Slab (Interior)					
	AVE HEAT LOSS	PEAK HEAT LOSS	AVE HEAT GAIN	PEAK HEAT GAIN	HRS HT ON	HRS CL ON	AVE HEAT LOSS	PEAK HEAT LOSS	AVE HEAT GAIN	PEAK HEAT GAIN	HRS HT ON	HRS CL ON
JAN	0 85	0 89	0 00	0 00	744	0	0 38	0 39	0 00	0 00	744	0
FEB	0 90	0 94	0 00	0 00	672	0	0 40	0 42	0 00	0 00	672	0
MAR	0 93	0 95	0 00	0 00	744	0	0 43	0 44	0 00	0 00	744	0
APR	0 96	1 06	0 00	0 00	720	0	0 56	0 62	0 00	0 00	720	0
MAY	1 06	1 27	0 00	0 00	744	0	0 76	0 91	0 00	0 00	744	0
JUN	0 98	1 13	0 00	0 00	720	0	0 80	0 89	0 00	0 00	720	0
JUL	0 90	1 06	0 00	0 00	744	0	0 84	0 94	0 00	0 00	744	0
AUG	0 76	0 81	0 00	0 00	744	0	0 76	0 79	0 00	0 00	744	0
SEP	0 51	0 56	0 00	0 00	720	0	0 49	0 51	0 00	0 00	720	0
OCT	0 39	0 48	0 00	0 00	744	0	0 29	0 34	0 00	0 00	744	0
NOV	0 46	0 57	0 00	0 00	720	0	0 24	0 28	0 00	0 00	720	0
DEC	0 66	0 76	0 00	0 00	744	0	0 30	0 31	0 00	0 00	744	0

INSULATION DEPTH

Full Berm, 3 Story	8 0 ft
Full Berm, 1 Story	8 0 ft
Half Berm	5 0 ft
Slab on Grade	4 0 ft

Heat Flux Expressed in Btu/h ft²

To obtain W/m² units, multiply Btu/h ft² units by 3 156

TABLE 33

Monthly Average and Peak Heat Fluxes for Phoenix with R-10.4 Insulation

Phoenix R-10.4 Insulation

Full Berm 3-Story

	Slab (Perimeter)						Slab (Interior)						Wall (Level-1)						Wall (Level-2)						Wall (Level-3)					
	AVE HEAT LOSS	PEAK HEAT LOSS	AVE HEAT GAIN	PEAK HEAT GAIN	HRS HT ON	HRS CL ON	AVE HEAT LOSS	PEAK HEAT LOSS	AVE HEAT GAIN	PEAK HEAT GAIN	HRS HT ON	HRS CL ON	AVE HEAT LOSS	PEAK HEAT LOSS	AVE HEAT GAIN	PEAK HEAT GAIN	HRS HT ON	HRS CL ON	AVE HEAT LOSS	PEAK HEAT LOSS	AVE HEAT GAIN	PEAK HEAT GAIN	HRS HT ON	HRS CL ON	AVE HEAT LOSS	PEAK HEAT LOSS	AVE HEAT GAIN	PEAK HEAT GAIN	HRS HT ON	HRS CL ON
JAN	0 00	0 00	0 30	0 78	0	744	0 00	0 00	0 30	0 79	0	744	0 75	0 82	0 00	0 00	744	0	0 00	0 00	0 20	0 64	0	744	0 00	0 00	0 31	0 67	0	744
FEB	0 11	0 30	0 00	0 00	672	0	0 11	0 30	0 00	0 00	672	0	1 09	1 81	0 00	0 00	672	0	0 42	1 00	0 00	0 00	672	0	0 20	0 81	0 00	0 00	672	0
MAR	0 15	0 24	0 00	0 00	744	0	0 15	0 24	0 00	0 00	744	0	1 04	1 51	0 00	0 00	744	0	0 54	0 86	0 00	0 00	744	0	0 28	0 62	0 00	0 00	744	0
APR	0 22	0 30	0 00	0 00	720	0	0 20	0 29	0 00	0 00	720	0	0 80	1 34	0 00	0 00	720	0	0 64	1 00	0 00	0 00	720	0	0 40	0 74	0 00	0 00	720	0
MAY	0 27	0 36	0 00	0 00	744	0	0 26	0 35	0 00	0 00	744	0	0 55	1 12	0 00	0 00	744	0	0 68	1 06	0 00	0 00	744	0	0 51	0 85	0 00	0 00	744	0
JUN	0 22	0 24	0 00	0 00	720	0	0 20	0 22	0 00	0 00	720	0	0 13	0 30	0 06	0 12	447	273	0 48	0 56	0 00	0 00	720	0	0 41	0 43	0 00	0 00	720	0
JUL	0 20	0 21	0 00	0 00	744	0	0 17	0 19	0 00	0 00	744	0	0 00	0 00	0 26	0 42	0	744	0 33	0 41	0 00	0 00	744	0	0 37	0 39	0 00	0 00	744	0
AUG	0 18	0 19	0 00	0 00	744	0	0 15	0 16	0 00	0 00	744	0	0 00	0 00	0 46	0 48	0	744	0 19	0 26	0 00	0 00	744	0	0 33	0 35	0 00	0 00	744	0
SEP	0 16	0 17	0 00	0 00	720	0	0 14	0 14	0 00	0 00	720	0	0 00	0 00	0 41	0 46	0	720	0 08	0 12	0 00	0 00	720	0	0 29	0 31	0 00	0 00	720	0
OCT	0 00	0 00	0 10	0 72	0	744	0 00	0 00	0 14	0 82	0	744	0 00	0 00	0 38	1 33	0	744	0 00	0 00	0 33	0 99	0	744	0 00	0 00	0 13	0 76	0	744
NOV	0 00	0 00	0 19	0 66	0	720	0 00	0 00	0 21	0 69	0	720	0 13	0 31	0 18	0 60	455	265	0 00	0 00	0 32	0 71	0	720	0 00	0 00	0 20	0 54	0	720
DEC	0 00	0 00	0 28	0 75	0	744	0 00	0 00	0 28	0 77	0	744	0 43	0 77	0 08	0 21	715	29	0 00	0 00	0 30	0 71	0	744	0 00	0 00	0 29	0 64	0	744

Full Berm 1-Story

	Slab (Perimeter)						Slab (Interior)						Wall (Level-1)					
	AVE HEAT LOSS	PEAK HEAT LOSS	AVE HEAT GAIN	PEAK HEAT GAIN	HRS HT ON	HRS CL ON	AVE HEAT LOSS	PEAK HEAT LOSS	AVE HEAT GAIN	PEAK HEAT GAIN	HRS HT ON	HRS CL ON	AVE HEAT LOSS	PEAK HEAT LOSS	AVE HEAT GAIN	PEAK HEAT GAIN	HRS HT ON	HRS CL ON
JAN	0 00	0 00	0 23	0 73	0	744	0 00	0 00	0 24	0 73	0	744	0 59	0 66	0 00	0 00	744	0
FEB	0 17	0 33	0 00	0 00	672	0	0 14	0 32	0 00	0 00	672	0	1 13	1 68	0 00	0 00	672	0
MAR	0 24	0 31	0 00	0 00	744	0	0 18	0 28	0 00	0 00	744	0	1 22	1 60	0 00	0 00	744	0
APR	0 31	0 39	0 00	0 00	720	0	0 24	0 33	0 00	0 00	720	0	1 12	1 56	0 00	0 00	720	0
MAY	0 38	0 46	0 00	0 00	744	0	0 30	0 39	0 00	0 00	744	0	0 98	1 47	0 00	0 00	744	0
JUN	0 32	0 34	0 00	0 00	720	0	0 24	0 26	0 00	0 00	720	0	0 51	0 75	0 00	0 00	720	0
JUL	0 28	0 30	0 00	0 00	744	0	0 22	0 23	0 00	0 00	744	0	0 17	0 33	0 00	0 00	737	7
AUG	0 24	0 26	0 00	0 00	744	0	0 20	0 21	0 00	0 00	744	0	0 00	0 00	0 07	0 09	0	744
SEP	0 21	0 23	0 00	0 00	720	0	0 19	0 20	0 00	0 00	720	0	0 00	0 00	0 09	0 12	0	720
OCT	0 01	0 02	0 08	0 57	314	430	0 01	0 02	0 09	0 63	308	436	0 00	0 00	0 27	1 05	0	744
NOV	0 00	0 00	0 14	0 59	0	720	0 00	0 00	0 12	0 58	0	720	0 11	0 26	0 14	0 56	350	370
DEC	0 00	0 00	0 21	0 70	0	744	0 00	0 00	0 20	0 69	0	744	0 32	0 64	0 09	0 26	690	54

Half Berm

	Slab (Perimeter)						Slab (Interior)						Wall (Level-1)					
	AVE HEAT LOSS	PEAK HEAT LOSS	AVE HEAT GAIN	PEAK HEAT GAIN	HRS HT ON	HRS CL ON	AVE HEAT LOSS	PEAK HEAT LOSS	AVE HEAT GAIN	PEAK HEAT GAIN	HRS HT ON	HRS CL ON	AVE HEAT LOSS	PEAK HEAT LOSS	AVE HEAT GAIN	PEAK HEAT GAIN	HRS HT ON	HRS CL ON
JAN	0 05	0 08	0 11	0 54	390	354	0 00	0 00	0 24	0 73	0	744	1 00	1 15	0 00	0 00	744	0
FEB	0 28	0 38	0 00	0 00	672	0	0 15	0 32	0 00	0 00	672	0	1 08	1 86	0 00	0 00	672	0
MAR	0 36	0 44	0 00	0 00	744	0	0 20	0 28	0 00	0 00	744	0	0 91	1 49	0 00	0 00	744	0
APR	0 44	0 52	0 00	0 00	720	0	0 26	0 34	0 00	0 00	720	0	0 55	1 09	0 00	0 00	720	0
MAY	0 48	0 58	0 00	0 00	744	0	0 32	0 41	0 00	0 00	744	0	0 29	0 91	0 01	0 02	726	18
JUN	0 40	0 44	0 00	0 00	720	0	0 27	0 29	0 00	0 00	720	0	0 00	0 00	0 25	0 36	0	720
JUL	0 34	0 37	0 00	0 00	744	0	0 25	0 26	0 00	0 00	744	0	0 00	0 00	0 46	0 64	0	744
AUG	0 29	0 31	0 00	0 00	744	0	0 23	0 24	0 00	0 00	744	0	0 00	0 00	0 54	0 67	0	744
SEP	0 25	0.27	0 00	0 00	720	0	0 21	0 22	0 00	0 00	720	0	0 00	0 00	0 36	0 48	0	720
OCT	0 05	0 08	0 10	0 41	626	118	0 02	0 03	0 10	0 55	453	291	0 15	0 19	0 27	1 18	327	417
NOV	0 06	0 10	0 08	0 38	460	260	0 00	0 00	0 11	0 53	0	720	0 40	0 70	0 13	0 32	665	55
DEC	0 04	0 09	0 11	0 54	348	396	0 00	0 00	0 19	0 67	0	744	0 80	1 21	0 00	0 00	744	0

Slab On Grade

	Slab (Perimeter)						Slab (Interior)					
	AVE HEAT LOSS	PEAK HEAT LOSS	AVE HEAT GAIN	PEAK HEAT GAIN	HRS HT ON	HRS CL ON	AVE HEAT LOSS	PEAK HEAT LOSS	AVE HEAT GAIN	PEAK HEAT GAIN	HRS HT ON	HRS CL ON
JAN	0 15	0 33	0 04	0 07	739	5	0 00	0 00	0 16	0 78	0	744
FEB	0 42	0 52	0 00	0 00	672	0	0 15	0 27	0 00	0 00	672	0
MAR	0 50	0 58	0 00	0 00	744	0	0 22	0 29	0 00	0 00	744	0
APR	0 51	0 62	0 00	0 00	720	0	0 30	0 38	0 00	0 00	720	0
MAY	0 49	0 62	0 00	0 00	744	0	0 37	0 45	0 00	0 00	744	0
JUN	0 32	0 40	0 00	0 00	720	0	0 32	0 34	0 00	0 00	720	0
JUL	0 19	0 25	0 00	0 00	744	0	0 29	0 31	0 00	0 00	744	0
AUG	0 11	0 14	0 00	0 00	744	0	0 26	0 28	0 00	0 00	744	0
SEP	0 08	0 09	0 00	0 00	720	0	0 24	0 25	0 00	0 00	720	0
OCT	0 01	0 03	0 19	0 99	55	689	0 04	0 06	0 09	0 49	493	251
NOV	0 10	0 22	0 14	0 62	350	370	0 00	0 00	0 12	0 58	0	720
DEC	0 14	0 29	0 09	0 42	579	165	0 00	0 00	0 17	0 69	0	744

INSULATION DEPTH
- Full Berm, 3 Story 8 0 ft
- Full Berm, 1 Story 8 0 ft
- Half Berm 5 0 ft
- Slab on Grade 4 0 ft

Heat Flux Expressed in Btu/h ft²

To obtain W/m² units, multiply Btu/h ft² units by 3 156

TABLE 34

Monthly Average and Peak Heat Fluxes for Miami with R-10.4 Insulation

Miami R-10.4 Insulation

Full Berm 3-Story

	Slab (Perimeter)						Slab (Interior)						Wall (Level-1)						Wall (Level-2)						Wall (Level-3)					
	AVE HEAT LOSS	PEAK HEAT LOSS	AVE HEAT GAIN	PEAK HEAT GAIN	HRS HT ON	HRS CL ON	AVE HEAT LOSS	PEAK HEAT LOSS	AVE HEAT GAIN	PEAK HEAT GAIN	HRS HT ON	HRS CL ON	AVE HEAT LOSS	PEAK HEAT LOSS	AVE HEAT GAIN	PEAK HEAT GAIN	HRS HT ON	HRS CL ON	AVE HEAT LOSS	PEAK HEAT LOSS	AVE HEAT GAIN	PEAK HEAT GAIN	HRS HT ON	HRS CL ON	AVE HEAT LOSS	PEAK HEAT LOSS	AVE HEAT GAIN	PEAK HEAT GAIN	HRS HT ON	HRS CL ON
JAN	0 00	0 00	0 41	0 89	0	744	0 00	0 00	0 36	0 85	0	744	0 04	0 08	0 14	0 42	616	128	0 00	0 00	0 41	0 81	0	744	0 00	0 00	0 46	0 82	0	744
FEB	0 03	0 09	0 12	0 15	72	600	0 04	0 11	0 08	0 10	115	557	0 20	0 56	0 00	0 00	672	0	0 07	0 23	0 07	0 08	79	593	0 05	0 14	0 16	0 19	24	648
MAR	0 03	0 14	0 01	0 02	587	157	0 05	0 15	0 00	0 00	744	0	0 25	0 72	0 00	0 00	744	0	0 11	0 43	0 00	0 00	744	0	0 07	0 33	0 03	0 05	262	482
APR	0 09	0 18	0 00	0 00	720	0	0 11	0 19	0 00	0 00	720	0	0 23	0 69	0 00	0 00	720	0	0 22	0 56	0 00	0 00	720	0	0 13	0 46	0 00	0 00	720	0
MAY	0 16	0 24	0 00	0 00	744	0	0 17	0 26	0 00	0 00	744	0	0 14	0 63	0 00	0 00	744	0	0 29	0 66	0 00	0 00	744	0	0 24	0 58	0 00	0 00	744	0
JUN	0 11	0 13	0 00	0 00	720	0	0 12	0 13	0 00	0 00	720	0	0 01	0 02	0 08	0 15	55	665	0 15	0 20	0 00	0 00	720	0	0 14	0 17	0 00	0 00	720	0
JUL	0 09	0 10	0 00	0 00	744	0	0 10	0 11	0 00	0 00	744	0	0 00	0 00	0 20	0 25	0	744	0 08	0 11	0 00	0 00	744	0	0 12	0 13	0 00	0 00	744	0
AUG	0 08	0 09	0 00	0 00	744	0	0 08	0 09	0 00	0 00	744	0	0 00	0 00	0 28	0 30	0	744	0 02	0 04	0 01	0 01	591	153	0 09	0 10	0 00	0 00	744	0
SEP	0 07	0 08	0 00	0 00	720	0	0 07	0 08	0 00	0 00	720	0	0 00	0 00	0 31	0 32	0	720	0 00	0 00	0 03	0 05	0	720	0 08	0 08	0 00	0 00	720	0
OCT	0 00	0 00	0 07	0 40	0	744	0 00	0 00	0 07	0 40	0	744	0 00	0 00	0 37	0 82	0	744	0 00	0 00	0 24	0 56	0	744	0 00	0 00	0 13	0 45	0	744
NOV	0 00	0 00	0 21	0 67	0	720	0 00	0 00	0 19	0 66	0	720	0 00	0 00	0 26	0 73	0	720	0 00	0 00	0 34	0 70	0	720	0 00	0 00	0 25	0 59	0	720
DEC	0 00	0 00	0 30	0 78	0	744	0 00	0 00	0 27	0 76	0	744	0 09	0 12	0 17	0 61	381	363	0 00	0 00	0 39	0 77	0	744	0 00	0 00	0 35	0 70	0	744

Full Berm 1-Story

	Slab (Perimeter)						Slab (Interior)						Wall (Level-1)					
	AVE HEAT LOSS	PEAK HEAT LOSS	AVE HEAT GAIN	PEAK HEAT GAIN	HRS HT ON	HRS CL ON	AVE HEAT LOSS	PEAK HEAT LOSS	AVE HEAT GAIN	PEAK HEAT GAIN	HRS HT ON	HRS CL ON	AVE HEAT LOSS	PEAK HEAT LOSS	AVE HEAT GAIN	PEAK HEAT GAIN	HRS HT ON	HRS CL ON
JAN	0 00	0 00	0 46	0 95	0	744	0 00	0 00	0 39	0 88	0	744	0 00	0 00	0 26	0 66	0	744
FEB	0 03	0 08	0 15	0 17	49	623	0 04	0 10	0 11	0 14	83	589	0 04	0 31	0 02	0 03	472	200
MAR	0 04	0 13	0 03	0 04	395	349	0 04	0 14	0 01	0 03	508	236	0 16	0 56	0 00	0 00	744	0
APR	0 08	0 17	0 00	0 00	720	0	0 08	0 17	0 00	0 00	720	0	0 23	0 63	0 00	0 00	720	0
MAY	0 14	0 23	0 00	0 00	744	0	0 14	0 23	0 00	0 00	744	0	0 24	0 65	0 00	0 00	744	0
JUN	0 09	0 11	0 00	0 00	720	0	0 09	0 10	0 00	0 00	720	0	0 06	0 12	0 02	0 05	459	261
JUL	0 07	0 08	0 00	0 00	744	0	0 06	0 07	0 00	0 00	744	0	0 00	0 00	0 11	0 17	0	744
AUG	0 05	0 06	0 00	0 00	744	0	0 05	0 06	0 00	0 00	744	0	0 00	0 00	0 21	0 24	0	744
SEP	0 03	0 04	0 00	0 00	720	0	0 04	0 05	0 00	0 00	720	0	0 00	0 00	0 26	0 27	0	720
OCT	0 00	0 00	0 16	0 55	0	744	0 00	0 00	0 11	0 50	0	744	0 00	0 00	0 41	0 78	0	744
NOV	0 00	0 00	0 28	0 75	0	720	0 00	0 00	0 22	0 70	0	720	0 00	0 00	0 39	0 79	0	720
DEC	0 00	0 00	0 37	0 85	0	744	0 00	0 00	0 30	0 79	0	744	0 00	0 00	0 25	0 75	0	744

Half Berm

	Slab (Perimeter)						Slab (Interior)						Wall (Level-1)					
	AVE HEAT LOSS	PEAK HEAT LOSS	AVE HEAT GAIN	PEAK HEAT GAIN	HRS HT ON	HRS CL ON	AVE HEAT LOSS	PEAK HEAT LOSS	AVE HEAT GAIN	PEAK HEAT GAIN	HRS HT ON	HRS CL ON	AVE HEAT LOSS	PEAK HEAT LOSS	AVE HEAT GAIN	PEAK HEAT GAIN	HRS HT ON	HRS CL ON
JAN	0 00	0 00	0 40	0 88	0	744	0 00	0 00	0 41	0 90	0	744	0 15	0 27	0 07	0 30	641	103
FEB	0 03	0 09	0 09	0 11	87	585	0 03	0 09	0 11	0 14	80	592	0 23	0 62	0 00	0 00	672	0
MAR	0 05	0 15	0 00	0 00	744	0	0 04	0 14	0 01	0 04	536	208	0 19	0 81	0 00	0 00	744	0
APR	0 11	0 19	0 00	0 00	720	0	0 08	0 17	0 00	0 00	720	0	0 13	0 69	0 01	0 02	654	66
MAY	0 17	0 26	0 00	0 00	744	0	0 14	0 23	0 00	0 00	744	0	0 10	0 53	0 07	0 11	300	444
JUN	0 12	0 14	0 00	0 00	720	0	0 09	0 11	0 00	0 00	720	0	0 00	0 00	0 19	0 25	0	720
JUL	0 08	0 10	0 00	0 00	744	0	0 07	0 08	0 00	0 00	744	0	0 00	0 00	0 28	0 31	0	744
AUG	0 06	0 07	0 00	0 00	744	0	0 05	0 06	0 00	0 00	744	0	0 00	0 00	0 32	0 34	0	744
SEP	0 04	0 05	0 00	0 00	720	0	0 04	0 05	0 00	0 00	720	0	0 00	0 00	0 32	0 36	0	720
OCT	0 00	0 00	0 14	0 53	0	744	0 00	0 00	0 12	0 51	0	744	0 00	0 00	0 27	0 80	0	744
NOV	0 00	0 00	0 26	0 73	0	720	0 00	0 00	0 24	0 71	0	720	0 06	0 12	0 12	0 56	111	609
DEC	0 00	0 00	0 32	0 82	0	744	0 00	0 00	0 32	0 80	0	744	0 28	0 45	0 14	0 40	632	112

Slab On Grade

	Slab (Perimeter)						Slab (Interior)					
	AVE HEAT LOSS	PEAK HEAT LOSS	AVE HEAT GAIN	PEAK HEAT GAIN	HRS HT ON	HRS CL ON	AVE HEAT LOSS	PEAK HEAT LOSS	AVE HEAT GAIN	PEAK HEAT GAIN	HRS HT ON	HRS CL ON
JAN	0 00	0 00	0 33	0 83	0	744	0 00	0 00	0 43	0 92	0	744
FEB	0 03	0 12	0 03	0 06	359	313	0 03	0 09	0 12	0 16	65	607
MAR	0 07	0 18	0 00	0 00	744	0	0 04	0 14	0 01	0 01	500	244
APR	0 12	0 22	0 00	0 00	720	0	0 08	0 17	0 00	0 00	720	0
MAY	0 15	0 25	0 00	0 00	744	0	0 15	0 23	0 00	0 00	744	0
JUN	0 06	0 09	0 00	0 00	720	0	0 10	0 11	0 00	0 00	720	0
JUL	0 01	0 03	0 02	0 06	477	267	0 09	0 24	0 00	0 00	744	0
AUG	0 00	0 00	0 10	0 13	0	744	0 16	0 23	0 00	0 00	744	0
SEP	0 00	0 00	0 13	0 14	0	720	0 05	0 07	0 00	0 00	720	0
OCT	0 00	0 00	0 31	0 78	0	744	0 00	0 00	0 16	0 60	0	744
NOV	0 00	0 00	0 34	0 83	0	720	0 00	0 00	0 27	0 73	0	720
DEC	0 00	0 00	0 28	0 84	0	744	0 00	0 00	0 35	0 83	0	744

INSULATION DEPTH

Full Berm, 3 Story	8 0 ft
Full Berm, 1 Story	8 0 ft
Half Berm	5 0 ft
Slab on Grade	4 0 ft

Heat Flux Expressed in Btu/h ft²

To obtain W/m² units, multiply Btu/h ft² units by 3 156

TABLE 35

Monthly Average and Peak Heat Fluxes for Minneapolis with R-20.8 Insulation

Minneapolis R-20.8 Insulation

Full Berm 3-Story

	Slab (Perimeter)						Slab (Interior)						Wall (Level-1)						Wall (Level-2)						Wall (Level-3)					
	AVE HEAT LOSS	PEAK HEAT LOSS	AVE HEAT GAIN	PEAK HEAT GAIN	HRS HT ON	HRS CL ON	AVE HEAT LOSS	PEAK HEAT LOSS	AVE HEAT GAIN	PEAK HEAT GAIN	HRS HT ON	HRS CL ON	AVE HEAT LOSS	PEAK HEAT LOSS	AVE HEAT GAIN	PEAK HEAT GAIN	HRS HT ON	HRS CL ON	AVE HEAT LOSS	PEAK HEAT LOSS	AVE HEAT GAIN	PEAK HEAT GAIN	HRS HT ON	HRS CL ON	AVE HEAT LOSS	PEAK HEAT LOSS	AVE HEAT GAIN	PEAK HEAT GAIN	HRS HT ON	HRS CL ON
JAN	0 63	0.65	0.00	0 00	744	0	0 48	0 49	0 00	0 00	744	0	1 55	1 62	0 00	0 00	744	0	1 10	1.14	0 00	0 00	744	0	1 09	1.11	0 00	0 00	744	0
FEB	0.59	0.61	0 00	0 00	672	0	0 44	0 46	0 00	0 00	672	0	1 66	1 67	0 00	0 00	672	0	1 18	1.21	0 00	0 00	672	0	1.07	1 08	0 00	0.00	672	0
MAR	0 56	0 57	0 00	0 00	744	0	0 41	0 43	0 00	0 00	744	0	1 63	1 66	0 00	0 00	744	0	1 23	1.25	0 00	0 00	744	0	1.05	1 06	0 00	0 00	744	0
APR	0 63	0.72	0.00	0 00	720	0	0.49	0 58	0 00	0 00	720	0	1.59	2 11	0 00	0 00	720	0	1 43	1.76	0 00	0.00	720	0	1.22	1 55	0.00	0 00	720	0
MAY	0 68	0 77	0.00	0 00	744	0	0 54	0 63	0 00	0 00	744	0	1 39	1 94	0 00	0 00	744	0	1.48	1.86	0 00	0 00	744	0	1 31	1.66	0 00	0 00	744	0
JUN	0.71	0.81	0.00	0 00	720	0	0 57	0 67	0 00	0 00	720	0	1 06	1 65	0 00	0 00	720	0	1 47	1 88	0 00	0 00	720	0	1.37	1 74	0 00	0 00	720	0
JUL	0.85	1.03	0.00	0.00	744	0	0 71	0 89	0 00	0 00	744	0	0 86	1 81	0 00	0 00	744	0	1.56	2.34	0.00	0 00	744	0	1.60	2.31	0.00	0 00	744	0
AUG	0 63	0.64	0 00	0.00	744	0	0 49	0 50	0 00	0 00	744	0	0 42	0 47	0 00	0 00	744	0	1 01	1.05	0 00	0.00	744	0	1 17	1 20	0 00	0 00	744	0
SEP	0 38	0 41	0.00	0 00	720	0	0 25	0 28	0 00	0 00	720	0	0 29	0 50	0 19	0 57	647	73	0 55	0 62	0 03	0 06	717	3	0 75	0 84	0 00	0 00	720	0
OCT	0 19	0.23	0.00	0.00	744	0	0 08	0 12	0 03	0 08	678	66	0 53	0 71	0 20	0 53	697	47	0 32	0 46	0 14	0 41	695	49	0 47	0 57	0 08	0 19	732	12
NOV	0 24	0.25	0 00	0 00	720	0	0 13	0 14	0 00	0 00	720	0	0 95	1 23	0 00	0 00	720	0	0 53	0 60	0 00	0 00	720	0	0 58	0 59	0 00	0 00	720	0
DEC	0 25	0.25	0 00	0 00	744	0	0 14	0 14	0 00	0 00	744	0	1 35	1 41	0 00	0 00	744	0	0 69	0.77	0 00	0 00	744	0	0 59	0.59	0 00	0 00	744	0

Full Berm 1-Story

	Slab (Perimeter)						Slab (Interior)						Wall (Level-1)					
	AVE HEAT LOSS	PEAK HEAT LOSS	AVE HEAT GAIN	PEAK HEAT GAIN	HRS HT ON	HRS CL ON	AVE HEAT LOSS	PEAK HEAT LOSS	AVE HEAT GAIN	PEAK HEAT GAIN	HRS HT ON	HRS CL ON	AVE HEAT LOSS	PEAK HEAT LOSS	AVE HEAT GAIN	PEAK HEAT GAIN	HRS HT ON	HRS CL ON
JAN	0.98	0 98	0 00	0 00	744	0	0 78	0 79	0 00	0 00	744	0	2 85	2 94	0 00	0 00	744	0
FEB	0 99	0.99	0 00	0 00	672	0	0 77	0 78	0 00	0 00	672	0	3 00	3 02	0 00	0 00	672	0
MAR	0 99	0 99	0 00	0 00	744	0	0 75	0.76	0 00	0 00	744	0	3 02	3 03	0.00	0 00	744	0
APR	1 09	1 17	0 00	0 00	720	0	0 84	0 92	0 00	0 00	720	0	3 07	3 52	0 00	0 00	720	0
MAY	1 15	1 24	0.00	0 00	744	0	0 90	0 99	0 00	0 00	744	0	2 95	3 43	0 00	0 00	744	0
JUN	1 20	1.30	0 00	0 00	720	0	0 94	1 04	0 00	0 00	720	0	2 69	3 21	0 00	0 00	720	0
JUL	1.33	1 51	0 00	0 00	744	0	1 09	1 26	0 00	0 00	744	0	2 59	3 43	0.00	0 00	744	0
AUG	1 10	1 11	0 00	0 00	744	0	0 87	0.88	0 00	0 00	744	0	2 01	2 07	0.00	0 00	744	0
SEP	0 83	0 85	0 00	0 00	720	0	0 64	0 67	0 00	0 00	720	0	1 56	1 74	0 00	0 00	720	0
OCT	0 64	0 68	0 00	0 00	744	0	0 47	0 52	0 00	0 00	744	0	1 52	1 70	0 00	0 00	744	0
NOV	0 69	0 70	0 00	0 00	720	0	0 53	0 54	0 00	0 00	720	0	1 93	2 19	0 00	0 00	720	0
DEC	0 72	0 73	0 00	0 00	744	0	0 54	0 55	0 00	0 00	744	0	2 32	2 42	0 00	0 00	744	0

Half Berm

	Slab (Perimeter)						Slab (Interior)						Wall (Level-1)					
	AVE HEAT LOSS	PEAK HEAT LOSS	AVE HEAT GAIN	PEAK HEAT GAIN	HRS HT ON	HRS CL ON	AVE HEAT LOSS	PEAK HEAT LOSS	AVE HEAT GAIN	PEAK HEAT GAIN	HRS HT ON	HRS CL ON	AVE HEAT LOSS	PEAK HEAT LOSS	AVE HEAT GAIN	PEAK HEAT GAIN	HRS HT ON	HRS CL ON
JAN	1 34	1 35	0 00	0 00	744	0	0 82	0 82	0 00	0 00	744	0	1 25	1 31	0.00	0 00	744	0
FEB	1 36	1 37	0 00	0 00	672	0	0 81	0 82	0 00	0 00	672	0	1 30	1 32	0 00	0.00	672	0
MAR	1 38	1 38	0 00	0 00	744	0	0.81	0 81	0 00	0 00	744	0	1 22	1 26	0.00	0 00	744	0
APR	1 49	1 56	0 00	0 00	720	0	0 91	0 99	0 00	0 00	720	0	1 04	1 69	0 00	0 00	720	0
MAY	1 54	1 64	0.00	0 00	744	0	0 98	1 07	0 00	0 00	744	0	0 81	1 38	0 00	0 00	744	0
JUN	1 56	1 67	0 00	0 00	720	0	1 04	1 13	0 00	0 00	720	0	0 51	1 05	0 00	0 00	720	0
JUL	1 68	1.86	0 00	0.00	744	0	1 19	1 36	0 00	0 00	744	0	0 39	1.30	0 00	0 00	744	0
AUG	1.43	1.44	0 00	0 00	744	0	0 98	0 99	0 00	0 00	744	0	0 25	0 38	0 10	0 26	695	49
SEP	1.16	1 19	0 00	0 00	720	0	0 74	0 78	0 00	0 00	720	0	0 35	0 60	0 20	0 66	612	108
OCT	0 99	1 05	0 00	0 00	744	0	0 57	0 62	0 00	0 00	744	0	0 64	0 76	0 18	0 43	715	29
NOV	1 08	1 11	0 00	0 00	720	0	0 64	0 65	0 00	0 00	720	0	0 96	1 23	0 00	0 00	720	0
DEC	1.14	1 17	0 00	0 00	744	0	0 65	0 66	0 00	0 00	744	0	1 18	1 24	0 00	0 00	744	0

Slab On Grade

	Slab (Perimeter)						Slab (Interior)					
	AVE HEAT LOSS	PEAK HEAT LOSS	AVE HEAT GAIN	PEAK HEAT GAIN	HRS HT ON	HRS CL ON	AVE HEAT LOSS	PEAK HEAT LOSS	AVE HEAT GAIN	PEAK HEAT GAIN	HRS HT ON	HRS CL ON
JAN	1 30	1.33	0 00	0 00	744	0	0 88	0 88	0.00	0.00	744	0
FEB	1 36	1.37	0.00	0.00	672	0	0 88	0 88	0 00	0 00	672	0
MAR	1 38	1 38	0 00	0 00	744	0	0 89	0 89	0 00	0 00	744	0
APR	1 45	1 56	0 00	0 00	720	0	1 01	1 07	0 00	0 00	720	0
MAY	1 45	1 58	0 00	ɔ 00	744	0	1 09	1 17	0 00	0 00	744	0
JUN	1 39	1 54	0 00	0 00	720	0	1 15	1.24	0 00	0 00	720	0
JUL	1 42	1.64	0 00	0 00	744	0	1 30	1 47	0 00	0 00	744	0
AUG	1 13	1 15	0 00	0 00	744	0	1 08	1 09	0 00	0 00	744	0
SEP	0 86	0 92	0.00	0 00	720	0	0 82	0.85	0 00	0 00	720	0
OCT	0 76	0.84	'0 00	0 00	744	0	0 63	0 67	0 00	0 00	744	0
NOV	0.92	1.01	0.00	0 00	720	0	0 68	0 69	0 00	0 00	720	0
DEC	1 07	1 11	0 00	0 00	744	0	0 70	0.71	0 00	0 00	744	0

INSUALTION DEPTH

Full Berm, 3 Story	8 0 ft
Full Berm, 1 Story	8 0 ft
Half Berm	5 0 ft
Slab on Grade	4 0 ft

Heat Flux Expressed in Btu/h ft²

To obtain W/m² units, multiply Btu/h ft² units by 3 156

TABLE 36
Monthly Average and Peak Heat Fluxes for Boston with R-20.8 Insulation

Boston R-20.8 Insulation

Full Berm 3-Story

	Slab (Perimeter)						Slab (Interior)						Wall (Level-1)						Wall (Level-2)						Wall (Level-3)					
	AVE HEAT LOSS	PEAK HEAT LOSS	AVE HEAT GAIN	PEAK HEAT GAIN	HRS HT ON	HRS CL ON	AVE HEAT LOSS	PEAK HEAT LOSS	AVE HEAT GAIN	PEAK HEAT GAIN	HRS HT ON	HRS CL ON	AVE HEAT LOSS	PEAK HEAT LOSS	AVE HEAT GAIN	PEAK HEAT GAIN	HRS HT ON	HRS CL ON	AVE HEAT LOSS	PEAK HEAT LOSS	AVE HEAT GAIN	PEAK HEAT GAIN	HRS HT ON	HRS CL ON	AVE HEAT LOSS	PEAK HEAT LOSS	AVE HEAT GAIN	PEAK HEAT GAIN	HRS HT ON	HRS CL ON
JAN	0 41	0 43	0 00	0 00	744	0	0 31	0 32	0 00	0 00	744	0	1 19	1 23	0 00	0 00	744	0	0 82	0 85	0 00	0 00	744	0	0 76	0 76	0 00	0 00	744	0
FEB	0 39	0 40	0 00	0 00	672	0	0 29	0 30	0 00	0 00	672	0	1 25	1 26	0 00	0 00	672	0	0 87	0 90	0 00	0 00	672	0	0 75	0 75	0 00	0 00	672	0
MAR	0 37	0 38	0 00	0 00	744	0	0 27	0 28	0 00	0 00	744	0	1 23	1 24	0 00	0 00	744	0	0 91	0 92	0 00	0 00	744	0	0 74	0 74	0 00	0 00	744	0
APR	0 46	0 54	0 00	0 00	720	0	0 36	0 44	0 00	0 00	720	0	1 33	1 73	0 00	0 00	720	0	1 11	1 44	0 00	0 00	720	0	0 92	1 25	0 00	0 00	720	0
MAY	0 51	0 60	0 00	0 00	744	0	0 41	0 51	0 00	0 00	744	0	1 28	1 74	0 00	0 00	744	0	1 21	1 56	0 00	0 00	744	0	1 01	1 36	0 00	0 00	744	0
JUN	0 56	0 65	0 00	0 00	720	0	0 46	0 56	0 00	0 00	720	0	0 99	1 58	0 00	0 00	720	0	1 24	1 63	0 00	0 00	720	0	1 09	1 45	0 00	0 00	720	0
JUL	0 71	0 88	0 00	0 00	744	0	0 60	0 78	0 00	0 00	744	0	0 86	1 80	0 00	0 00	744	0	1 37	2 14	0 00	0 00	744	0	1 34	2 04	0 00	0 00	744	0
AUG	0 50	0 50	0 00	0 00	744	0	0 39	0 40	0 00	0 00	744	0	0 44	0 49	0 00	0 00	744	0	0 87	0 90	0 00	0 00	744	0	0 93	0 95	0 00	0 00	744	0
SEP	0 26	0 29	0 00	0 00	720	0	0 16	0 19	0 00	0 00	720	0	0 32	0 47	0 20	0 58	654	66	0 43	0 52	0 08	0 19	708	12	0 53	0 62	0 05	0 09	716	4
OCT	0 08	0 12	0 09	0 27	699	45	0 02	0 04	0 07	0 60	401	343	0 42	0 61	0 20	0 56	685	59	0 23	0 34	0 15	0 51	654	90	0 28	0 37	0 14	0 40	694	50
NOV	0 13	0 14	0 00	0 00	720	0	0 05	0 06	0 00	0 00	720	0	0 76	0 97	0 00	0 00	720	0	0 40	0 46	0 00	0 00	720	0	0 39	0 40	0 00	0 00	720	0
DEC	0 14	0 15	0 00	0 00	744	0	0 07	0 07	0 00	0 00	744	0	1 06	1 11	0 00	0 00	744	0	0 53	0 60	0 00	0 00	744	0	0 41	0 42	0 00	0 00	744	0

Full Berm 1-Story

	Slab (Perimeter)						Slab (Interior)						Wall (Level-1)					
	AVE HEAT LOSS	PEAK HEAT LOSS	AVE HEAT GAIN	PEAK HEAT GAIN	HRS HT ON	HRS CL ON	AVE HEAT LOSS	PEAK HEAT LOSS	AVE HEAT GAIN	PEAK HEAT GAIN	HRS HT ON	HRS CL ON	AVE HEAT LOSS	PEAK HEAT LOSS	AVE HEAT GAIN	PEAK HEAT GAIN	HRS HT ON	HRS CL ON
JAN	0 68	0 68	0 00	0 00	744	0	0 53	0 53	0 00	0 00	744	0	2 11	2 17	0 00	0 00	744	0
FEB	0 69	0 69	0 00	0 00	672	0	0 52	0 52	0 00	0 00	672	0	2 20	2 21	0 00	0 00	672	0
MAR	0 69	0 69	0 00	0 00	744	0	0 51	0 52	0 00	0 00	744	0	2 21	2 22	0 00	0 00	744	0
APR	0 80	0 87	0 00	0 00	720	0	0 61	0 69	0 00	0 00	720	0	2 39	2 73	0 00	0 00	720	0
MAY	0 87	0 96	0 00	0 00	744	0	0 67	0 76	0 00	0 00	744	0	2 42	2 81	0 00	0 00	744	0
JUN	0 93	1 02	0 00	0 00	720	0	0 72	0 82	0 00	0 00	720	0	2 23	2 74	0 00	0 00	720	0
JUL	1 07	1 25	0 00	0 00	744	0	0 87	1 05	0 00	0 00	744	0	2 22	3 04	0 00	0 00	744	0
AUG	0 85	0 86	0 00	0 00	744	0	0 67	0 67	0 00	0 00	744	0	1 68	1 74	0 00	0 00	744	0
SEP	0 60	0 63	0 00	0 00	720	0	0 44	0 48	0 00	0 00	720	0	1 27	1 40	0 00	0 00	720	0
OCT	0 41	0 46	0 00	0 00	744	0	0 27	0 32	0 00	0 00	744	0	1 12	1 33	0 00	0 00	744	0
NOV	0 47	0 49	0 00	0 00	720	0	0 34	0 36	0 00	0 00	720	0	1 46	1 65	0 00	0 00	720	0
DEC	0 50	0 51	0 00	0 00	744	0	0 36	0 37	0 00	0 00	744	0	1 76	1 84	0 00	0 00	744	0

Half Berm

	Slab (Perimeter)						Slab (Interior)						Wall (Level-1)					
	AVE HEAT LOSS	PEAK HEAT LOSS	AVE HEAT GAIN	PEAK HEAT GAIN	HRS HT ON	HRS CL ON	AVE HEAT LOSS	PEAK HEAT LOSS	AVE HEAT GAIN	PEAK HEAT GAIN	HRS HT ON	HRS CL ON	AVE HEAT LOSS	PEAK HEAT LOSS	AVE HEAT GAIN	PEAK HEAT GAIN	HRS HT ON	HRS CL ON
JAN	0 93	0 94	0 00	0 00	744	0	0 56	0 56	0 00	0 00	744	0	0 93	0 97	0 00	0 00	744	0
FEB	0 95	0 96	0 00	0 00	672	0	0 56	0 56	0 00	0 00	672	0	0 96	0 98	0 00	0 00	672	0
MAR	0 96	0 97	0 00	0 00	744	0	0 56	0 56	0 00	0 00	744	0	0 91	0 93	0 00	0 00	744	0
APR	1 08	1 15	0 00	0 00	720	0	0 66	0 74	0 00	0 00	720	0	0 92	1 41	0 00	0 00	720	0
MAY	1 15	1 24	0 00	0 00	744	0	0 74	0 82	0 00	0 00	744	0	0 79	1 30	0 00	0 00	744	0
JUN	1 19	1 30	0 00	0 00	720	0	0 80	0 88	0 00	0 00	720	0	0 48	1 03	0 00	0 00	720	0
JUL	1 32	1 50	0 00	0 00	744	0	0 95	1 12	0 00	0 00	744	0	0 42	1 35	0 00	0 00	744	0
AUG	1 09	1 10	0 00	0 00	744	0	0 75	0 75	0 00	0 00	744	0	0 23	0 31	0 09	0 22	707	37
SEP	0 83	0 87	0 00	0 00	720	0	0 52	0 55	0 00	0 00	720	0	0 33	0 50	0 26	0 72	639	81
OCT	0 66	0 71	0 00	0 00	744	0	0 35	0 40	0 00	0 00	744	0	0 48	0 67	0 20	0 53	695	49
NOV	0 75	0 77	0 00	0 00	720	0	0 42	0 43	0 00	0 00	720	0	0 75	0 95	0 00	0 00	720	0
DEC	0 80	0 82	0 00	0 00	744	0	0 44	0 45	0 00	0 00	744	0	0 92	0 98	0 00	0 00	744	0

Slab On Grade

	Slab (Perimeter)						Slab (Interior)					
	AVE HEAT LOSS	PEAK HEAT LOSS	AVE HEAT GAIN	PEAK HEAT GAIN	HRS HT ON	HRS CL ON	AVE HEAT LOSS	PEAK HEAT LOSS	AVE HEAT GAIN	PEAK HEAT GAIN	HRS HT ON	HRS CL ON
JAN	0 94	0 97	0 00	0 00	744	0	0 61	0 61	0 00	0 00	744	0
FEB	0 98	0 99	0 00	0 00	672	0	0 61	0 62	0 00	0 00	672	0
MAR	1 00	1 00	0 00	0 00	744	0	0 62	0 63	0 00	0 00	744	0
APR	1 11	1 18	0 00	0 00	720	0	0 74	0 81	0 00	0 00	720	0
MAY	1 16	1 25	0 00	0 00	744	0	0 82	0 90	0 00	0 00	744	0
JUN	1 13	1 26	0 00	0 00	720	0	0 89	0 97	0 00	0 00	720	0
JUL	1 19	1 40	0 00	0 00	744	0	1 05	1 22	0 00	0 00	744	0
AUG	0 92	0 94	0 00	0 00	744	0	0 84	0 85	0 00	0 00	744	0
SEP	0 67	0 72	0 00	0 00	720	0	0 59	0 62	0 00	0 00	720	0
OCT	0 54	0 62	0 00	0 00	744	0	0 41	0 45	0 00	0 00	744	0
NOV	0 68	0 75	0 00	0 00	720	0	0 47	0 48	0 00	0 00	720	0
DEC	0 79	0 83	0 00	0 00	744	0	0 49	0 50	0 00	0 00	744	0

INSULATION DEPTH

Full Berm, 3 Story	8 0 ft
Full Berm, 1 Story	8 0 ft
Half Berm	5 0 ft
Slab on Grade	4 0 ft

Heat Flux Expressed in Btu/h ft²

To obtain W/m² units, multiply Btu/h ft² units by 3 156

TABLE 37

Monthly Average and Peak Heat Fluxes for Seattle with R-20.8 Insulation

Seattle R-20.8 Insulation

	Slab (Perimeter)						Slab (Interior)					
	AVE HEAT LOSS	PEAK HEAT LOSS	AVE HEAT GAIN	PEAK HEAT GAIN	HRS HT ON	HRS CL ON	AVE HEAT LOSS	PEAK HEAT LOSS	AVE HEAT GAIN	PEAK HEAT GAIN	HRS HT ON	HRS CL ON
Full Berm 3-Story												
JAN	0.37	0.38	0.00	0.00	744	0	0.26	0.27	0.00	0.00	744	0
FEB	0.35	0.36	0.00	0.00	672	0	0.25	0.26	0.00	0.00	672	0
MAR	0.33	0.34	0.00	0.00	744	0	0.24	0.24	0.00	0.00	744	0
APR	0.43	0.51	0.00	0.00	720	0	0.33	0.41	0.00	0.00	720	0
MAY	0.59	0.76	0.00	0.00	744	0	0.49	0.66	0.00	0.00	744	0
JUN	0.61	0.71	0.00	0.00	720	0	0.51	0.61	0.00	0.00	720	0
JUL	0.64	0.74	0.00	0.00	744	0	0.54	0.64	0.00	0.00	744	0
AUG	0.57	0.59	0.00	0.00	744	0	0.46	0.49	0.00	0.00	744	0
SEP	0.30	0.33	0.00	0.00	720	0	0.20	0.23	0.00	0.00	720	0
OCT	0.12	0.16	0.05	0.12	731	13	0.04	0.07	0.06	0.47	562	182
NOV	0.06	0.09	0.03	0.09	700	20	0.00	0.01	0.04	0.40	159	561
DEC	0.10	0.11	0.00	0.00	744	0	0.02	0.03	0.00	0.00	744	0
Full Berm 1-Story												
JAN	0.66	0.67	0.00	0.00	744	0	0.49	0.49	0.00	0.00	744	0
FEB	0.67	0.68	0.00	0.00	672	0	0.49	0.49	0.00	0.00	672	0
MAR	0.68	0.68	0.00	0.00	744	0	0.49	0.49	0.00	0.00	744	0
APR	0.79	0.86	0.00	0.00	720	0	0.59	0.67	0.00	0.00	720	0
MAY	0.97	1.13	0.00	0.00	744	0	0.76	0.92	0.00	0.00	744	0
JUN	1.00	1.10	0.00	0.00	720	0	0.78	0.88	0.00	0.00	720	0
JUL	1.03	1.13	0.00	0.00	744	0	0.81	0.91	0.00	0.00	744	0
AUG	0.95	0.98	0.00	0.00	744	0	0.74	0.77	0.00	0.00	744	0
SEP	0.68	0.70	0.00	0.00	720	0	0.49	0.52	0.00	0.00	720	0
OCT	0.48	0.53	0.00	0.00	744	0	0.31	0.36	0.00	0.00	744	0
NOV	0.43	0.47	0.00	0.00	720	0	0.28	0.31	0.00	0.00	720	0
DEC	0.49	0.50	0.00	0.00	744	0	0.33	0.34	0.00	0.00	744	0
Half Berm												
JAN	0.91	0.93	0.00	0.00	744	0	0.53	0.53	0.00	0.00	744	0
FEB	0.94	0.95	0.00	0.00	672	0	0.54	0.54	0.00	0.00	672	0
MAR	0.96	0.96	0.00	0.00	744	0	0.54	0.54	0.00	0.00	744	0
APR	1.07	1.14	0.00	0.00	720	0	0.65	0.72	0.00	0.00	720	0
MAY	1.25	1.41	0.00	0.00	744	0	0.83	0.99	0.00	0.00	744	0
JUN	1.27	1.37	0.00	0.00	720	0	0.86	0.96	0.00	0.00	720	0
JUL	1.29	1.40	0.00	0.00	744	0	0.90	1.00	0.00	0.00	744	0
AUG	1.20	1.24	0.00	0.00	744	0	0.83	0.85	0.00	0.00	744	0
SEP	0.93	0.96	0.00	0.00	720	0	0.57	0.61	0.00	0.00	720	0
OCT	0.74	0.80	0.00	0.00	744	0	0.40	0.45	0.00	0.00	744	0
NOV	0.71	0.75	0.00	0.00	720	0	0.36	0.40	0.00	0.00	720	0
DEC	0.78	0.81	0.00	0.00	744	0	0.41	0.43	0.00	0.00	744	0
Slab On Grade												
JAN	0.96	1.00	0.00	0.00	744	0	0.59	0.60	0.00	0.00	744	0
FEB	1.01	1.03	0.00	0.00	672	0	0.60	0.61	0.00	0.00	672	0
MAR	1.03	1.04	0.00	0.00	744	0	0.62	0.62	0.00	0.00	744	0
APR	1.12	1.20	0.00	0.00	720	0	0.74	0.80	0.00	0.00	720	0
MAY	1.27	1.45	0.00	0.00	744	0	0.93	1.08	0.00	0.00	744	0
JUN	1.24	1.36	0.00	0.00	720	0	0.97	1.06	0.00	0.00	720	0
JUL	1.23	1.34	0.00	0.00	744	0	1.01	1.11	0.00	0.00	744	0
AUG	1.10	1.15	0.00	0.00	744	0	0.94	0.96	0.00	0.00	744	0
SEP	0.82	0.86	0.00	0.00	720	0	0.67	0.70	0.00	0.00	720	0
OCT	0.67	0.74	0.00	0.00	744	0	0.48	0.52	0.00	0.00	744	0
NOV	0.68	0.74	0.00	0.00	720	0	0.43	0.47	0.00	0.00	720	0
DEC	0.80	0.85	0.00	0.00	744	0	0.48	0.50	0.00	0.00	744	0

	Wall (Level-1)						Wall (Level-2)						Wall (Level-3)					
	AVE HEAT LOSS	PEAK HEAT LOSS	AVE HEAT GAIN	PEAK HEAT GAIN	HRS HT ON	HRS CL ON	AVE HEAT LOSS	PEAK HEAT LOSS	AVE HEAT GAIN	PEAK HEAT GAIN	HRS HT ON	HRS CL ON	AVE HEAT LOSS	PEAK HEAT LOSS	AVE HEAT GAIN	PEAK HEAT GAIN	HRS HT ON	HRS CL ON
Full Berm 3-Story																		
JAN	1.31	1.40	0.00	0.00	744	0	0.83	0.88	0.00	0.00	744	0	0.73	0.73	0.00	0.00	744	0
FEB	1.41	1.43	0.00	0.00	672	0	0.92	0.95	0.00	0.00	672	0	0.73	0.73	0.00	0.00	672	0
MAR	1.41	1.44	0.00	0.00	744	0	0.98	1.00	0.00	0.00	744	0	0.72	0.72	0.00	0.00	744	0
APR	1.40	1.86	0.00	0.00	720	0	1.18	1.51	0.00	0.00	720	0	0.91	1.24	0.00	0.00	720	0
MAY	1.42	2.32	0.00	0.00	744	0	1.44	2.14	0.00	0.00	744	0	1.20	1.88	0.00	0.00	744	0
JUN	1.18	1.70	0.00	0.00	720	0	1.38	1.79	0.00	0.00	720	0	1.20	1.58	0.00	0.00	720	0
JUL	1.04	1.53	0.00	0.00	744	0	1.34	1.76	0.00	0.00	744	0	1.25	1.62	0.00	0.00	744	0
AUG	0.81	0.86	0.00	0.00	744	0	1.12	1.20	0.00	0.00	744	0	1.10	1.15	0.00	0.00	744	0
SEP	0.56	0.69	0.11	0.27	703	17	0.62	0.71	0.00	0.00	720	0	0.64	0.73	0.00	0.00	720	0
OCT	0.58	0.77	0.13	0.34	719	25	0.38	0.51	0.12	0.32	716	28	0.37	0.47	0.11	0.30	717	27
NOV	0.77	0.93	0.00	0.00	720	0	0.38	0.50	0.01	0.01	719	1	0.30	0.37	0.02	0.05	716	4
DEC	1.07	1.18	0.00	0.00	744	0	0.56	0.62	0.00	0.00	744	0	0.39	0.41	0.00	0.00	744	0
Full Berm 1-Story																		
JAN	2.19	2.29	0.00	0.00	744	0												
FEB	2.31	2.33	0.00	0.00	672	0												
MAR	2.34	2.37	0.00	0.00	744	0												
APR	2.44	2.81	0.00	0.00	720	0												
MAY	2.63	3.37	0.00	0.00	744	0												
JUN	2.48	2.93	0.00	0.00	720	0												
JUL	2.39	2.82	0.00	0.00	744	0												
AUG	2.11	2.20	0.00	0.00	744	0												
SEP	1.63	1.72	0.00	0.00	720	0												
OCT	1.41	1.58	0.00	0.00	744	0												
NOV	1.50	1.65	0.00	0.00	720	0												
DEC	1.79	1.91	0.00	0.00	744	0												
Half Berm																		
JAN	1.03	1.10	0.00	0.00	744	0												
FEB	1.02	1.10	0.00	0.00	672	0												
MAR	0.97	1.09	0.00	0.00	744	0												
APR	0.90	1.38	0.00	0.00	720	0												
MAY	0.85	1.81	0.00	0.00	744	0												
JUN	0.65	1.15	0.00	0.00	720	0												
JUL	0.59	1.07	0.00	0.00	744	0												
AUG	0.46	0.49	0.00	0.00	744	0												
SEP	0.45	0.56	0.22	0.56	672	48												
OCT	0.56	0.71	0.19	0.47	706	38												
NOV	0.70	0.80	0.00	0.00	720	0												
DEC	0.91	1.00	0.00	0.00	744	0												

INSULATION DEPTH

Full Berm, 3 Story	8.0 ft
Full Berm, 1 Story	8.0 ft
Half Berm	5.0 ft
Slab on Grade	4.0 ft

Heat Flux Expressed in Btu/h ft²

To obtain W/m² units, multiply Btu/h ft² units by 3.156

TABLE 38

Monthly Average and Peak Heat Fluxes for Kansas City with R-20.8 Insulation

Kansas City R-20.8 Insulation

Full Berm 3-Story

	Slab (Perimeter)						Slab (Interior)						Wall (Level-1)						Wall (Level-2)						Wall (Level-3)					
	AVE HEAT LOSS	PEAK HEAT LOSS	AVE HEAT GAIN	PEAK HEAT GAIN	HRS HT ON	HRS CL ON	AVE HEAT LOSS	PEAK HEAT LOSS	AVE HEAT GAIN	PEAK HEAT GAIN	HRS HT ON	HRS CL ON	AVE HEAT LOSS	PEAK HEAT LOSS	AVE HEAT GAIN	PEAK HEAT GAIN	HRS HT ON	HRS CL ON	AVE HEAT LOSS	PEAK HEAT LOSS	AVE HEAT GAIN	PEAK HEAT GAIN	HRS HT ON	HRS CL ON	AVE HEAT LOSS	PEAK HEAT LOSS	AVE HEAT GAIN	PEAK HEAT GAIN	HRS HT ON	HRS CL ON
JAN	0 31	0 32	0 00	0 00	744	0	0 22	0 22	0 00	0 00	744	0	1 42	1 58	0 00	0 00	744	0	0 71	0 78	0 00	0 00	744	0	0 58	0 58	0 00	0 00	744	0
FEB	0 30	0 30	0 00	0 00	672	0	0 21	0 21	0 00	0 00	672	0	1 65	1 70	0 00	0 00	672	0	0 85	0 92	0 00	0 00	672	0	0 59	0 60	0 00	0 00	672	0
MAR	0 28	0 29	0 00	0 00	744	0	0 20	0 20	0 00	0 00	744	0	1 57	1 65	0 00	0 00	744	0	0 97	1 00	0 00	0 00	744	0	0 60	0 61	0 00	0 00	744	0
APR	0 38	0 46	0 00	0 00	720	0	0 29	0 37	0 00	0 00	720	0	1 44	1 98	0 00	0 00	720	0	1 18	1 51	0 00	0 00	720	0	0 81	1 13	0 00	0 00	720	0
MAY	0 55	0 71	0 00	0 00	744	0	0 46	0 63	0 00	0 00	744	0	1 16	2 18	0 00	0 00	744	0	1 40	2 14	0 00	0 00	744	0	1 12	1 78	0 00	0 00	744	0
JUN	0 57	0 67	0 00	0 00	720	0	0 48	0 58	0 00	0 00	720	0	0 80	1 37	0 00	0 00	720	0	1 27	1 71	0 00	0 00	720	0	1 13	1 50	0 00	0 00	720	0
JUL	0 60	0 70	0 00	0 00	744	0	0 51	0 61	0 00	0 00	744	0	0 49	1 06	0 00	0 00	744	0	1 16	1 62	0 00	0 00	744	0	1 17	1 54	0 00	0 00	744	0
AUG	0 53	0 55	0 00	0 00	744	0	0 43	0 46	0 00	0 00	744	0	0 29	0 35	0 00	0 00	744	0	0 87	0 98	0 00	0 00	744	0	1 00	1 06	0 00	0 00	744	0
SEP	0 26	0 29	0 00	0 00	720	0	0 17	0 20	0 00	0 00	720	0	0 16	0 30	0 23	0 76	596	124	0 37	0 44	0 10	0 25	701	19	0 54	0 62	0 05	0 08	717	3
OCT	0 09	0 12	0 08	0 24	705	39	0 03	0 04	0 07	0 56	462	282	0 33	0 47	0 25	0 73	650	94	0 16	0 26	0 16	0 59	599	145	0 27	0 35	0 14	0 41	691	53
NOV	0 04	0 06	0 06	0 21	631	89	0 00	0 00	0 06	0 48	0	720	0 53	0 80	0 02	0 05	716	4	0 17	0 27	0 08	0 25	645	75	0 19	0 25	0 06	0 17	689	31
DEC	0.07	0 08	0 00	0 00	744	0	0 01	0 01	0 01	0 01	471	273	1 11	1 30	0 00	0 00	744	0	0 37	0 47	0 00	0 00	744	0	0 27	0 29	0 00	0 00	744	0

Full Berm 1-Story

	Slab (Perimeter)						Slab (Interior)						Wall (Level-1)					
	AVE HEAT LOSS	PEAK HEAT LOSS	AVE HEAT GAIN	PEAK HEAT GAIN	HRS HT ON	HRS CL ON	AVE HEAT LOSS	PEAK HEAT LOSS	AVE HEAT GAIN	PEAK HEAT GAIN	HRS HT ON	HRS CL ON	AVE HEAT LOSS	PEAK HEAT LOSS	AVE HEAT GAIN	PEAK HEAT GAIN	HRS HT ON	HRS CL ON
JAN	0 55	0 56	0 00	0 00	744	0	0 41	0 41	0 00	0 00	744	0	2 10	2.28	0 00	0 00	744	0
FEB	0 57	0 58	0 00	0 00	672	0	0 41	0 41	0 00	0 00	672	0	2 37	2 42	0 00	0 00	672	0
MAR	0 59	0 60	0 00	0 00	744	0	0 41	0 41	0 00	0 00	744	0	2 37	2.42	0 00	0 00	744	0
APR	0 71	0 78	0 00	0 00	720	0	0 51	0.59	0 00	0 00	720	0	2 38	2 82	0 00	0 00	720	0
MAY	0 89	1 05	0 00	0 00	744	0	0 69	0 85	0 00	0 00	744	0	2 31	3 16	0 00	0 00	744	0
JUN	0 91	1 02	0 00	0 00	720	0	0 71	0 81	0 00	0 00	720	0	2 02	2 54	0 00	0 00	720	0
JUL	0 93	1 04	0 00	0 00	744	0	0 75	0 85	0 00	0 00	744	0	1 74	2 27	0 00	0 00	744	0
AUG	0 84	0 88	0 00	0 00	744	0	0 67	0 70	0 00	0 00	744	0	1 46	1 57	0 00	0 00	744	0
SEP	0 56	0 59	0 00	0 00	720	0	0 42	0 46	0 00	0 00	720	0	1 02	1 16	0 00	0 00	720	0
OCT	0 37	0 41	0 00	0 00	744	0	0 25	0 30	0 00	0 00	744	0	0 93	1 09	0 00	0 00	744	0
NOV	0 32	0 35	0 00	0 00	720	0	0 21	0 25	0 00	0 00	720	0	1 08	1 33	0 00	0 00	720	0
DEC	0 37	0 39	0 00	0 00	744	0	0 26	0 27	0 00	0 00	744	0	1 61	1 82	0 00	0 00	744	0

Half Berm

	Slab (Perimeter)						Slab (Interior)						Wall (Level-1)					
	AVE HEAT LOSS	PEAK HEAT LOSS	AVE HEAT GAIN	PEAK HEAT GAIN	HRS HT ON	HRS CL ON	AVE HEAT LOSS	PEAK HEAT LOSS	AVE HEAT GAIN	PEAK HEAT GAIN	HRS HT ON	HRS CL ON	AVE HEAT LOSS	PEAK HEAT LOSS	AVE HEAT GAIN	PEAK HEAT GAIN	HRS HT ON	HRS CL ON
JAN	0 81	0 84	0 00	0 00	744	0	0 44	0 44	0 00	0 00	744	0	1 13	1 33	0 00	0 00	744	0
FEB	0 86	0 88	0 00	0 00	672	0	0 44	0 45	0 00	0 00	672	0	1 27	1 37	0 00	0 00	672	0
MAR	0 89	0 90	0.00	0 00	744	0	0 45	0 45	0 00	0 00	744	0	1 02	1 16	0 00	0 00	744	0
APR	1 00	1 08	0 00	0 00	720	0	0 57	0 64	0 00	0 00	720	0	0 83	1 44	0 00	0 00	720	0
MAY	1 16	1 33	0 00	0 00	744	0	0 75	0 91	0 00	0 00	744	0	0 53	1 50	0 00	0 00	744	0
JUN	1 16	1 27	0 00	0 00	720	0	0 79	0 88	0 00	0 00	720	0	0 32	0 87	0 00	0 00	720	0
JUL	1 15	1 27	0 00	0 00	744	0	0 83	0 92	0 00	0 00	744	0	0 13	0 64	0 00	0 00	744	0
AUG	1 05	1 09	0 00	0 00	744	0	0 76	0 78	0 00	0 00	744	0	0 13	0 22	0 00	0 00	744	0
SEP	0 77	0 80	0 00	0 00	720	0	0 50	0 53	0 00	0 00	720	0	0 21	0 41	0 31	0 88	618	102
OCT	0 59	0 65	0 00	0 00	744	0	0 32	0 37	0 00	0 00	744	0	0 45	0 56	0 25	0 62	692	52
NOV	0 57	0 62	0 00	0 00	720	0	0 28	0 31	0 00	0 00	720	0	0 60	0 91	0 03	0 08	712	8
DEC	0 67	0 71	0 00	0 00	744	0	0 33	0 35	0 00	0 00	744	0	1 10	1 21	0 00	0 00	744	0

Slab On Grade

	Slab (Perimeter)						Slab (Interior)					
	AVE HEAT LOSS	PEAK HEAT LOSS	AVE HEAT GAIN	PEAK HEAT GAIN	HRS HT ON	HRS CL ON	AVE HEAT LOSS	PEAK HEAT LOSS	AVE HEAT GAIN	PEAK HEAT GAIN	HRS HT ON	HRS CL ON
JAN	0 88	0 95	0 00	0 00	744	0	0 48	0 49	0 00	0 00	744	0
FEB	0 98	1 00	0 00	0 00	672	0	0 50	0 51	0 00	0 00	672	0
MAR	1 00	1 01	0 00	0 00	744	0	0 52	0 53	0 00	0 00	744	0
APR	1 07	1 17	0 00	0 00	720	0	0 65	0 71	0 00	0 00	720	0
MAY	1 14	1 35	0 00	0 00	744	0	0 85	1 00	0 00	0 00	744	0
JUN	1 06	1 20	0 00	0 00	720	0	0 89	0 98	0 00	0 00	720	0
JUL	0 98	1 14	0 00	0 00	744	0	0 93	1 03	0 00	0 00	744	0
AUG	0 84	0 90	0 00	0 00	744	0	0 85	0 88	0.00	0 00	744	0
SEP	0 57	0 62	0 00	0 00	720	0	0 57	0 59	0.00	0 00	720	0
OCT	0 45	0 53	0 00	0 00	744	0	0 37	0 41	0.00	0 00	744	0
NOV	0 48	0 58	0 00	0 00	720	0	0 32	0 35	0 00	0 00	720	0
DEC	0 68	U 76	0.00	0 00	744	0	0 37	0 38	0 00	0 00	744	0

INSULATION DEPTH
- Full Berm, 3 Story 8 0 ft
- Full Berm, 1 Story 8 0 ft
- Half Berm 5 0 ft
- Slab on Grade 4 0 ft

Heat Flux Expressed in Btu/h ft²

To obtain W/m² units, multiply Btu/h ft² units by 3 156

TABLE 39

Monthly Average and Peak Heat Fluxes for Minneapolis with R-10.4 (Deep) Insulation

Minneapolis R-10.4 Deep Insulation

Full Berm 3-Story

	Slab (Perimeter)						Slab (Interior)						Wall (Level-1)						Wall (Level-2)						Wall (Level-3)					
	AVE HEAT LOSS	PEAK HEAT LOSS	AVE HEAT GAIN	PEAK HEAT GAIN	HRS HT ON	HRS CL ON	AVE HEAT LOSS	PEAK HEAT LOSS	AVE HEAT GAIN	PEAK HEAT GAIN	HRS HT ON	HRS CL ON	AVE HEAT LOSS	PEAK HEAT LOSS	AVE HEAT GAIN	PEAK HEAT GAIN	HRS HT ON	HRS CL ON	AVE HEAT LOSS	PEAK HEAT LOSS	AVE HEAT GAIN	PEAK HEAT GAIN	HRS HT ON	HRS CL ON	AVE HEAT LOSS	PEAK HEAT LOSS	AVE HEAT GAIN	PEAK HEAT GAIN	HRS HT ON	HRS CL ON
JAN	0.63	0.65	0.00	0.00	744	0	0.48	0.49	0.00	0.00	744	0	1.36	1.43	0.00	0.00	744	0	1.11	1.15	0.00	0.00	744	0	1.12	1.14	0.00	0.00	744	0
FEB	0.59	0.61	0.00	0.00	672	0	0.44	0.46	0.00	0.00	672	0	1.44	1.46	0.00	0.00	672	0	1.19	1.23	0.00	0.00	672	0	1.10	1.11	0.00	0.00	672	0
MAR	0.56	0.58	0.00	0.00	744	0	0.41	0.43	0.00	0.00	744	0	1.40	1.43	0.00	0.00	744	0	1.26	1.28	0.00	0.00	744	0	1.08	1.09	0.00	0.00	744	0
APR	0.63	0.72	0.00	0.00	720	0	0.49	0.58	0.00	0.00	720	0	1.30	1.87	0.00	0.00	720	0	1.46	1.79	0.00	0.00	720	0	1.26	1.59	0.00	0.00	720	0
MAY	0.68	0.78	0.00	0.00	744	0	0.54	0.64	0.00	0.00	744	0	1.10	1.66	0.00	0.00	744	0	1.52	1.90	0.00	0.00	744	0	1.35	1.70	0.00	0.00	744	0
JUN	0.71	0.81	0.00	0.00	720	0	0.57	0.67	0.00	0.00	720	0	0.77	1.34	0.00	0.00	720	0	1.51	1.93	0.00	0.00	720	0	1.42	1.78	0.00	0.00	720	0
JUL	0.85	1.03	0.00	0.00	744	0	0.71	0.89	0.00	0.00	744	0	0.59	1.54	0.00	0.00	744	0	1.60	2.40	0.00	0.00	744	0	1.64	2.35	0.00	0.00	744	0
AUG	0.63	0.64	0.00	0.00	744	0	0.49	0.50	0.00	0.00	744	0	0.30	0.39	0.04	0.11	732	12	1.07	1.12	0.00	0.00	744	0	1.22	1.24	0.00	0.00	744	0
SEP	0.38	0.41	0.00	0.00	720	0	0.25	0.28	0.00	0.00	720	0	0.29	0.51	0.20	0.65	624	96	0.61	0.67	0.01	0.01	719	1	0.79	0.88	0.00	0.00	720	0
OCT	0.20	0.24	0.00	0.00	744	0	0.08	0.12	0.03	0.08	680	64	0.55	0.68	0.21	0.52	705	39	0.36	0.49	0.13	0.36	709	35	0.50	0.60	0.07	0.15	736	8
NOV	0.25	0.25	0.00	0.00	720	0	0.13	0.14	0.00	0.00	720	0	0.92	1.21	0.00	0.00	720	0	0.55	0.62	0.00	0.00	720	0	0.61	0.62	0.00	0.00	720	0
DEC	0.25	0.25	0.00	0.00	744	0	0.14	0.15	0.00	0.00	744	0	1.23	1.25	0.00	0.00	744	0	0.70	0.78	0.00	0.00	744	0	0.62	0.62	0.00	0.00	744	0

Full Berm 1-Story

	Slab (Perimeter)						Slab (Interior)						Wall (Level-1)					
	AVE HEAT LOSS	PEAK HEAT LOSS	AVE HEAT GAIN	PEAK HEAT GAIN	HRS HT ON	HRS CL ON	AVE HEAT LOSS	PEAK HEAT LOSS	AVE HEAT GAIN	PEAK HEAT GAIN	HRS HT ON	HRS CL ON	AVE HEAT LOSS	PEAK HEAT LOSS	AVE HEAT GAIN	PEAK HEAT GAIN	HRS HT ON	HRS CL ON
JAN	1.05	1.05	0.00	0.00	744	0	0.80	0.81	0.00	0.00	744	0	2.93	3.03	0.00	0.00	744	0
FEB	1.06	1.06	0.00	0.00	672	0	0.78	0.79	0.00	0.00	672	0	3.08	3.10	0.00	0.00	672	0
MAR	1.06	1.06	0.00	0.00	744	0	0.77	0.78	0.00	0.00	744	0	3.10	3.12	0.00	0.00	744	0
APR	1.17	1.25	0.00	0.00	720	0	0.86	0.94	0.00	0.00	720	0	3.11	3.60	0.00	0.00	720	0
MAY	1.24	1.33	0.00	0.00	744	0	0.92	1.01	0.00	0.00	744	0	2.97	3.46	0.00	0.00	744	0
JUN	1.28	1.38	0.00	0.00	720	0	0.96	1.06	0.00	0.00	720	0	2.68	3.21	0.00	0.00	720	0
JUL	1.41	1.60	0.00	0.00	744	0	1.11	1.28	0.00	0.00	744	0	2.58	3.41	0.00	0.00	744	0
AUG	1.18	1.19	0.00	0.00	744	0	0.90	0.90	0.00	0.00	744	0	2.05	2.12	0.00	0.00	744	0
SEP	0.90	0.92	0.00	0.00	720	0	0.66	0.70	0.00	0.00	720	0	1.66	1.86	0.00	0.00	720	0
OCT	0.70	0.74	0.00	0.00	744	0	0.49	0.54	0.00	0.00	744	0	1.66	1.80	0.00	0.00	744	0
NOV	0.76	0.77	0.00	0.00	720	0	0.56	0.57	0.00	0.00	720	0	2.06	2.34	0.00	0.00	720	0
DEC	0.78	0.80	0.00	0.00	744	0	0.57	0.57	0.00	0.00	744	0	2.42	2.52	0.00	0.00	744	0

INSULATION DEPTH
Full Berm, 3 Story 16.0 ft
Full Berm, 1 Story 13.0 ft

Heat Flux Expressed in Btu/h ft²

To obtain W/m² units, multiply Btu/h ft² units by 3.156

TABLE 40

Monthly Average and Peak Heat Fluxes for Boston with R-10.4 (Deep) Insulation

Boston R-10.4 Deep Insulation

Full Berm 3-Story

	Slab (Perimeter)						Slab (Interior)						Wall (Level-1)						Wall (Level-2)						Wall (Level-3)					
	AVE HEAT LOSS	PEAK HEAT LOSS	AVE HEAT GAIN	PEAK HEAT GAIN	HRS HT ON	HRS CL ON	AVE HEAT LOSS	PEAK HEAT LOSS	AVE HEAT GAIN	PEAK HEAT GAIN	HRS HT ON	HRS CL ON	AVE HEAT LOSS	PEAK HEAT LOSS	AVE HEAT GAIN	PEAK HEAT GAIN	HRS HT ON	HRS CL ON	AVE HEAT LOSS	PEAK HEAT LOSS	AVE HEAT GAIN	PEAK HEAT GAIN	HRS HT ON	HRS CL ON	AVE HEAT LOSS	PEAK HEAT LOSS	AVE HEAT GAIN	PEAK HEAT GAIN	HRS HT ON	HRS CL ON
JAN	0.42	0.43	0.00	0.00	744	0	0.31	0.32	0.00	0.00	744	0	1.04	1.08	0.00	0.00	744	0	0.82	0.86	0.00	0.00	744	0	0.78	0.79	0.00	0.00	744	0
FEB	0.39	0.40	0.00	0.00	672	0	0.29	0.30	0.00	0.00	672	0	1.09	1.10	0.00	0.00	672	0	0.89	0.91	0.00	0.00	672	0	0.77	0.77	0.00	0.00	672	0
MAR	0.37	0.38	0.00	0.00	744	0	0.27	0.28	0.00	0.00	744	0	1.05	1.07	0.00	0.00	744	0	0.93	0.95	0.00	0.00	744	0	0.76	0.77	0.00	0.00	744	0
APR	0.46	0.54	0.00	0.00	720	0	0.36	0.44	0.00	0.00	720	0	1.12	1.55	0.00	0.00	720	0	1.13	1.46	0.00	0.00	720	0	0.94	1.27	0.00	0.00	720	0
MAY	0.51	0.61	0.00	0.00	744	0	0.41	0.51	0.00	0.00	744	0	1.04	1.53	0.00	0.00	744	0	1.23	1.59	0.00	0.00	744	0	1.04	1.39	0.00	0.00	744	0
JUN	0.56	0.66	0.00	0.00	720	0	0.46	0.56	0.00	0.00	720	0	0.74	1.31	0.00	0.00	720	0	1.27	1.66	0.00	0.00	720	0	1.13	1.48	0.00	0.00	720	0
JUL	0.71	0.88	0.00	0.00	744	0	0.60	0.78	0.00	0.00	744	0	0.62	1.57	0.00	0.00	744	0	1.40	2.18	0.00	0.00	744	0	1.37	2.07	0.00	0.00	744	0
AUG	0.50	0.50	0.00	0.00	744	0	0.39	0.40	0.00	0.00	744	0	0.32	0.36	0.03	0.06	739	5	0.92	0.95	0.00	0.00	744	0	0.96	0.99	0.00	0.00	744	0
SEP	0.26	0.29	0.00	0.00	720	0	0.16	0.19	0.00	0.00	720	0	0.31	0.46	0.24	0.67	646	74	0.49	0.56	0.07	0.14	713	7	0.57	0.66	0.03	0.06	717	3
OCT	0.09	0.12	0.12	0.25	703	41	0.02	0.04	0.07	0.60	407	337	0.43	0.62	0.22	0.57	690	54	0.27	0.37	0.15	0.47	680	64	0.31	0.40	0.13	0.37	702	42
NOV	0.14	0.14	0.00	0.00	720	0	0.05	0.06	0.00	0.00	720	0	0.74	0.95	0.00	0.00	720	0	0.42	0.48	0.00	0.00	720	0	0.42	0.43	0.00	0.00	720	0
DEC	0.15	0.15	0.00	0.00	744	0	0.07	0.07	0.00	0.00	744	0	0.97	0.98	0.00	0.00	744	0	0.54	0.60	0.00	0.00	744	0	0.44	0.44	0.00	0.00	744	0

Full Berm 1-Story

	Slab (Perimeter)						Slab (Interior)						Wall (Level-1)					
	AVE HEAT LOSS	PEAK HEAT LOSS	AVE HEAT GAIN	PEAK HEAT GAIN	HRS HT ON	HRS CL ON	AVE HEAT LOSS	PEAK HEAT LOSS	AVE HEAT GAIN	PEAK HEAT GAIN	HRS HT ON	HRS CL ON	AVE HEAT LOSS	PEAK HEAT LOSS	AVE HEAT GAIN	PEAK HEAT GAIN	HRS HT ON	HRS CL ON
JAN	0.73	0.74	0.00	0.00	744	0	0.54	0.54	0.00	0.00	744	0	2.17	2.24	0.00	0.00	744	0
FEB	0.74	0.75	0.00	0.00	672	0	0.53	0.54	0.00	0.00	672	0	2.27	2.28	0.00	0.00	672	0
MAR	0.75	0.75	0.00	0.00	744	0	0.52	0.53	0.00	0.00	744	0	2.28	2.30	0.00	0.00	744	0
APR	0.86	0.93	0.00	0.00	720	0	0.62	0.70	0.00	0.00	720	0	2.44	2.79	0.00	0.00	720	0
MAY	0.94	1.02	0.00	0.00	744	0	0.69	0.78	0.00	0.00	744	0	2.44	2.85	0.00	0.00	744	0
JUN	0.99	1.09	0.00	0.00	720	0	0.74	0.83	0.00	0.00	720	0	2.22	2.72	0.00	0.00	720	0
JUL	1.14	1.31	0.00	0.00	744	0	0.89	1.06	0.00	0.00	744	0	2.21	3.03	0.00	0.00	744	0
AUG	0.92	0.93	0.00	0.00	744	0	0.68	0.69	0.00	0.00	744	0	1.72	1.79	0.00	0.00	744	0
SEP	0.66	0.68	0.00	0.00	720	0	0.46	0.49	0.00	0.00	720	0	1.36	1.50	0.00	0.00	720	0
OCT	0.47	0.51	0.00	0.00	744	0	0.29	0.34	0.00	0.00	744	0	1.24	1.45	0.00	0.00	744	0
NOV	0.53	0.54	0.00	0.00	720	0	0.36	0.37	0.00	0.00	720	0	1.57	1.78	0.00	0.00	720	0
DEC	0.55	0.57	0.00	0.00	744	0	0.38	0.39	0.00	0.00	744	0	1.85	1.91	0.00	0.00	744	0

INSULATION DEPTH

Full Berm, 3 Story 16.0 ft

Full Berm, 1 Story 13.0 ft.

Heat Flux Expressed in Btu/h ft²

To obtain W/m² units, multiply Btu/h ft² units by 3.156

TABLE 41

Monthly Average and Peak Heat Fluxes for Seattle with R-10.4 (Deep) Insulation

Seattle R-10.4 Deep Insulation

Full Berm 3-Story

Slab (Perimeter)

	AVE HEAT LOSS	PEAK HEAT LOSS	AVE HEAT GAIN	PEAK HEAT GAIN	HRS HT ON	HRS CL ON
JAN	0.37	0.38	0.00	0.00	744	0
FEB	0.35	0.36	0.00	0.00	672	0
MAR	0.34	0.35	0.00	0.00	744	0
APR	0.43	0.51	0.00	0.00	720	0
MAY	0.60	0.76	0.00	0.00	744	0
JUN	0.61	0.72	0.00	0.00	720	0
JUL	0.65	0.75	0.00	0.00	744	0
AUG	0.57	0.60	0.00	0.00	744	0
SEP	0.31	0.33	0.00	0.00	720	0
OCT	0.12	0.16	0.04	0.10	733	11
NOV	0.07	0.09	0.03	0.07	704	16
DEC	0.10	0.11	0.00	0.00	744	0

Slab (Interior)

	AVE HEAT LOSS	PEAK HEAT LOSS	AVE HEAT GAIN	PEAK HEAT GAIN	HRS HT ON	HRS CL ON
JAN	0.26	0.27	0.00	0.00	744	0
FEB	0.25	0.26	0.00	0.00	672	0
MAR	0.24	0.24	0.00	0.00	744	0
APR	0.33	0.41	0.00	0.00	720	0
MAY	0.49	0.66	0.00	0.00	744	0
JUN	0.51	0.61	0.00	0.00	720	0
JUL	0.54	0.64	0.00	0.00	744	0
AUG	0.46	0.49	0.00	0.00	744	0
SEP	0.20	0.23	0.00	0.00	720	0
OCT	0.04	0.07	0.06	0.47	566	178
NOV	0.01	0.01	0.04	0.40	175	545
DEC	0.02	0.03	0.00	0.00	744	0

Wall (Level-1)

	AVE HEAT LOSS	PEAK HEAT LOSS	AVE HEAT GAIN	PEAK HEAT GAIN	HRS HT ON	HRS CL ON
JAN	1.18	1.26	0.00	0.00	744	0
FEB	1.22	1.27	0.00	0.00	672	0
MAR	1.21	1.28	0.00	0.00	744	0
APR	1.16	1.63	0.00	0.00	720	0
MAY	1.14	2.08	0.00	0.00	744	0
JUN	0.92	1.43	0.00	0.00	720	0
JUL	0.81	1.30	0.00	0.00	744	0
AUG	0.63	0.66	0.00	0.00	744	0
SEP	0.50	0.61	0.17	0.42	688	32
OCT	0.56	0.71	0.17	0.42	712	32
NOV	0.73	0.86	0.00	0.00	720	0
DEC	0.99	1.08	0.00	0.00	744	0

Wall (Level-2)

	AVE HEAT LOSS	PEAK HEAT LOSS	AVE HEAT GAIN	PEAK HEAT GAIN	HRS HT ON	HRS CL ON
JAN	0.85	0.90	0.00	0.00	744	0
FEB	0.93	0.97	0.00	0.00	672	0
MAR	1.00	1.02	0.00	0.00	744	0
APR	1.20	1.54	0.00	0.00	720	0
MAY	1.45	2.17	0.00	0.00	744	0
JUN	1.41	1.82	0.00	0.00	720	0
JUL	1.37	1.80	0.00	0.00	744	0
AUG	1.16	1.24	0.00	0.00	744	0
SEP	0.68	0.76	0.00	0.00	720	0
OCT	0.43	0.54	0.11	0.27	724	20
NOV	0.42	0.52	0.00	0.00	720	0
DEC	0.58	0.64	0.00	0.00	744	0

Wall (Level-3)

	AVE HEAT LOSS	PEAK HEAT LOSS	AVE HEAT GAIN	PEAK HEAT GAIN	HRS HT ON	HRS CL ON
JAN	0.76	0.76	0.00	0.00	744	0
FEB	0.75	0.75	0.00	0.00	672	0
MAR	0.75	0.75	0.00	0.00	744	0
APR	0.94	1.27	0.00	0.00	720	0
MAY	1.23	1.91	0.00	0.00	744	0
JUN	1.24	1.61	0.00	0.00	720	0
JUL	1.28	1.66	0.00	0.00	744	0
AUG	1.13	1.18	0.00	0.00	744	0
SEP	0.67	0.76	0.00	0.00	720	0
OCT	0.40	0.50	0.10	0.27	723	21
NOV	0.33	0.40	0.01	0.01	719	1
DEC	0.42	0.44	0.00	0.00	744	0

Full Berm 1-Story

Slab (Perimeter)

	AVE HEAT LOSS	PEAK HEAT LOSS	AVE HEAT GAIN	PEAK HEAT GAIN	HRS HT ON	HRS CL ON
JAN	0.71	0.72	0.00	0.00	744	0
FEB	0.73	0.74	0.00	0.00	672	0
MAR	0.74	0.75	0.00	0.00	744	0
APR	0.86	0.93	0.00	0.00	720	0
MAY	1.04	1.20	0.00	0.00	744	0
JUN	1.07	1.17	0.00	0.00	720	0
JUL	1.10	1.20	0.00	0.00	744	0
AUG	1.02	1.05	0.00	0.00	744	0
SEP	0.74	0.77	0.00	0.00	720	0
OCT	0.55	0.59	0.00	0.00	744	0
NOV	0.49	0.53	0.00	0.00	720	0
DEC	0.54	0.56	0.00	0.00	744	0

Slab (Interior)

	AVE HEAT LOSS	PEAK HEAT LOSS	AVE HEAT GAIN	PEAK HEAT GAIN	HRS HT ON	HRS CL ON
JAN	0.51	0.51	0.00	0.00	744	0
FEB	0.51	0.51	0.00	0.00	672	0
MAR	0.50	0.50	0.00	0.00	744	0
APR	0.60	0.68	0.00	0.00	720	0
MAY	0.77	0.94	0.00	0.00	744	0
JUN	0.80	0.90	0.00	0.00	720	0
JUL	0.83	0.93	0.00	0.00	744	0
AUG	0.76	0.79	0.00	0.00	744	0
SEP	0.51	0.54	0.00	0.00	720	0
OCT	0.34	0.39	0.00	0.00	744	0
NOV	0.30	0.33	0.00	0.00	720	0
DEC	0.35	0.36	0.00	0.00	744	0

Wall (Level-1)

	AVE HEAT LOSS	PEAK HEAT LOSS	AVE HEAT GAIN	PEAK HEAT GAIN	HRS HT ON	HRS CL ON
JAN	2.27	2.38	0.00	0.00	744	0
FEB	2.37	2.42	0.00	0.00	672	0
MAR	2.40	2.46	0.00	0.00	744	0
APR	2.48	2.85	0.00	0.00	720	0
MAY	2.64	3.41	0.00	0.00	744	0
JUN	2.49	2.95	0.00	0.00	720	0
JUL	2.40	2.85	0.00	0.00	744	0
AUG	2.15	2.21	0.00	0.00	744	0
SEP	1.72	1.81	0.00	0.00	720	0
OCT	1.53	1.68	0.00	0.00	744	0
NOV	1.62	1.75	0.00	0.00	720	0
DEC	1.89	2.02	0.00	0.00	744	0

INSULATION DEPTH
Full Berm, 3 Story 16.0 ft
Full Berm, 1 Story 13.0 ft

Heat Flux Expressed in Btu/h ft²

To obtain W/m² units, multiply Btu/h ft² units by 3.156

TABLE 42

Monthly Average and Peak Heat Fluxes for Kansas City with R-10.4 (Deep) Insulation

Kansas City R-10.4 Deep Insulation

Full Berm 3-Story

	Slab (Perimeter)						Slab (Interior)					
	AVE HEAT LOSS	PEAK HEAT LOSS	AVE HEAT GAIN	PEAK HEAT GAIN	HRS HT ON	HRS CL ON	AVE HEAT LOSS	PEAK HEAT LOSS	AVE HEAT GAIN	PEAK HEAT GAIN	HRS HT ON	HRS CL ON
JAN	0 31	0 32	0 00	0 00	744	0	0 22	0 22	0 00	0 00	744	0
FEB	0 30	0 31	0 00	0 00	672	0	0 21	0 21	0 00	0 00	672	0
MAR	0 28	0 29	0 00	0 00	744	0	0 20	0 20	0 00	0 00	744	0
APR	0 38	0 46	0 00	0 00	720	0	0 29	0 37	0 00	0 00	720	0
MAY	0 55	0 71	0 00	0 00	744	0	0 46	0 63	0 00	0 00	744	0
JUN	0 57	0 67	0 00	0 00	720	0	0 48	0 58	0 00	0 00	720	0
JUL	0 60	0 70	0 00	0 00	744	0	0 51	0 61	0 00	0 00	744	0
AUG	0 53	0 56	0 00	0 00	744	0	0 43	0 46	0 00	0 00	744	0
SEP	0 27	0 29	0 00	0 00	720	0	0 17	0 20	0 00	0 00	720	0
OCT	0 09	0 12	0 08	0 23	708	36	0 03	0 05	0 07	0 56	466	278
NOV	0 04	0 06	0 06	0 20	640	80	0 00	0 00	0 06	0 48	0	720
DEC	0 07	0 08	0 00	0 00	744	0	0 01	0 01	0 00	0 01	501	243

	Wall (Level-1)						Wall (Level-2)						Wall (Level-3)					
	AVE HEAT LOSS	PEAK HEAT LOSS	AVE HEAT GAIN	PEAK HEAT GAIN	HRS HT ON	HRS CL ON	AVE HEAT LOSS	PEAK HEAT LOSS	AVE HEAT GAIN	PEAK HEAT GAIN	HRS HT ON	HRS CL ON	AVE HEAT LOSS	PEAK HEAT LOSS	AVE HEAT GAIN	PEAK HEAT GAIN	HRS HT ON	HRS CL ON
JAN	1 28	1 47	0 00	0 00	744	0	0 70	0 78	0 00	0 00	744	0	0 60	0 60	0 00	0 00	744	0
FEB	1 48	1 55	0 00	0 00	672	0	0 85	0 92	0 00	0 00	672	0	0 61	0 62	0 00	0 00	672	0
MAR	1 32	1 43	0 00	0 00	744	0	0 98	1 02	0 00	0 00	744	0	0 63	0 64	0 00	0 00	744	0
APR	1 15	1 74	0 00	0 00	720	0	1 20	1 53	0 00	0 00	720	0	0 84	1 15	0 00	0 00	720	0
MAY	0 85	1 86	0 00	0 00	744	0	1 43	2 17	0 00	0 00	744	0	1 15	1 81	0 00	0 00	744	0
JUN	0 56	1 12	0 00	0 00	720	0	1 31	1 76	0 00	0.00	720	0	1 16	1 53	0 00	0 00	720	0
JUL	0 30	0 84	0 00	0 00	744	0	1 20	1 67	0 00	0 00	744	0	1 20	1 58	0.00	0 00	744	0
AUG	0 19	0 25	0 00	0 00	744	0	0 92	1 03	0 00	0.00	744	0	1.04	1 09	0 00	0 00	744	0
SEP	0 16	0 33	0 28	0 84	609	111	0 42	0 48	0 09	0 21	707	13	0 56	0 64	0 03	0.05	718	2
OCT	0 37	0 48	0 27	0 70	677	67	0 19	0 28	0 16	0.55	643	101	0 29	0 37	0 13	0 39	699	45
NOV	0 55	0 84	0 04	0 09	711	9	0 19	0 29	0 08	0.23	665	55	0 21	0 27	0 05	0 14	697	23
DEC	1 09	1 23	0 00	0 00	744	0	0 37	0 47	0 00	0 00	744	0	0 29	0 31	0 00	0 00	744	0

Full Berm 1-Story

	Slab (Perimeter)						Slab (Interior)						Wall (Level-1)					
	AVE HEAT LOSS	PEAK HEAT LOSS	AVE HEAT GAIN	PEAK HEAT GAIN	HRS HT ON	HRS CL ON	AVE HEAT LOSS	PEAK HEAT LOSS	AVE HEAT GAIN	PEAK HEAT GAIN	HRS HT ON	HRS CL ON	AVE HEAT LOSS	PEAK HEAT LOSS	AVE HEAT GAIN	PEAK HEAT GAIN	HRS HT ON	HRS CL ON
JAN	0 59	0 61	0 00	0 00	744	0	0 42	0 42	0 00	0 00	744	0	2 17	2 41	0 00	0.00	744	0
FEB	0 62	0 63	0 00	0 00	672	0	0 42	0 42	0 00	0 00	672	0	2 46	2 52	0 00	0.00	672	0
MAR	0 65	0 66	0.00	0 00	744	0	0 42	0 42	0 00	0 00	744	0	2 41	2 49	0 00	0.00	744	0
APR	0 77	0 84	0.00	0 00	720	0	0 52	0 60	0 00	0 00	720	0	2 39	2 87	0 00	0 00	720	0
MAY	0 96	1 11	0 00	0 00	744	0	0 70	0 86	0 00	0 00	744	0	2 28	3 12	0 00	0 00	744	0
JUN	0 98	1 08	0 00	0 00	720	0	0 72	0 82	0 00	0 00	720	0	2 01	2 54	0 00	0 00	720	0
JUL	1 00	1 11	0 00	0 00	744	0	0 76	0 86	0 00	0 00	744	0	1 73	2 26	0 00	0 00	744	0
AUG	0 90	0 94	0 00	0 00	744	0	0 69	0 72	0 00	0 00	744	0	1 48	1 61	0 00	0 00	744	0
SEP	0 61	0 63	0 00	0 00	720	0	0.44	0.47	0.00	0 00	720	0	1.10	1 27	0 00	0 00	720	0
OCT	0 41	0 46	0 00	0 00	744	0	0 27	0 32	0 00	0 00	744	0	1 04	1 19	0 00	0 00	744	0
NOV	0 36	0 40	0 00	0 00	720	0	0 23	0 26	0 00	0 00	720	0	1 19	1 48	0 00	0 00	720	0
DEC	0 42	0 44	0 00	0 00	744	0	0 28	0 29	0 00	0 00	744	0	1 75	1 92	0 00	0 00	744	0

INSULATION DEPTH
Full Berm, 3 Story 16 0 ft
Full Berm, 1 Story 13 0 ft

Heat Flux Expressed in Btu/h ft²

To obtain W/m² units, multiply Btu/h ft² units by 3 156

APPENDIX B

BOUNDARY CONDITIONS AT GROUND SURFACES

Transient heat flow at the ground surface is calculated with an hourly time step using TMY (see Reference 21) solar meteorological data inputs. An overview of the calculation procedure for each component of the surface boundary condition follows.

The energy balance equation at the ground air interface may be written,

$$R = H + LE + G \tag{B-1}$$

WHERE:

- R = net radiation (solar and longwave)
- H = sensible heat flux between the surface and the air (positive when directed upward)
- LE = latent heat of evaporation between the ground surface and air (positive when surface is losing water by evaporation)
- G = sensible heat flux conducted into the ground (positive when the ground is losing heat)

NET RADIATION, R. The net radiation, R, is defined as the difference between the incoming and outgoing streams of solar and longwave sky radiation. Thus,

$$R = Q(1-\alpha) + I\downarrow - \epsilon\ \sigma\ T_s^4 \tag{B-2}$$

WHERE:

- Q = incoming solar radiation
- α = surface albedo or reflectivity
- $I\downarrow$ = longwave radiation emitted downward by the atmosphere and absorbed by the ground.
- ϵ = surface infrared emissivity
- σ = Stephan-Boltzman constant
- T_s= surface temperature.

Infrared radiation heat exchange between the ground surface and the sky is estimated using the statistical methods developed by Clark and Allen (See Reference 24, The Estimation of Atmospheric Radiation for Clear and Cloudy Skies, by Clark). This analysis includes the effects of clouds and humidity on the apparent emissivity of the sky. The incoming solar radiation (Q) incident on the ground surface is read directly from the weather tapes.

SENSIBLE HEAT FLUX, H. The sensible heat flux between the ground surface and the air due to surface convection is estimated with the equation,

$$H = h_{cos}\ (T_s - T_a) \tag{B-3}$$

WHERE:

- h_{cos} = surface convection heat transfer coefficent (Btu/h ft^2°F)
- T_s = ground surface temperataure (°F)
- T_a = ambient dry-bulb air temperature (°F)

The surface convection heat transfer coefficient, h_{cos}, is estimated using the empirical model developed by Kreith and Sellers (See Reference 25, General Principles of Natural Evaporation, by Kreith) and is expressed as,

$$h_{cos} = \quad C_p \, D_h \tag{B-4}$$

where,

= air density (lbs/ft^3)
C_p = specific heat (Btu/lb°F)
D_h = nebulous transfer coefficient (ft/h)

The nebulous transfer coefficient, D_h, is estimated by the relationship,

$$D_h = 0.016 \; u \; \ln \; \frac{L}{L_o}^{-2} \tag{B-5}$$

WHERE:

u = horizontal wind speed (ft/h)
L = height at which the horizontal wind speed is measured (inches)
L_o = roughness length (inches)

In equation (B-5), D_h must be corrected when the air temperature varies with height. For further explanation, refer to Kreith and Sellers (Reference 23).

LATENT HEAT FLUX, LE. The latent heat flux depends on the vapor pressure difference between the ground surface and the air. The equations used in the computer model are based on Speltz and Meixel (see Reference 26, A Computer Simulation of the Thermal Perfomance of Earth Covered Roofs, by Speltz). In the arid southwest region of the country, such as Phoenix, evapotransporation is maintained at a level high enough to support the growth of plant cover.

HEAT CONDUCTION INTO THE GROUND, G. The transient heat conduction into the ground is modeled by use of a two-dimensional explicit finite-differencing routine. The following difference equation is used to calculate heat conduction from the upper boundary nodes to the ground surface:

$$\underline{1} \; \frac{T_{m,n}^{p+1} - T_{m,n}^{p}}{t} =$$

$$\frac{(T_{m+1,n}^{p}) + (T_{m-1,n}^{p}) - 2T_{m,n}^{p}}{(\quad x)^2} + \frac{T_s + T_{m,n-1}^{p} - 2T_{m,n}^{p}}{(\quad y)^2} \tag{B-6}$$

T represents the temperature and the subscripts 'm' and 'n' denotes the x and y position of the node in relation to the temperature field, the superscript 'p' designates the time increment, x and y represent the node spacing, is the thermal diffusivity and t represents the time increment of the calculation. T_s represents the ground surface temperature.

APPENDIX C

CALCULATION WORKSHEETS

This Appendix contains worksheets that may be used in calculating heating and cooling loads for earth-sheltered buildings zones. Refer to Section 18, Energy Calculations, Paragraph 5, for instructions.

Bin Number ________ Job Location ____________

Room Number ______ Room Name ____________

Month: J F M A M J J A S O N D

Days: M T W T F S S M-S

Building Temperature Set Points (Thermostat)

	01-08	09-16	17-24
Dry Bulb	___	___	___
Wet Bulb	___	___	___
Enthalpy	___	___	___

CLIMATOLOGICAL DATA				OBSERVATION HOUR GAP																							
				01-08									09-16								17-24						
				Conditions Above Set Point					Conditions Below Set Point				Conditions Above Set Point				Conditions Below Set Point				Conditions Above Set Point				Conditions Below Set Point		
Temp. Range	Mean Bin Temp. (Dry Bulb)	MCWB (1)	Enthalpy Outside Air(2)	O h (3)	ΔT (4)	Deg h(5)	ΔH (6)	Btuh/lb(7)	O h (3)	ΔT (8)	Deg h(9)	O h (3)	ΔT (4)	Deg h(9)	ΔH (6)	Btuh/lb(7)	O h (3)	ΔT (8)	Deg h(9)	O h (3)	ΔT (4)	Deg h(5)	ΔH (6)	Btuh/lb(7)	O h (3)	ΔT (8)	Deg h(9)
110/114	112																										
105/109	107																										
100/104	102																										
95/99	97																										
90/94	92																										
85/89	87																										
80/84	82																										
75/79	77																										
70/74	72																										
65/69	67																										
60/64	62																										
55/59	57																										
50/54	52																										
45/41	47																										
40/44	42																										
35/39	37																										
30/34	32																										
25/29	27																										
20/24	22																										
15/19	17																										
10/14	12																										
5/9	7																										
0/4	2																										
-5/-1	-3																										
-10/-6	-8																										
-15/-11	-13																										
-20/-16	-18																										
-25/-21	-23																										
-30/-26	-28																										
-35/-31	-33																										
TOTALS																											
Adjusted Total for Days in Week																											

(1) Mean coincident wet bulb temperature from NAVFAC P-89, Engineering Weather Data)

(2) Enthalpy of outside air for each temperature bin (based on mean coincident wet bulb temperature)

(3) Observed Hours (from NAVFAC P-89, Engineering Weather Data)

(4)(8) ΔT = building dry bulb temperature less mean bin dry bulb temperature

(5)(9) degree hours = observed hours X ΔT

(6) ΔH = enthalpy of outside air less enthalpy of the building inside air

(7) Btu hours/pound = observed hours X ΔH

WORKSHEET B

HEATING/COOLING LOAD CALCULATIONS

(Integrated Values)

Bin Number ________ Job Location ____________

Room Number ______ Room Name ____________

	J	F	M	A	M	J	J	A	S	O	N	D
Month												

	M	T	W	T	F	S	S	M-S
Days								

	01-08	09-16	17-24	01-24
Hours				

Degree Hours Above Set Point _______ Degree Hours Below Set Point _______

Enthalpy Hours Above Set Point _______ Hours Below Set Point _____

Hours Above Set Point ________

Transmission Sensible Heat Above Grade Surfaces	Area (ft²)	"U" Factor	Conditions Above Building Set Point Temperature: Degree Hours	Sensible Ht. Gain			Conditions Below Building Set Point Temperature: Degree Hours		Sensibl Ht. Los
North Exterior Wall									
East Exterior Wall									
South Exterior Wall									
West Exterior Wall									
Roof									
Exterior Glazing									

Transmission Sensible Heat Earth Contact Surfaces	Area (ft²)	Monthly Avg. Heat Loss/Gain	Hours Above Set Point	Sensible Ht. Gain		Sensible Ht. Loss	Hours Below Set Point	Sensible Ht. Gain	Sensibl Ht. Los
North Wall									
East Wall									
South Wall									
West Wall									
Slab Perimeter									
Slab Interior									

Transmission Sensible Heat Between Building Zones	Area (ft²)	"U" Factor	Temp. Diff.	Hours Abv Set Point	Sensible Ht. Gain		Sensible Ht. Loss	Hours Blw Set Point	Sensible Ht. Gain	Sensibl Ht. Los
Floors										
Ceilings										
Partitions										

Transmission Solar Heat Gain	Area (ft²)	Solar Fac. (Btu/h ft²)	Hours Abv Set Point	Sensible Ht. Gain			Hours Blw Set Point	Sensible Ht. Gain	
North Exterior Glass									
East Exterior Glass									
South Exterior Glass									
West Exterior Glass									

Body Heat Gain	No. People	Rate Gain (Btu/h)	Hours Abv Set Point	Sensible Ht. Gain	Latent Load		Hours Blw Set Point	Sensible Ht. Gain	
Sensible									
Latent									

Equipment Heat Gain	Conv. Factor	Area (ft²)	W/ft²	Hours Abv Set Point	Sensible Ht. Gain			Hours Blw Set Point	Sensible Ht. Gain	
Lights	3.412									
Electric Equipment	3.412									
Miscellaneous	3.412									

Infiltration	Enthalpy Hours (Btu h/lb)	Constant	Air Quantity (ft³/min)	Total Heat	Degree Hours	Sensible Ht. Gain	Latent Load		Degree Hours		Sensibl Ht. Lo:
Total		4.5									
Sensible		1.085									
TOTALS											

Ventilation	Enthalpy Hours (Btu h/lb)	Constant	Air Quantity (ft³/min)	Total Heat	Degree Hours	Sensible Ht. Gain	Latent Load		Degree Hours		Sensibl Ht. Lo:
Total		4.5									
Sensible		1.085									
TOTALS											

WORKSHEET C

HEATING/COOLING LOAD CALCULATIONS

(Design Peak)

Room No. ____________________

Room Name ____________________

Job No. __________ __________

Job Location ____________________

SPACE CONDITIONS

Summer	Outside	Inside	Difference
Dry-Bulb	F	F	
Wet-Bulb	F	F	
Dew Point	F	F	
Relative Humidity	%	%	
Winter	F	F	

COOLING / HEATING

Transmission Sensible Heat Above Grade Surfaces	Area (ft²)	Temp. Diff.	"U" Factor	Sensible Btu/h	Heating: Temp. Diff.	Heating: "U" Factor	Heating: Sensible Btu/h
North Exterior Wall							
East Exterior Wall							
South Exterior Wall							
West Exterior Wall							
Roof							
Transmission Sensible Heat Earth Contact Surfaces	**Area (ft²)**	**BTU/ hr ft²**			**BTU/ hr ft²**		
North Wall							
East Wall							
South Wall							
West Wall							
Slab Perimeter							
Slab Interior							
Transmission Sensible Heat Between Building Zones	**Area (ft²)**	**Temp. Diff.**	**"U" Factor**		**Temp. Diff.**	**"U" Factor**	
Floor							
Ceiling							
Partitions							
Glass Summary Calculations	**Area (ft²)**	**Solar Fac. (Btu/h ft²)**			**Temp. Diff.**	**"U" Factor**	
North Exterior Glass							
East Exterior Glass							
South Exterior Glass							
West Exterior Glass							
Total Transmission & Solar Heat Gain 1					Heat Loss 5		

Infiltration	Air Change/h	Volume (ft³)	Conv. Factor	ΔT or ΔH	Sensible Btu/h	Latent Btu/h	Heating: Air Change/h	Heating: Temp. Diff.	Heating: Sensible Btu/h
Sensible			0.018						
Latent			0.018						
Total Infiltration Heat Gain 2							Total Infiltr. Heat Loss 6		

Body Heat Gains	Number People	Rate of Heat Gain	Sensible Btu/h	Latent Btu/h
Sensible				
Latent				
Total Body Heat Gains 3				

Equipment Heat Gains	Conv. Factor	Watts	Sensible Btu/h
Electric Lights	3.412		
Electric Equipment	3.412		
Miscellaneous	3.412		
Total Equipment Gains 4			

	Sensible	Latent
TOTAL HEAT GAINS (1+2+3+4)		

TOTAL HEAT LOSSES (5+6) ______

APPENDIX D

EARTH THERMAL CONDUCTIVITY FACTORS[1]
in Btu/hr ft °F

Moisture Content of Soil	Type of Soil: Sand	Silt	Clay
Low (less than 4% by weight)	0.167 (0.29)[2]	0.083 (0.14)	0.083 (0.14)
Medium (from 4% to 20% by weight)	1.083 (1.87)	0.750 (1.30)	0.583 (1.01)
High (greater than 20% by weight)	1.250 (2.16)	1.250 (2.16)	1.250 (2.16)

NOTE: Dry Soil is exceedingly rare in most parts of the United States, and a low moisture content should be assumed only if the assumption can be proven valid.

1 Source: National Bureau of Standards, 1982.

2 Values inserted in parenthesis are W/m °C

CRITERIA SOURCES
(Publications Containing Criteria Cited in this Manual)

DOD Construction Manual (DOD 4270.1-M), Department of Defense, The Pentagon, Washington, DC 20301.

NAVFAC DESIGN MANUALS AND P-PUBLICATIONS

DM-2.2	Loads
DM-3.1	Plumbing Systems
DM-3.3	Heating, Ventilating, Air-conditioning and Dehumidifying Systems
DM-4.2	Electric Power Distribution Systems
DM-4.4	Electric Utilization Systems
DM-4.7	Wire Communication and Signal Systems
DM-5.11	Soil Conservation
DM-7.1	Soil Mechanics
DM-7.2	Foundations and Earth Structures
DM-7.3	Soil Dynamics, Deep Stabilization and Special Geotechnical Construction
DM-8	Fire Protection Engineering
P-89	Engineering Weather Date
P-355	Seismic Design of Buildings
P-397	Explosive Storage
P-422	Life-Cycle Costing
P-960	Installation Design

Department of Defense activities may obtain copies of Design Manuals and P-Publications from the Commanding Officer, Naval Publications and Forms Center, 5801 Tabor Avenue, Philadelphia, PA 19120. Department of Defense activities must use the Military Standard Requisitioning and Issue Procedure (MILSTRIP), using the stock control number obtained from NAVSUP Publication 2002.

Other Government Agencies and commercial organizations may procure Design Manuals and P-Publications from the Superintendent of Documents, U.S. Government Printing Office, Washington, DC. 20420.

National Fire Protection Association, Quincy, MA 02210.

70 National Electrical Code 1981
101 Life Safety Code
Fire Protection Handbook.

Uniform Building Code, International Conference of Building Officials, 5360 S. Workman Mill Road, Whittier, CA 90601.

INDEX

A — Page

Access 48,49
Air-Handling Systems . . . 56,119
Automatic Fire Extinguishing
  System 57

B

Backfilling 37
Below-Grade Heat Flux 89
Bentonite Clay 73
Brick Paved Roof 95
Building Envelope 88
Built-Up Asphalt 62
Building Use Parameters . . . 79

C

Cementitious Materials . . . 63
Climate Evaluation 3
  Infiltration Loads 3
  Cooling Loads 3
  Internal Heat Gains 3
  Ground Temperatures 4
Climatic Regions 4, 5, 6
Communications 123
Compaction 37
Computer Model. 129
Containers 46
Cooling Loads 3
Corner Site 30

D

Dampproofing 60
Daylighting 54
Design Examples 155
Dewatering 36
Drain Paved Roof. 95
Drainage 33, 57

E

Elevators 58
Emergency Power 122
Energy Calculations 124
Erosion Control 46
Excavation 11, 36
Exits 49, 58
Expansion Joint 103

F — Page

Fire Alarm 123
Fire Department Access . . . 57
Fire Protection 56
  Air Handling Systems . . . 56
  Automatic Fire-Extinguishing
    System 57
  Elevators 58
  Exits 58
  Fire-Fighting Access . . 49, 57
  Life-Safety 15, 58
  Refuge Areas 58
  Smoke and Heat Venting . . 56
Flat Site 24

G

Gain System 55
Ground Conditions 10
Ground Moisture 86
Ground Temperatures 4
Ground-Water Conditions . . 10, 78
Ground-Water Drainage 33

H

Heat Gains 3
Heat Loads 12
Heat Transfer Principles . . 124

I

Infiltration-Dominant
  Buildings 13
Infiltration Loads 3
Insulation 85
  Building Envelope 88
  Climate 86
  Ground Moisture 86
  Installation 94
  Internal Heat 86
  Passive Solar Grain 91
  Thermal Breaks 92
  Thermal Performance 85
  Waterproofing 91
Internal Heat 86
Internal Heat Gains 3
Irrigation 43
  Manual 43
  Spray Heads 44
  Trickle Systems 44

L — Page

Life-Cycle Cost Evaluation . . 16
Life-Safety 15, 58
Lighting 120
Liquid-Applied Systems 65

M

Manual Irrigation Systems . . 43
Masonry Cavity Wall . . . 104, 111
Mechanical and Electrical Systems 119
  Air Handling Systems 119
  Communications 123
  Emergency Power 122
  Fire Alarm 123
  Lighting 120
  Sanitary Systems 120
  Service Location 122
  Transformer Location 121
  Ventilation Requirements . . 119
  Water Systems 120
Mechanical Zoning 49
Modified Bitumens 68
Moisture Sources 60
Mowing 47

N

Natural Lighting, Skylights and Passive Solar 54
  Daylighting 54
  Gain Systems 55
  Passive Solar Applications . 54
Natural Light 50, 52

P

Parapet 107
Passive Solar Applications . . 54
Passive Solar Gain 91
Perimeter 15
Physical Characteristics . . . 52
Pipe Penetration 102
Pitch Membranes 62
Planting and Irrigation . . . 39
  Irrigation Systems 43
  Plant Materials 39
  Plant Species 40
Plastic Sheets 70

R — Page

Refuge Areas 58
Resistances - Insulation . . . 87
Roof Drains 102

S

Sanitary and Storm Sewers. . 10,120
Service Location 122
Site Context 20
  Coherence 20
  Corner Site 30
  Identity 20
  Flat Site 24
  Nature 20
  Reference 20
  Shore Front 28
  Territoriality 22
  Tradition 21
  Unwooded Suburban Site . . 24
  Wooded Site 22
Site Evaluation 7
  Access to Building Services 9
  Excavation Costs 11
  Ground Conditions 10
  Ground Water Conditions . . 10
  Isolation and Protection . 10
  Sanitary and Storm Sewers . 10
  Site Records 7
  Topography 10
  Utilities 8
Site Records 7
Shore Front 28
Skylights 54
Smoke and Heat Venting . . . 56
Sod Placement 47
SOLMET 131, 132
Spray Heads 44
Stability - Excavation - Dewatering 36
Structural Bay. 49
Structural Deck 101
Structural Systems 171
Structure 116
  Blast Loads 118
  Design Loads 116
  Roof Loads 116
  Seismic Loads 117
  Wall Loads 116
Surface-to-Volume Ratios . . 169

T — Page

Territoriality 22
Thermal Breaks 92
Thermal Conductivity 124
Thermal Lag 170
Thermal Performance . . . 85, 151
Topography 10
Transformer Location 121
Trickle Systems 44
Typical Meterological Year
(TMY) 132

U

Unwooded Site 24
Utilities 8

V

Ventilation 13, 119
Vulcanized Materials 70

W — Page

Water Systems 120

Waterproofing 60
- Bentonite Clay 73
- Building Use Parameters . . 79
- Built-Up Asphalt 62
- Cementitious Materials . . . 63
- Dampproofing 60
- Ground-Water Conditions . . 78
- Liquid Applied Systems . . . 65
- Modified Bitumens 68
- Moisture Sources 60
- Pitch Membranes 62
- Plastic Sheets 70
- Vulcanized Materials 70

Weeding 47
Window Sill 108

Wooded Site 23
Work Conditions 53

www.ingramcontent.com/pod-product-compliance
Ingram Content Group UK Ltd.
Pitfield, Milton Keynes, MK11 3LW, UK
UKHW051459290926
15471UKWH00008B/1299

9 781589 639683